CAMBRIDGE TRACTS IN MATHEMATICS

65. *Normal topological spaces*

RICHARD A. ALÒ
Associate Professor of Mathematics
Carnegie-Mellon University

AND

HARVEY L. SHAPIRO
Associate Professor of Mathematics
Northern Illinois University

Normal topological spaces

CAMBRIDGE UNIVERSITY PRESS

CAMBRIDGE UNIVERSITY PRESS
Cambridge, New York, Melbourne, Madrid, Cape Town, Singapore, São Paulo, Delhi

Cambridge University Press
The Edinburgh Building, Cambridge CB2 8RU, UK

Published in the United States of America by Cambridge University Press, New York

www.cambridge.org
Information on this title: www.cambridge.org/9780521202718

First published 1974
This digitally printed version 2008

A catalogue record for this publication is available from the British Library

Library of Congress Catalogue Card Number: 73-79304

ISBN 978-0-521-20271-8 hardback
ISBN 978-0-521-09530-3 paperback

Contents

Introduction *page* vii

I. Preliminaries 1

1 Notation and terminology 1

2 Continuous functions and pseudometrics 19

3 Uniform spaces 32

4 $\mathscr{Z}$-filters and convergence 41

5 Compactifications and realcompletions 55

II. Normality and real-valued continuous functions 70

6 Normal spaces and C- and C^*-embedding 70

7 Normal spaces and z-embedding 77

8 Pseudometrics and normal sequences of covers 85

9 Counter-examples 91

III. Normality and normal covers 113

10 Definitions and elementary properties 113

11 Normal spaces and open covers 126

12 Collectionwise normal spaces and open covers 134

13 Uniform spaces, normal covers, and pseudometrics 142

IV. Normality and pseudometrics 163

14 P^{γ}-embedding and equivalent formulations 163

15 Extending continuous pseudometrics and other topological concepts 185

16 Extending $\aleph_0$-separable continuous pseudometrics and real-valued continuous functions *page* 197
17 Extending totally bounded continuous pseudometrics and uniformly continuous functions 202

V. Normality and uniformities 210
18 Uniformities generated by γ-separable continuous pseudometrics 210
19 Extending uniformities generated by continuous functions 218
20 Some applications 225

Appendix 241
21 Countably paracompact spaces and related questions 244
22 Generalized metric spaces and related results 261

References 281

Guide to the References 299

Index 301

CORRIGENDA

Page 66 (lines 13 and 16): FOR $Y-f(x)$ READ $Z-f(x)$
Page 66 (line 14): FOR image of X READ image of βX
Page 67 (line 17): FOR rings of sets READ strong delta normal bases
Page 68 (line 21): DELETE if and
Page 68 (line 23): FOR $\mathrm{cl}_{\upsilon X} S$ READ $\mathrm{cl}_{\upsilon Y} S$
Page 147 (4th line from foot): FOR For every f is READ For every f in

Introduction

The origins of general set theoretical topology go back to the works of Alexandroff, Fréchet, Hausdorff, and Urysohn, to mention just a few of the more important names. It interacts with various other branches of mathematics such as functional analysis, modern algebra, the theory of partial differential equations, algebraic topology, differential topology and differential geometry. It was influenced strongly by algebraic geometry, functional analysis, and the theories of real and complex variables (the last as commenced by the early works of Weierstrass). With these various elements interacting, new advances in general topology have not in general been under the influence of the more powerful developments in other areas of mathematics. Hence one of the objectives of general topology has been to find and develop a unified system and unified approach to this area. Terms and ideas which appear dissimilar at first should be analyzed and new terms and ideas invented to encompass these. Considerable progress has been made by the Soviet mathematicians in this area. Behind much of this work is found the notions (and variations of these notions) of locally finite covers, star-finite refinements, normal sequences of covers as well as alterations of the notion of pseudometrics and concepts from dimension theory.

Completely regular spaces have been shown to be natural generalizations of metric spaces since they are precisely the ones that have Hausdorff compactifications, they are the spaces that are embeddable in cubes, and they are the spaces which are associated with uniformities where the metric space notions of completeness and Cauchy sequences have their most general setting. (Hence the class of uniform spaces are important spaces for consideration.) However a slightly stronger condition than completely regular has been found to be more influential in the development of a unifying theory (however, by no means completely satisfactory, yet more satisfactory than completely regular spaces). This is the notion of a normal space. Even though the class of normal spaces is neither hereditary nor productive (as the completely regular spaces are), they do have significant properties. They are, for example, characterized among the spaces that are uniformizable as those spaces for which

(1) every closed subspace is C-embedded, and

(2) every open point-finite cover has associated with it an open cover with the property that the closures of its members refines the original cover.

This gives information as to when a cover of a normal space is a normal cover, one of the important notions mentioned above. Hence it is worthwhile to investigate characterizations of normal spaces in the light of recent developments and to utilize these notions to arrive at again new concepts useful in obtaining a unified theory.

Paracompact spaces have also had their impact on general topology. It is in the theory of these spaces that there is couched the solution to the metrization problem. Again the notion lends itself readily to the notions of normal covers. For purposes of dimension theory there is some doubt whether general paracompact spaces or general normal spaces are adequate for extending the results of the already well-established theory for metric spaces. As mentioned above new ideas have evolved which along with paracompactness and normality have assisted in giving meaningful results. For example such a notion would be that of a paracompact M-space (this last having been introduced by K. Morita). Every metric space and every compact space are paracompact M-spaces and the countable product of paracompact M-spaces is of the same type.

Thus we arrive at some of the reasons for the present text. It is clear that today in the field of general set theoretical topology a text is needed to discuss material beyond the usual basic general topology course. However we feel that this 'second text' should serve a number of roles which, as they turn out, prove to be closely related, one to the others. It is desirable to present to the student recent developments in general topology. It is desirable to endow him with the impetus, and the inspiration to carry on mathematical research. For this it is hoped that the material presently at hand will give the necessary new ideas, the necessary techniques, and the necessary goals for achieving meaningful research in set theoretical topology. But what precisely does this phrase, 'meaningful research in set theoretical topology', mean? We have pointed out some of the areas that have influenced it and those that it has inspired. We have also discussed the need for defining a methodology and for implementing new ideas to increase the advances in this area as well as to further inspire other areas.

Here then, we discuss in great detail the notion of normal cover and the related uniform concepts. We look at many characterizations of normal spaces and their interrelations with paracompact spaces and the various ways of embedding subspaces. We also look at the new concepts that have made such a recent impact (such as M-spaces) and consider their relationship to the above concepts.

With this material then and with the many research problems suggested by it, we hope to assist in this last aim. A prerequisite for the text is of course a basic general topology course which may have hinted at the notions of paracompactness and uniform spaces (but it is not absolutely necessary that they did). We utilize much of the theory of *Rings of Continuous Functions* as developed in [XI]. However, the necessary tools for this work are developed here, leaving for additional leisure reading only the notions whose techniques are not pertinent to the ensuing discussion. Bringing in this area (*Rings of Continuous Functions*) also assists in the interaction of general topology with other branches of mathematics.

As to the structure of the text, first of all, we have made an attempt to gather many of the characterizations of both a normal space and a collectionwise normal space into one text. Around this basic theme, the following course has been developed.

Chapter I contains basic preliminary results. All of these

results (along with some in Chapter II) except those on uniform spaces, should be familiar to anyone with a basic course in topology. In Chapter II, the structural relationship between normal spaces and the real-valued continuous functions defined on them, is considered. Here also are stated and proved the known characterizations of normality in terms of C-embedding and C^*-embedding. Characterizations of normal spaces and of collectionwise normal spaces in terms of various open covers are given in Chapter III. Then a list of equivalent conditions for a topological space to be normal are stated and, although many of these conditions have appeared in one place or another in the literature, there has been no attempt to organize them and to assimilate the ideas which they encompass. These then are used as a constructing frame on which is built the entire course. To assist in this task, many theorems about refinement of covers that are part of the folklore of this subject but never seem to be published are also shown here. Then a proof of Katětov's theorem characterizing collectionwise normal spaces is given.

In Chapter IV the extension of pseudometrics is discussed. Using various classes of pseudometrics, normal spaces and collectionwise normal spaces are characterized. It is shown that, roughly speaking, extending continuous pseudometries is to collectionwise normal spaces as extending real-valued continuous functions is to normal spaces.

In Chapter V a study is made of extending uniformities. These concepts are then used to give additional characterizations of normal and collectionwise normal spaces. Also in this chapter we make a point to emphasize some additional applications of the new results brought forth in the text.

In the Appendix, we present no proofs of theorems but only a survey of some ideas presently under consideration in this fast developing field of general topology. We hope that this will give the reader some idea of the many problems presently being researched and at the same time encourage him to search the literature for the proofs and additional ideas. In a book such as this, it is impossible to include many of the ideas that seem to be worthwhile. We have tried to select some which appear pertinent to the results in the text.

It is difficult to recognize all those who have contributed, in one way or another, to the development of this book. We have taught the material to several classes of graduate students and to all of them we owe a debt of thanks for their patience, their corrections and their suggestions. Their number prohibits us from naming them all. On the other hand, we must name and thank the following students who worked more closely with us – Jogendra Kumar Kohli, Arvind Kumar Misra, Greg Naber, Linn Imler Sennott, and Maury Weir. They read parts of the manuscript at various stages and some contributed either new results or assisted in the exposition of some of the areas. We would also like to thank Professors Robert Heath, Jun-iti Nagata and Akihiro Okuyama for their many valuable comments, suggestions, and assistance in the preparation of Section 22 in the Appendix. Professor Mary Ellen Rudin was very kind to permit us to use a pre-print of her paper (see [333] and [417]) where she solves the long unsolved and difficult Dowker problem. Eddie Lamarr Buckner and Janice Frederick did a wonderful job in typing much of the manuscript. Professor C. T. C. Wall was very helpful with the editing. We wish to thank our respective departments of mathematics for providing research support for the completion of this project. Finally the second named author must thank the Mathematics Department at Carnegie-Mellon University for providing a two year visiting appointment which enabled the commencement of this work.

September 1973

RICHARD A. ALÒ
Carnegie-Mellon University

HARVEY L. SHAPIRO
Northern Illinois University

This book is dedicated to
Elena Chimenti, Giuseppe Alò
Serafina Torchia
Melanie Jeanne, Kenneth Mark, Lili Rosanne
Michelina Luisa
and to
Flora and Maurice Shapiro
the many abstractions I have known
the memory of the Kent State massacre

I. *Preliminaries*

1 Notation and terminology

To establish a common foundation on which to build it is necessary to make a few comments about terminology and notation. This section is devoted to this and also to the recollecting (for some) of concepts from set theoretical topology. A quick perusal by the informed, then, should be useful.

The set of real numbers is denoted by **R**. Its subset of non-negative real numbers is denoted by $\mathbf{R}^+$, whereas its subset of positive integers is denoted by **N**. The set of real numbers x for which $0 \leqslant x \leqslant 1$ is denoted by E. These are the most commonly cited subsets of **R** used in this book. From time to time, other subsets are considered. However they will be defined when needed.

If X is a set, 2^X denotes the collection of all subsets of X and it is called the *power set of* X. A distinction is made between a collection $\mathscr{U}$ of subsets of a set X and a family $\mathscr{U} = (U_\alpha)_{\alpha \in I}$ of subsets of a set X as follows. By a *family* $\mathscr{U} = (U_\alpha)_{\alpha \in I}$ of subsets of X is meant that there exists a set I, called the *index set* of the family, and a function f mapping I into 2^X such that $f(\alpha) = U_\alpha$. Loosely speaking, a family of subsets is a collection of subsets in which each member has been labeled.

In the sequel definitions are given in terms of either families or collections and it is left to the reader to formulate the implicit definition using the other concept. In the text, free use is made of whichever concept seems the most natural.

Let $\mathscr{T}$ denote the collection of all open subsets of a topological space X. When it is desired to call particular attention to the topology $\mathscr{T}$ of X, or when the underlying point-set is to be provided with more than one topology, reference shall be made to X as '*the topological space* $(X,\mathscr{T})$'.

The concepts of a subspace and relative topologies (as in [XVIII]) are concerned with the reaction of a collection of subsets of a topological space on a distinguished subset. To be able to use the idea more fully the following definitions are made.

1.1 DEFINITION If A is a subset of the set X, and if $\mathscr{U} = (U_\alpha)_{\alpha \in I}$ is a family of subsets of X, then by the *trace of* $\mathscr{U}$ on A is meant the family $(U_\alpha \cap A)_{\alpha \in I}$. The trace of $\mathscr{U}$ on A is denoted by $\mathscr{U}|A$. If $\mathscr{S}$ is a collection of subsets of X, then by the trace of $\mathscr{S}$ on A is meant the collection $\{S \cap A : S \in \mathscr{S}\}$ and it also is denoted by $\mathscr{S}|A$. Conversely if $\mathscr{S}$ is a collection of subsets of A then $\mathscr{U}$ is an extension of $\mathscr{S}$ if for each $S \in \mathscr{S}$ there is a $U \in \mathscr{U}$ such that $U \cap A = S$. (Thus the topology $\mathscr{T}$ on a set X is an extension of the relative topology on any of its subsets, whereas the relative topology on a subset A of X is just the trace of $\mathscr{T}$ on A.)

The *closure* of a subset A of X will be denoted by $\operatorname{cl} A$, or, when there is possibility of confusion, by $\operatorname{cl}_X A$. Analogously, the *interior* of a subset A of X will be denoted by $\operatorname{int} A$ or $\operatorname{int}_X A$. A subset G is *regular open* if $G = \operatorname{int} \operatorname{cl} G$ and a set F is *regular closed* if $F = \operatorname{cl} \operatorname{int} F$. The *complement* of a subset B of X will be denoted by $X - B$. If A and B are contained in X, then by $A - B$ is meant $\{x \in X : x \in A \text{ and } x \notin B\}$. The cardinal of a set X is denoted by $|X|$. The collection of all finite subsets of the set X will be denoted by $[X]$.

Often for collections of subsets of a set X, it is necessary to look at intersections of finite subcollections. More precisely, if $\mathscr{U} = (U_\alpha)_{\alpha \in I}$ is a family of subsets of a set X, then by $\mathscr{U}^F$ is meant the family of subsets of X obtained by taking for each $J \in [I]$, the set $\cap \{U_\alpha : \alpha \in J\}$.

Many times in general topology a particular property which is defined on the entire space will have worthwhile implications when it is localized to certain open bases for points of the space. For example such has been the case with compactness and connectedness, being utilized in the motivation of local compactness and local connectedness. Dieudonné (see [95]) was the first to significantly localize the notion of compactness in yet another way called paracompactness. To pave the way for this definition it is necessary to look at properties of open covers of the whole space which also are local assertions. It is interesting to see that these notions which lead to Dieudonné's form of 'compactness' also lead to several characterizations of normality.

1.2 DEFINITION Let x be an element in the topological space X, and let $\mathscr{U} = (U_\alpha)_{\alpha \in I}$ be a family of subsets of X. The family $(U_\alpha)_{\alpha \in I}$ is *point-finite at x* if there exists a finite subset K of I such that $x \notin U_\alpha$ for every $\alpha \notin K$. The family $(U_\alpha)_{\alpha \in I}$ is *locally finite at x* if there exist a neighborhood G of x and a finite subset J of I such that $G \cap U_\alpha = \varnothing$ for every $\alpha \notin J$. The family $(U_\alpha)_{\alpha \in I}$ is *discrete at x* if there exist a neighborhood H of x and a subset K of I with $|K| \leqslant 1$ such that $H \cap U_\alpha = \varnothing$ for every $\alpha \notin K$. The family $(U_\alpha)_{\alpha \in I}$ is *point-finite* (respectively, *locally finite*, respectively, *discrete*) if $(U_\alpha)_{\alpha \in I}$ is point-finite (respectively, locally finite, respectively, discrete) at every point $x \in X$. It will be said that $\mathscr{U}$ is *σ-locally finite* (respectively, *σ-point-finite*, respectively, *σ-discrete*) if $\mathscr{U}$ is a union of a countable number of families each of which is locally finite (respectively, point-finite, respectively, discrete). It is *closure preserving* if for every subset J of I

$$\operatorname{cl}\Big(\bigcup_{\alpha \in J} U_\alpha\Big) = \bigcup_{\alpha \in J} \operatorname{cl} U_\alpha.$$

A set G in a topological space $(X, \mathscr{T})$ is called a *G_δ-set* if G can be written as a countable intersection of open sets (that is, if there exists a sequence $(G_n)_{n \in \mathbf{N}} \subset \mathscr{T}$ such that $G = \bigcap_{n \in \mathbf{N}} G_n$). The set is called an *F_σ-set* if it can be written as a countable union of closed sets.

These concepts are applied as qualifiers to the concept of an open cover which is next to be defined.

1.3 DEFINITION Let $\mathscr{U} = (U_\alpha)_{\alpha \in I}$ be a family of subsets of a set X, and let γ be a cardinal number. The family $\mathscr{U}$ is a *cover* of X in case $I \neq \varnothing$ and $X = \bigcup_{\alpha \in I} U_\alpha$. The family $(U_\alpha)_{\alpha \in I}$ has *power at most γ* in case $|I| \leqslant \gamma$. If $\mathscr{U}$ is a finite cover we say that $\mathscr{U}$ has *power less than $\aleph_0$*. The family is said to be *open* (respectively, *closed*) in case U_α is open (respectively, closed) for every $\alpha \in I$. When precision is needed, it shall be said that $(U_\alpha)_{\alpha \in I}$ is *open in* X (respectively, *closed in* X).

Now if S is a subset of X and if $(V_\beta)_{\beta \in J}$ is a family of subsets of S, then, by the above, $(V_\beta)_{\beta \in J}$ is an open (respectively, closed) cover of S if $J \neq \varnothing$, if $S = \bigcup_{\beta \in J} V_\beta$, and if, for each $\beta \in J$, V_β is an open (respectively, closed) subset of S in its relative topology. Thus if $(V_\beta)_{\beta \in J}$ is an open cover of S, then $V_\beta \subset S$ for each $\beta \in J$.

As was mentioned above, the closure preserving property of a family of subsets of a set is very useful. For finite families of subsets this property always holds. Along with several other useful facts the following proposition establishes this fact for locally finite families of sets.

1.4 PROPOSITION *Suppose that* $\mathscr{U} = (U_\alpha)_{\alpha \in I}$ *is a family of subsets of a topological space* X.

(1) *The family* $(U_\alpha)_{\alpha \in I}$ *is locally finite if and only if* $(\mathrm{cl}\, U_\alpha)_{\alpha \in I}$ *is locally finite.*

(2) *If* $\mathscr{U}$ *is locally finite then* $\mathscr{U}$ *is closure preserving.*

(3) *The family* $\mathscr{U}$ *is discrete if and only if* $\mathscr{U}$ *is closure preserving and the elements of* $\mathscr{U}$ *have pairwise disjoint closures.*

(4) *If* $\mathscr{U} = (U_\alpha)_{\alpha \in I}$ *is locally finite, then the collection* $\mathscr{U}^F$ *is locally finite.*

Proof. The facts here are basic and their proofs are straightforward. Consequently many of them are left as exercises. Here statement (2) is shown as a pattern for the proofs. It is clear that

$$\bigcup_{\alpha \in I} \mathrm{cl}\, U_\alpha \subset \mathrm{cl}\,(\bigcup_{\alpha \in I} U_\alpha).$$

Conversely suppose that $x \notin \bigcup_{\alpha \in I} \mathrm{cl}\, U_\alpha$. Since $\mathscr{U}$ is locally finite, there exist a neighborhood W of x and a finite subset K of I such that $W \cap U_\alpha = \varnothing$ if $\alpha \notin K$. Since $x \notin \mathrm{cl}\, U_\alpha$ for all $\alpha \in I$, certainly, for each $\alpha \in K$, there exists a neighborhood G_α of x such that $G_\alpha \cap U_\alpha = \varnothing$. Let $G = W \cap (\bigcap_{\alpha \in K} G_\alpha)$ and note that G is a neighborhood of x whose intersection with $\bigcup_{\alpha \in I} U_\alpha$ is empty. Hence $x \notin \mathrm{cl}\,(\bigcup_{\alpha \in I} U_\alpha)$. □

In considering the collection of covers on a given space, it is seen that there is a natural preorder to be defined on it. But first let us recall the definition of a preorder.

1.5 DEFINITION A binary relation $*$ on a set A is a *preorder* if it is *reflexive* ($a * a$ for all a in A) and *transitive* ($a * b$ and $b * c$ implies $a * c$ for all a, b and c in A). A set with a preorder is called a *preordered set*. Usually a preorder $*$ is denoted by $\leqslant$. A preorder on A is called a *partial order* if for a and b in A, $a \leqslant b$ and $b \geqslant a$ implies $a = b$. We say $a < b$ if $a \leqslant b$ and $a \neq b$.

A subset B of a preordered set A is a *chain* if for every pair a and b of elements in B either $a \leqslant b$ or $b < a$. A partially ordered set that is also a chain is called a *totally ordered set*. A partially ordered set W is called *well ordered* if every non-empty subset of W has a first element (that is, for each $B \subset W$ there exists $\beta \in B$ such that $\beta \leqslant b$ for all $b \in B$). A subset B of A is *bounded below* (respectively, *above*) if there is an $\alpha \in A$ such that for each $b \in B$ either $\alpha < b$ (respectively, $b < \alpha$) or $\alpha = b$. The element α is called a *lower* (respectively, *upper*) *bound* for B. An element $\alpha \in A$ is *maximal* (respectively, *minimal*) *in* A if there is no member a in A such that $\alpha < a$ (respectively, $a < \alpha$).

One of the fundamental theorems of set theory is the following.

1.6 THEOREM *The following statements are equivalent.*

(1) (*Axiom of choice*) *Given any non-empty family* $(X_\alpha)_{\alpha \in I}$ *of non-empty pairwise disjoint sets there exists a set* S *consisting of exactly one element from each* X_α.

(2) (*Zermelo's theorem*) *Every set can be well ordered.*

(3) (*Zorn's lemma*) *If each chain in a partially ordered set has an upper bound, then there is a maximal element of the set.*

1.7 DEFINITION Suppose that X is a non-empty set. If $\mathscr{U} = (U_\alpha)_{\alpha \in I}$ and $\mathscr{V} = (V_\beta)_{\beta \in J}$ are two families of subsets of X, then $\mathscr{U}$ is a *refinement of* $\mathscr{V}$ (or $\mathscr{U}$ *refines* $\mathscr{V}$), written $\mathscr{U} < \mathscr{V}$, if $I \neq \varnothing$, if $\bigcup_{\alpha \in I} U_\alpha = \bigcup_{\beta \in J} V_\beta$ and if each element of $\mathscr{U}$ is a subset of some element of $\mathscr{V}$. It is said that $\mathscr{U}$ is a *screen of* $\mathscr{V}$ (or that $\mathscr{U}$ *screens* $\mathscr{V}$ in case $I = J$ and $U_\alpha \subset V_\alpha$ for all $\alpha \in I$). If X is now a topological space then the family $\mathscr{U}$ is a *strong screen* of $\mathscr{V}$ if $J = I$ and if $\operatorname{cl} U_\alpha \subset V_\alpha$ for all $\alpha \in I$. If $\mathscr{U}$ is a screen of $\mathscr{V}$ then we say that $\mathscr{V}$ is an *expansion* of $\mathscr{U}$. The family $(U_\alpha)_{\alpha \in I}$ is said to have the *finite intersection property* (respectively, *countable intersection property*) if for every finite (respectively, countable) subset F of I, $\bigcap_{\alpha \in F} U_\alpha \neq \varnothing$. If $n \in \mathbf{N}$, we say that $\mathscr{U}$ has *finite order n* if whenever J is a subset of I such that $\bigcap_{\alpha \in J} U_\alpha \neq \varnothing$ then $|J| \leqslant n$.

Thus a refinement of a cover is itself a cover. Notice also that a refinement of a cover may contain more sets than the given cover. The following lemma demonstrates that when this happens the indexing sets may be made more precise.

1.8 LEMMA *Suppose that S is a subset of a topological space X and that $\mathscr{V} = (V_\beta)_{\beta \in J}$ is a locally finite open cover of X whose trace on S refines the open cover $\mathscr{U} = (U_\alpha)_{\alpha \in I}$ of S. Then there exists a locally finite open cover $\mathscr{W}$ of X such that $\mathscr{W}|S$ is a screen of $\mathscr{U}$.*

Proof. Let $\pi \colon J \to I$ be a function such that $V_\beta \cap S \subset U_{\pi(\beta)}$ for each $\beta \in J$. For each $\alpha \in I$, let

$$W_\alpha = \bigcup_{\beta \in \pi^{-1}(\alpha)} V_\beta.$$

Observe that $\mathscr{W} = (W_\alpha)_{\alpha \in I}$ is an open cover of X such that $W_\alpha \cap S \subset U_\alpha$ for each $\alpha \in I$. To see that $\mathscr{W}$ is locally finite, let x be any element of X. Since $\mathscr{V}$ is locally finite, there exists a neighborhood G of x and a finite subset K of I such that $G \cap V_\beta = \varnothing$ if $\beta \notin K$. Since π is a function, the set $\pi(K)$ is finite. Furthermore, if $\alpha \notin \pi(K)$, then $G \cap W_\alpha = \varnothing$ and it follows that $\mathscr{W}$ is locally finite. □

As hinted above, the important aspect of this lemma is that it will assist us in eliminating indexing sets that are not the 'right size'. In other words, if one has an open cover $\mathscr{U}$ of a subspace S of X and an open cover $\mathscr{V}$ of X such that $\mathscr{V}|S$ refines $\mathscr{U}$, then it is always possible to find another cover of X with the same indexing set as $\mathscr{U}$ which will do the same thing.

So far collections of subsets of a non-empty topological space X have been under consideration. Other collections which should be considered are collections of real-valued functions on X. These are playing an ever more important role in modern topology. For example they can be utilized to give characterizations of some of the topological separation axioms. More will be said about this in Section 2.

Let us record now for the sake of completeness and for reference the topological separation axioms. The T_1-separation axiom will not be a part of the definition of a completely regular space, normal space, etc. as required by some authors (for example, see [XI]).

1.9 DEFINITION A topological space X is a *T_1-space* if for each $x \in X$, $\{x\}$ is closed. A space X is a *Hausdorff space* if for each pair of distinct points x and y in X there are disjoint open sets U and V such that $x \in U$ and $y \in V$. It is a *regular space* if for each $x \in X$ and each closed set F with $x \notin F$, there are disjoint open sets U and V such that $x \in U$ and $F \subset V$. It is a *completely regular space* if for each $x \in X$ and each closed set F with $x \notin F$ there is a real-valued continuous function f on X such that $f(x) = 0$ and $f(y) = 1$ for every $y \in F$. We say that X is a *Tychonoff space* if X is a completely regular T_1-space. The space X is said to be a *normal space* if for each pair F_1 and F_2 of disjoint closed sets there are disjoint open sets U_1 and U_2 such that $F_i \subset U_i$ $(i = 1, 2)$. We say that X is *hereditarily normal* if every subset of X is normal. The space is *perfectly normal* if X is normal and if every closed subset of X is a G_δ.

Now suppose that X is a topological space and that γ is an infinite cardinal number. The space X is said to be *γ-collectionwise normal* if for every discrete family $(F_\alpha)_{\alpha \in I}$ of closed subsets of X of power at most γ there is a family $(G_\alpha)_{\alpha \in I}$ of pairwise disjoint open subsets of X such that $F_\alpha \subset G_\alpha$ for each $\alpha \in I$. The space X is said to be *collectionwise normal* if for every discrete family $\mathscr{F} = (F_\alpha)_{\alpha \in I}$ of closed subsets of X there is a family $\mathscr{G} = (G_\alpha)_{\alpha \in I}$ of pairwise disjoint open subsets of X such that $F_\alpha \subset G_\alpha$ for every $\alpha \in I$ (that is, $\mathscr{F}$ is screened by $\mathscr{G}$). We say that X is *hereditarily collectionwise normal* if every subset of X is collectionwise normal.

The various notions of 'compactness' described above may now be defined and some comment on the interrelationships between these concepts clarifies their status as useful generalizations.

1.10 DEFINITION Suppose that X is a topological space and that γ is an infinite cardinal number. It is said that X is a *compact space* if every open cover of X has a finite *subcover* (that is, if $(U_\alpha)_{\alpha \in I}$ is an open cover of X then there exists a finite set $F \subset I$ such that $(U_\alpha)_{\alpha \in F}$ covers X). By a *compactification*† of X we mean a compact Hausdorff space in which X is dense (up to homeomorphism). The space X is *countably compact* if every countable open cover of X has a finite subcover. We say that X is *locally compact* if every point of X has a compact neighborhood. The space is *zero-dimensional* if there is a base for the topology consisting of closed and open subsets of X. It is said to be a *Lindelöf* space if every open cover of X has a countable subcover. The space X is *γ-paracompact* if every open cover of X of power at most γ has a locally finite open refinement. The space X is *paracompact* if every open cover of X has a locally finite open refinement. If X is $\aleph_0$-paracompact then it is said that X is *countably paracompact*.

This definition of paracompactness is the one given by Kuratowski [xx]. It differs from the original definition given by Dieudonné [95] in that Dieudonné requires a paracompact space to be a Hausdorff space. Here Kuratowski's definition has been adopted because it is desired that a pseudometric space be paracompact. A proof of this was given by Stone in [359] and it can be found in Chapter 5 of Kelley's book (see [xviii]). There it is also shown that a paracompact regular space is normal and that a paracompact Hausdorff space is regular. Various characterizations of paracompactness have been given. Kelley mentions those due to E. Michael, A. H. Stone, J. S. Griffin and himself. These various characterizations will be used from time to time.

Finally let us mention here that it is this concept of paracompactness, more than any other, that led to the solution of

† We prefer to identify the space X with its homeomorphic copy as opposed to constantly referring to a compactification as an ordered pair (see [xviii]).

the metrizability problem, that is, to the necessary and sufficient conditions for a topological space to be metrizable (see [300], [347], [50]). In Chapter VI, we will see how this concept also leads to other interesting spaces.

Since the notions related to paracompactness and paracompactness itself are important for this treatise, it is important to mention the answers for paracompactness, to some of the questions usually asked about a topological property. Since pseudometric spaces are paracompact, metric spaces are paracompact Hausdorff spaces. Regular Lindelöf spaces are also paracompact. The real line with the topology defined by taking as a subbase for the open sets, the half open intervals, is a regular Lindelöf space. The product of this space with itself is not normal and hence not paracompact. Hence paracompactness is not preserved by taking topological products. Likewise subspaces and quotient spaces of paracompact spaces are not in general paracompact. In the case of subspaces, however, paracompactness is hereditary to closed subspaces. Another example of a non-paracompact space is the space of all ordinals less than the first uncountable ordinal with the order topology. (This example will be discussed in greater detail in Section 9.)

Probably the only concept in 1.10 that is not familiar to the reader is that of γ-collectionwise normality and collectionwise normality. The last concept was introduced by Bing in [50] when he gave his solution to the metrization problem. Later it will be shown that every paracompact space is collectionwise normal. Collectionwise normal spaces have become more interesting in recent years because of their relation to the extension theory of functions and pseudometrics. First then let us give a useful characterization of normality and then relate γ-collectionwise normality for $\gamma = \aleph_0$ to normality.

1.11 PROPOSITION *If X is a topological space, then the following statements are equivalent.*

(1) *The space X is normal.*

(2) *If F is a closed subset of X and if U is an open subset such that $F \subset U$, then there exists an open subset V of X such that $F \subset V \subset \mathrm{cl}\, V \subset U$.*

The easy proof is elementary but should be established by the reader.

The equivalence of $\aleph_0$-collectionwise normality to normality is given in the next theorem.

1.12 THEOREM *A topological space X is a normal space if and only if X is $\aleph_0$-collectionwise normal.*

Proof. Suppose that X is a normal space and let $(F_n)_{n\in\mathbf{N}}$ be a discrete countable family of closed subsets of X. Then $A_1 = \bigcup_{n=2}^{\infty} F_n$ is a closed set (see 1.4) and clearly F_1 is disjoint from it. By 1.11 there exists an open set G_1 of X such that

$$F_1 \subset G_1 \quad \text{and} \quad (\text{cl}\, G_1) \cap A_1 = \varnothing.$$

The family $(\text{cl}\, G_1, F_2, F_3, \ldots)$ is a countable discrete family of closed subsets of X. Hence $\text{cl}\, G_1 \cup \left(\bigcup_{n=3}^{\infty} F_n\right) = A_2$ is a closed set and there exists an open set G_2 of X such that $F_2 \subset G_2$ and $\text{cl}\, G_2 \cap A_2 = \varnothing$. Then $(\text{cl}\, G_1, \text{cl}\, G_2, F_3, F_4, \ldots)$ is a countable discrete family of closed sets. Appealing to finite induction a family $(G_n)_{n\in\mathbf{N}}$ of pairwise disjoint open sets is obtained such that $F_n \subset G_n$ for all $n \in \mathbf{N}$. Hence X is $\aleph_0$-collectionwise normal. Since the converse is obvious, the proof is now complete. □

An important concept to be defined later is that of a uniform space, the natural generalization of a metric space where 'uniform continuity' and 'completeness' may be discussed. Tukey in [XXVIII] gives a careful exposition of the concept by means of the concepts next to be defined. It is interesting to note that covers and refinements of covers play an important role in the theory of uniform spaces also. Hence the role of covers in topology appears as a good concept for unifying others.

1.13 DEFINITION Let $\mathscr{U} = (U_\alpha)_{\alpha\in I}$ and $\mathscr{V} = (V_\beta)_{\beta\in J}$ be two covers of a set X. If $A \subset X$, then by the *star of A with respect* to $\mathscr{U}$ is meant $\cup\{U_\alpha : \alpha \in I \text{ and } U_\alpha \cap A \neq \varnothing\}$. The star of A with respect to $\mathscr{U}$ is denoted by $\operatorname{st}(A, \mathscr{U})$. If $x \in X$, then we write $\operatorname{st}(x, \mathscr{U})$ for $\operatorname{st}(\{x\}, \mathscr{U})$. Iterated stars are defined by

$$\operatorname{st}^n(A, \mathscr{U}) = \operatorname{st}(\operatorname{st}^{n-1}(A, \mathscr{U}), \mathscr{U}).$$

Let $\mathscr{U}^* = (\operatorname{st}(U_\alpha, \mathscr{U}))_{\alpha \in I}$ and let $\mathscr{U}^\Delta = (\operatorname{st}(x, \mathscr{U}))_{x \in X}$. It is said that $\mathscr{U}$ is a *star-refinement* of $\mathscr{V}$ (and write $\mathscr{U} <^* \mathscr{V}$) if $\mathscr{U}^* < \mathscr{V}$. Similarly $\mathscr{U}$ is a *Δ-refinement of* $\mathscr{V}$ (and write $\mathscr{U} <^\Delta \mathscr{V}$) if $\mathscr{U}^\Delta < \mathscr{V}$. Finally $\mathscr{U} = (U_\alpha)_{\alpha \in I}$ is *star-finite* if for every $\alpha \in I$ there exists a finite subset K of I such that $U_\alpha \cap U_\beta = \varnothing$ if $\beta \notin K$.

If $\mathscr{U}$ is a cover of a set X, then it is easy to verify that $\mathscr{U}^*$ is a cover of X. If, in addition, X is not empty, then $\mathscr{U}^\Delta$ is a cover of X.

Also note that the relations 'star-refinement' and 'Δ-refinement' that were just defined are preorders on the collection of all covers of a non-empty set X.

Just as continuous real-valued functions play an important role in the general theory of topological spaces, collections of continuous pseudometrics will play an important role in the theory to be developed here. A pseudometric on the topological space X can be obtained from a cover of X if the cover has some additional structure. One structural concept that will be applicable is that of being able to obtain from the cover a sequence of covers which form a chain with respect to the preorder of star-refinement.

1.14 Definition A sequence $(\mathscr{U}_n)_{n \in \mathbf{N}}$ of open covers, $\mathscr{U}_n$, of a set X is said to be *normal* if $\mathscr{U}_{n+1} <^* \mathscr{U}_n$ for each $n \in \mathbf{N}$. A cover $\mathscr{U}$ of a topological space X is said to be *normal* in case there exists a normal sequence $\mathscr{U}^{\mathbf{N}} = (\mathscr{U}_n)_{n \in \mathbf{N}}$ of open covers of X such that $\mathscr{U}_1 < \mathscr{U}$.

The word normal as applied here will not be confusing as compared to the concept of a normal topological space since the context of the discussion will assist in the clarification. As an example consider the real line $\mathbf{R}$ and let $S(x, \delta)$ be an open interval centered at x with width 2δ and let $\mathscr{U}_n = (S(x, 1/3^n))_{x \in \mathbf{R}}$ for each

n in N. Then $\mathscr{U}_{n+1} <^* \mathscr{U}_n$ and $\mathscr{U}_1$ refines the normal cover $\mathscr{U} = (S(x, 1))_{x \in \mathbf{R}}$.

The following results are very easy to establish. These results can be found in Tukey's work (see [XXVIII]). They give us the important working relationships among refinements, star-refinements, and Δ-refinements. As a corollary to these results, it is seen that a sequence of Δ-refining covers of a set gives rise in a natural way to a normal sequence of covers. It is also interesting to see that normal (respectively, locally finite) open covers of the image space under a continuous map give rise to normal (respectively, locally finite) open covers of the pre-image. As a corollary to this, since the injection mapping is always continuous, it is obvious that normal or locally finite open covers are hereditary. These results should be established by the reader.

1.15 PROPOSITION (*Tukey*) *Suppose that $\mathscr{U}$ and $\mathscr{V}$ are two covers of a set X.*

(1) *If $A \subset B \subset X$, then* $\operatorname{st}(A, \mathscr{U}) \subset \operatorname{st}(B, \mathscr{U})$.

(2) *If $\mathscr{U} < \mathscr{V}$ and if $A \subset X$, then* $\operatorname{st}(A, \mathscr{U}) \subset \operatorname{st}(A, \mathscr{V})$.

1.16 COROLLARY *Suppose that $\mathscr{U}$ and $\mathscr{V}$ are two covers of a set X. If $A \subset B \subset X$ and if $\mathscr{U} < \mathscr{V}$, then* $\operatorname{st}(A, \mathscr{U}) \subset \operatorname{st}(B, \mathscr{V})$.

1.17 PROPOSITION (*Tukey*) *If $\mathscr{U} = (U_\alpha)_{\alpha \in I}$ is a cover of a set X, then the following statements are equivalent.*

(1) *$x, y \in U_\alpha$ for some $\alpha \in I$.*

(2) $x \in \operatorname{st}(y, \mathscr{U})$.

(3) $y \in \operatorname{st}(x, \mathscr{U})$.

1.18 PROPOSITION (*Tukey*) *If $\mathscr{U}$ and $\mathscr{V}$ are two covers of a set X and if $\mathscr{U} < \mathscr{V}$, then $\mathscr{U}^* < \mathscr{V}^*$. If, in addition, X is non-empty, then $\mathscr{U}^\Delta < \mathscr{V}^\Delta$.*

1.19 PROPOSITION (*Tukey*) *If $\mathscr{U}$ is a cover of a non-empty set X, then $\mathscr{U}^\Delta < \mathscr{U}^* < (\mathscr{U}^\Delta)^\Delta$.*

1.20 COROLLARY *If $(\mathscr{U}_n)_{n \in \mathbf{N}}$ is a sequence of covers of a set X such that $\mathscr{U}_{n+1} <^\Delta \mathscr{U}_n$ for all $n \in \mathbf{N}$, then $(\mathscr{U}_{2n})_{n \in \mathbf{N}}$ is a normal sequence.*

1.21 PROPOSITION *Suppose that f is a continuous map from a topological space X into a topological space Y. If $(U_\alpha)_{\alpha \in I}$ is a normal (respectively, locally finite) open cover of Y, then $(f^{-1}(U_\alpha))_{\alpha \in I}$ is a normal (respectively, locally finite) open cover of X.*

1.22 COROLLARY *If S is a subset of a topological space X and if $\mathscr{U}$ is a normal (respectively, locally finite) open cover of X, then $\mathscr{U}|S$ is a normal (respectively, locally finite) open cover of S.*

Proof. Since the canonical injection i from S into X is continuous, the result follows from 1.21. □

Given a finite sequence of covers of a set, it is often necessary to construct new covers from this sequence. One such construction would be to take the cover formed by considering the intersections of all members of the various covers as the following definition requires.

1.23 DEFINITION Suppose the n-tuple $(\mathscr{A}_1, \ldots, \mathscr{A}_n)$ is a finite sequence of covers of a set X and that $\mathscr{A}_i = (A_i(\alpha))_{\alpha \in J_i}$ for each $i = 1, \ldots, n$. Then by the *intersection cover*, $\bigwedge_{i=1}^{n} \mathscr{A}_i$, is meant the family

$$(A_1(\alpha_1) \cap \ldots \cap A_n(\alpha_n))_{(\alpha_1, \ldots, \alpha_n) \in J_1 \times \ldots \times J_n}.$$

These intersection covers have some natural relationships. If two finite sequences of covers of the same length are related by either star-refinement or by refinement between the corresponding elements of the finite tuples, then the corresponding intersection covers have the same properties. Also, any sequence of covers immediately generates a sequence of intersection covers.

From a normal cover then, intersection covers will be constructed and it is these which will be of assistance in obtaining a pseudometric from the cover.

1.24 PROPOSITION *If $(\mathscr{A}_1, \ldots, \mathscr{A}_n)$ and $(\mathscr{B}_1, \ldots, \mathscr{B}_n)$ are two finite sequences of covers of a set X and if $\mathscr{A}_i <^* \mathscr{B}_i$ (respectively, $\mathscr{A}_i < \mathscr{B}_i$) for each $i = 1, \ldots, n$, then*

$$\bigwedge_{i=1}^{n} \mathscr{A}_i <^* \bigwedge_{i=1}^{n} \mathscr{B}_i \quad \left(\textit{respectively} \bigwedge_{i=1}^{n} \mathscr{A}_i < \bigwedge_{i=1}^{n} \mathscr{B}_i\right).$$

Proof. For each $i = 1, \ldots, n$, let $\mathscr{A}_i = (A_i(\alpha))_{\alpha \in I_i}$ and $\mathscr{B}_i = (B_i(\beta))_{\beta \in J_i}$, and suppose that $(\alpha_1, \ldots, \alpha_n) \in I_1 \times \ldots \times I_n$. Since $\mathscr{A}_i <^* \mathscr{B}_i$ (respectively $\mathscr{A}_i < \mathscr{B}_i$), there is $\beta_i \in J_i$ so that $\mathrm{st}\,(A_i(\alpha_i), \mathscr{A}_i) \subset B_i(\beta_i)$ (respectively, $A_i(\alpha_i) \subset B_i(\beta_i)$). Then $B_1(\beta_1) \cap \ldots \cap B_n(\beta_n)$ is a member of the intersection cover $\bigwedge_{i=1}^{n} \mathscr{B}_i$ and one easily verifies that

$$\mathrm{st}\left(A_1(\alpha_1) \cap \ldots \cap A_n(\alpha_n), \bigwedge_{i=1}^{n} \mathscr{A}_i\right) \subset B_1(\beta_1) \cap \ldots \cap B_n(\beta_n)$$

(respectively, $A_1(\alpha_1) \cap \ldots \cap A_n(\alpha_n) \subset B_1(\beta_1) \cap \ldots \cap B_n(\beta_n)$). □

1.25 **Proposition** *A sequence of covers $(\mathscr{A}_i)_{i \in \mathbf{N}}$ of a set X generates a nested sequence of intersection covers. That is for each $n \in \mathbf{N}$,*

$$\bigwedge_{j=1}^{n+1} \mathscr{A}_j < \bigwedge_{j=1}^{n} \mathscr{A}_j.$$

Proof. For each $i \in \mathbf{N}$, let $\mathscr{A}_i = (A_i(\alpha))_{\alpha \in J_i}$. The result now follows from the fact that if $(\alpha_1, \ldots, \alpha_{n+1}) \in J_1 \times \ldots \times J_{n+1}$, then $\bigcap_{j=1}^{n+1} A_j(\alpha_j) \subset \bigcap_{j=1}^{n} A_j(\alpha_j)$. □

If $(X_\alpha, \mathscr{T}_\alpha)$, $\alpha \in I$, is a family of topological spaces, then a suitable topology $\mathscr{T}$ on the product set $X = \prod_{\alpha \in I} X_\alpha$ is obtained by utilizing the following base $\mathscr{B}$ for the open sets. Elements of $\mathscr{B}$ are product sets of the form $\prod_{\alpha \in I} B_\alpha$ where

(1) B_α is an element of a base $\mathscr{B}_\alpha \subset \mathscr{T}_\alpha$ for each $\alpha \in I$,
(2) for all but finitely many coordinates, $B_\alpha = X_\alpha$.

This topology is called the *Tychonoff product topology* for the product set X.

Let us label the points x in X as $x = (x_\alpha)_{\alpha \in I}$ where $x_\alpha \in X_\alpha$. For each α in I, a map pr_α from X into X_α may be defined as $\mathrm{pr}_\alpha(x) = x_\alpha$. This map is called the *α-projection map* of X onto X_α. Then the Tychonoff product topology is precisely that topology that has as a subbase the elements of the form $\mathrm{pr}_\alpha^{-1}(B_\alpha)$ where $\alpha \in I$ and $B_\alpha \in \mathscr{B}_\alpha$, a base for $\mathscr{T}_\alpha$. Thus the Tychonoff product topology on the product set X is the smallest topology on X for which the projection maps pr_α, $\alpha \in I$, are continuous.

This construction motivates an interesting generalization. If X is a set, if $((X_\alpha, \mathscr{T}_\alpha))_{\alpha \in I}$ is a family of topological spaces, and if for each α in I, f_α is a function from X into X_α then we put on X the smallest topology $\mathscr{T}$ such that each map f will be continuous. The topology $\mathscr{T}$ is called the *topology induced by the family* $(f_\alpha)_{\alpha \in I}$ or the *projective limit topology*. Thus the Tychonoff product topology on the set $X = \prod_{\alpha = I} X_\alpha$ is the topology induced by the family $(\mathrm{pr}_\alpha)_{\alpha \in I}$.

1.26 Exercises

(1) Prove 1.4, 1.11, 1.15, 1.16, 1.17, 1.18, 1.19, 1.20, and 1.21.

(2) Construct an example to show that a refinement of a cover may have more elements than the given cover.

(3) Show that a topological space X is collectionwise normal (respectively, paracompact) if and only if X is γ-collectionwise normal (respectively, γ-paracompact) for all infinite cardinal numbers γ.

(4) Prove that every zero-dimensional T_0-space is Tychonoff. A topological space is said to be *totally disconnected* if for each point $x \in X$, the union of all the connected subsets of X that contain x is $\{x\}$ itself. What implications hold between totally disconnected spaces and zero-dimensional spaces?

(5) If Y is locally compact and if X is an open subset of Y then X is locally compact. If X is dense in Y, then every compact neighborhood in X of a point $x \in X$ is a neighborhood in Y of x. Consequently, every isolated point in X is also isolated in Y. Moreover, if X is also locally compact, then X is open in Y.

(6) Every locally compact Hausdorff space X has a one-point compactification X^*. (Let X^* be X with one point adjoined to it, let X be an open subspace of X^* and let complements of compact subsets of X form a base for the neighborhoods for the new point.)

(7) If $\mathscr{U}$ is a cover of a set X, prove that $\mathscr{U}^*$ is a cover of X. If, in addition, X is non-empty, prove that $\mathscr{U}^\Delta$ is a cover of X.

(8) Prove that a topological space is compact if and only if every collection of closed subsets of X with the finite intersection

property is non-empty. Similarly, show that X is countably compact if and only if every countable collection of closed subsets with the finite intersection property is non-empty.

(9) Prove that every paracompact Hausdorff space is a regular space and that every paracompact regular space is a normal space. In fact, every paracompact Hausdorff space is a collectionwise normal space.

For exercises (10) through (13) we introduce the following terminology.

Let X be a totally ordered set with partial order $<$. Then X has a naturally defined topology on it. Specifically, for a and b in X, $a < b$, let

$$(a, b) = \{x \in X \colon a < x < b\}.$$

Call (a, b) an *open interval*. The collection of all open intervals (a, b) for a and b in X together with X form a base for the open sets of the *intrinsic topology* on X (see [395]).

(10) With the above topology the totally ordered space X is a hereditarily normal space. In fact, X is hereditarily collectionwise normal.

(11) If $(X, <)$ is a well ordered set, the collection of all subsets of X having one of the following forms constitutes a base for the open subsets of X.

(*a*) $\{x\}$ for x an element of X with immediate predecessor,
(*b*) $(y, x] = \{z \in X \colon y < z \leqslant x$ where x and y are in X, $y < x$ and x is an element with no immediate predecessor$\}$.

Thus in a well ordered set $(X, <)$, every element x with an immediate predecessor is isolated in its intrinsic topology (that is, $\{x\}$ is both open and closed). Consequently for well ordered sets, the intrinsic topology is zero-dimensional (see [395]).

(12) A totally ordered set in its intrinsic topology is countably paracompact (see Ball [39]).

(13) A totally ordered set is compact in its intrinsic topology if and only if it has both a least and a greatest element and it cannot be written as the union of two non-empty sets A and B

such that each element of A is less than each element of B, where A has no greatest member and B has no least member.

For exercises (14) through (18) we need the following terminology.

A *linear topological space* is a (real or complex) linear space L (vector space) with a topology such that addition and scalar multiplication are each continuous simultaneously in both variables, that is
(*a*) the map $+$ of $L \times L$ (with the product topology) into L which is given by $+(x, y) = x + y$ and
(*b*) the map $\cdot$ of $K \times L$ (with the product topology where K is the field of scalars) into L given by $\cdot(k, x) = k \cdot x$.
The topology of a linear topological space is called a *vector topology*.

(14) A *line segment* joining two points x and y of a linear space L is the set of all points of the form $ax + by$ with a and b non-negative real numbers such that $a + b = 1$. A subset A of L is *convex* if, whenever x and y are contained in L, also the line segment joining x and y is contained in L.
(*a*) Any subspace of a linear space is convex.
(*b*) If A is a subset of L, then the *translate* $x + A = \{x + a : a \in A\}$ for each $x \in L$ is convex whenever A is convex.
(*c*) Any finite linear combination of convex sets is convex.
(*d*) The intersection of the members of any collection of convex sets is convex.
(*e*) The union of the members of any collection of convex sets, *directed* by $\supset$ (that is, the union of any two members of the collection is contained in some third member) is convex.
(*f*) Every subset A of the linear space L is contained in a smallest convex set. This set is called the *convex hull*, *convex envelope*, or *convex extension* of A. The convex hull of A consists of all finite linear combinations $\Sigma\{a_i x_i : i = 1, \ldots, n\}$ where $x_i \in A$, $a_i \in \mathbf{R}$, $a_i \geqslant 0$, and $\sum_{i=1}^{n} a_i = 1$.

(15) A subset A of the linear space L is *circled* if $aA \subset A$ whenever $|a| \leqslant 1$.
(*a*) If the subset A is circled then $aA = A$ whenever $|a| = 1$.
(*b*) A circled subset A is *symmetric*, that is $A = -A$.
(*c*) A subset A is circled and convex if and only if A contains

$ax+by$ whenever x and y are in A and $|a|+|b| \leqslant 1$. If L is a real linear space then A is circled and convex if and only if A is convex and symmetric.

(*d*) The smallest circled set containing a subset A, called the *circled extension* of A, is the set $\cup\{aA: |a| \leqslant 1\}$. The smallest linear subspace containing a non-void convex circled set A is the set $\cup\{nA: n = 1, 2, \ldots\}$.

(*e*) The convex hull of a circled set is circled.

(*f*) The smallest convex circled set containing a set A is the convex hull of the circled extension of A, that is, precisely the set of linear combinations

$$\Sigma\{a_i x_i: i = 1, \ldots, n\},$$

where $x_i \in A$ and $\Sigma\{|a_i|: i = 1, \ldots, n\} \leqslant 1$.

(16) A set U is a neighborhood of a point $x \in L$ if and only if $-x+U$ is a neighborhood of $0 \in L$. Consequently a base for the neighborhood system of 0 is called a *local base* or a *local base for the topology* or a *system of nuclei*. Thus the topology on L is completely determined by any local base $\mathscr{U}$.

(17) As a result of (15) we can describe those collections of subsets of the linear space L that are local bases for some vector topology. In particular if $\mathscr{U}$ is a local base then the following statements hold.

(*a*) For each U and V in $\mathscr{U}$ there is a W in $\mathscr{U}$ such that $W \subset U \cap V$.

(*b*) For each U in $\mathscr{U}$ there is a member V in $\mathscr{U}$ such that $V+V \subset U$,

(*c*) For each U in $\mathscr{U}$ there is a member V in $\mathscr{U}$ such that $aV \subset U$ for each scalar a, $|a| \leqslant 1$, that is the circled extension of V is contained in U.

(*d*) For each x in L and U in $\mathscr{U}$ there is a scalar a such that $x \in aU$.

(*e*) For each U in $\mathscr{U}$ there is a member V of $\mathscr{U}$ and a circled set W such that $V \subset W \subset U$.

If in addition L is Hausdorff then $\cap\{U: U \in \mathscr{U}\} = \{0\}$. On the other hand, if $\mathscr{U}$ is a non-empty family of subsets of L that satisfy (*a*) through (*d*) and if $\mathscr{T}$ is the family of sets G in L such that for each $x \in G$ there is a $U \in \mathscr{U}$ with $x+U \subset G$, then $\mathscr{T}$ is a vector topology on L and $\mathscr{U}$ is a local base for this topology. If (*e*) also holds then $\mathscr{T}$ is a Hausdorff topology.

(18) Computations in linear topological spaces.

Let L be a linear topological space and let $\mathscr{U}$ be the family of neighborhoods of 0. Then the following statements hold.

(*a*) If A is a subset of L, if $x \in L$ and if a is a non-zero scalar then the closure of $x+A$ is $x+\operatorname{cl} A$ and the closure of aA is $a(\operatorname{cl} A)$.

(*b*) If A and B are subsets of L then $\operatorname{cl} A + \operatorname{cl} B \subset \operatorname{cl}(A+B)$.

(*c*) If U is an open subset of L then $A+U$ is open and

$$A + \operatorname{int} B \subset \operatorname{int}(A+B).$$

(*d*) If C and D are compact subsets, then $C+D$ is compact.

(*e*) The closure of the set A is $\bigcap \{A+U: U \in \mathscr{U}\}$.

(*f*) If C is compact and if U is open and contains C then there is a set $V \in \mathscr{U}$ such that $C+V \subset U$.

(*g*) If C is compact and if F is closed then $C+F$ is closed.

(*h*) The closure of a subspace is a subspace.

(*i*) The closure of a circled set is circled.

(*j*) If 0 is contained in the interior of a circled set A then $\operatorname{int} A$ is also circled.

(*k*) The collection of all closed circled neighborhoods of 0 is a local base (see 17(*c*)).

(*l*) Each convex neighborhood of 0 contains the closure of a convex circled neighborhood of 0.

(*m*) The closure of a convex set is convex.

2 Continuous functions and pseudometrics

Throughout this text the emphasis, as mentioned in the introduction will be placed on continuous functions, continuous pseudometrics and their interactions. One of the important considerations will be to discuss a form of embedding for a subspace which is stronger than C-embedding, where in the latter case every continuous real-valued function on the subspace must extend to the entire space. The notion of C-embedding has found its importance in the theory of realcompactifications which will also be discussed shortly. To begin, some of the terminology and elementary ideas associated with C-embedding and its closely related C^*-embedding are presented here.

2.1 Definition Suppose that S is a subset of a topological space X. Let $C(X)$ be the collection of all real-valued continuous functions on X. Let $C^*(X)$ be the subset of $C(X)$ consisting of all bounded functions in $C(X)$. It is said that S is *C-embedded* (respectively, *C^*-embedded*) *in* X if every function in $C(S)$ (respectively, $C^*(S)$) can be extended to a function in $C(X)$ (respectively, $C^*(X)$). That subcollection of $C^*(X)$ consisting of all functions f with the range of f contained in E is denoted by $C(X, E)$.

If X is a topological space and if $f \in C(X)$, set $Z(f)$ (or, when there is possibility of confusion, $Z_X(f)$) equal to the set $\{x \in X : f(x) = 0\}$. The set $Z(f)$ is called the *zero-set of* f. The complement of $Z(f)$ is called the *cozero-set of* f. If $S \subset X$, then S is a *zero-set* (respectively, *cozero-set*) if $S = Z(f)$ (respectively, $S = X - Z(f)$) for some $f \in C(X)$. Let $Z(X)$ denote the collection of all zero-sets in X.

Now let X be a topological space and let $(U_\alpha)_{\alpha \in I}$ be a family of subsets of X. The family $(U_\alpha)_{\alpha \in I}$ is a *cozero-set* (respectively, *zero-set*) *cover of* X if $(U_\alpha)_{\alpha \in I}$ is a cover of X and if U_α is a cozero-set (respectively, zero-set) for each $\alpha \in I$. By a *partition of unity* on a topological space X we mean a family $\Phi = (f_\alpha)_{\alpha \in I}$ of real-valued continuous functions on X such that $\sum_{\alpha \in I} f_\alpha(x) = 1$ for each $x \in X$.

Let $\mathbf{R}^X$ denote the collection of all real-valued functions on X. Then $\mathbf{R}^X$ is a commutative ring with unity element relative to the operations of pointwise addition and multiplication, and it is a lattice relative to the pointwise order $\leqslant$. Let us denote by $\mathbf{r}$ the real-valued function that takes the point x in X to the real number $r(x) = r$. Then the zero element of $\mathbf{R}^X$ is the constant function $\mathbf{0}$, the unity element is the constant function $\mathbf{1}$, the additive inverse $-f$ of f is given by the formula $(-f)(x) = -f(x)$ for all $x \in X$, and the least upper bound $f \vee g$ and the greatest lower bound $f \wedge g$ of f and g are given by the rules

$$(f \vee g)(x) = \max\{f(x), g(x)\} \quad \text{and} \quad (f \wedge g)(x) = \min\{f(x), g(x)\}$$

for all $x \in X$. The collection $C(X)$ is a subring and a sublattice of $\mathbf{R}^X$ and $C^*(X)$ is a subring and a sublattice of $C(X)$.

Now suppose that $\Phi = (f_\alpha)_{\alpha \in I}$ is a partition of unity on a topological space X and for each $\alpha \in I$ let $P_\alpha = X - Z(f_\alpha)$. We say that Φ is *point-finite* (respectively, *locally finite*) in case $(P_\alpha)_{\alpha \in I}$ is point-finite (respectively, locally finite). We say that Φ is *subordinate* to a given cover $\mathscr{U}$ in case $(P_\alpha)_{\alpha \in I}$ is a refinement of $\mathscr{U}$.

In Section 9 we will give an example of a C^*-embedded subset that is not C-embedded. The next result gives a sufficient condition for a subset S to be C-embedded or C^*-embedded in a topological space X.

2.2 PROPOSITION *Let S be a subset of a topological space X. If each non-negative function in $C(S)$ (respectively, $C^*(S)$) can be extended to a function in $C(X)$ then S is C-embedded in X (respectively, C^*-embedded in X).*

Proof. Let $f \in C(S)$. By hypothesis, $f \vee \mathbf{0}$ and $-(f \wedge \mathbf{0})$ have extensions g and h respectively, in $C(X)$. Then $g-h$ is a continuous extension of $(f \vee \mathbf{0}) - (-(f \wedge \mathbf{0})) = f$. The case for C^* is proved in a similar fashion. □

Ranges of continuous real-valued functions defined on compact subsets of a space are always bounded subsets of $\mathbf{R}$. However not all topological spaces X for which every $f \in C(X)$ is bounded, are compact. Consequently we look at the following definition.

2.3 DEFINITION A topological space X is *pseudocompact* if $C(X) = C^*(X)$.

Thus every compact topological space is pseudocompact. However a much stronger result is

2.4 PROPOSITION *If X is a countably compact space then X is pseudocompact.*

Proof. Suppose that $f \in C(X)$ and for each $n \in \mathbf{N}$, let $U_n = \{x \in X \colon |f(x)| < n\}$. Then $(U_n)_{n \in \mathbf{N}}$ is a countable open cover of X and hence has a finite subcover. But this implies that f is bounded and therefore X is pseudocompact. □

In general a pseudocompact space need not be countably compact (see Section 9). However, if this is a normal T_1-space, then the converse does hold as stated in the exercises.

Dowker [100] has observed that if X is a collectionwise normal space and if $(F_\alpha)_{\alpha \in I}$ is a discrete family of closed subsets of X, then there exists a locally finite family $(G_\alpha)_{\alpha \in I}$ of mutually disjoint open sets such that $F_\alpha \subset G_\alpha$ for every $\alpha \in I$.

Now Dowker's result can be obtained as a corollary in 2.7 after having proved a more general result in 2.6. First let us analyze the concept of a discrete family in terms of the points of X.

2.5 PROPOSITION *If $(A_\alpha)_{\alpha \in I}$ is a family of subsets of X, then the set $\{x \in X: (A_\alpha)_{\alpha \in I}$ is discrete at $x\}$ is open in X.*

Proof. If x is such that $(A_\alpha)_{\alpha \in I}$ is discrete at x, then there exists a neighborhood G of x such that $G \cap A_\alpha \neq \varnothing$ for at most one $\alpha \in I$. Moreover, if $y \in G$ then $(A_\alpha)_{\alpha \in I}$ is discrete at y and it follows that $G \subset \{x \in X: (A_\alpha)_{\alpha \in I}$ is discrete at $x\}$. □

In obtaining a discrete collection $(G_\alpha)_{\alpha \in I}$ of open subsets from a pairwise disjoint family $(H_\alpha)_{\alpha \in I}$ of open subsets which will cover the closed subsets $(B_\alpha)_{\alpha \in I}$ and such that $B_\alpha \subset G_\alpha$ for all $\alpha \in I$, the next proposition demonstrates the use of a weakened form of the concept of closure-preserving.

2.6 PROPOSITION *Let X be a normal space and suppose that $(B_\alpha)_{\alpha \in I}$ is a family of subsets of X such that $\bigcup_{\alpha \in I} B_\alpha$ is closed in X. If there exists a pairwise disjoint family $(H_\alpha)_{\alpha \in I}$ of open subsets of X such that $B_\alpha \subset H_\alpha$ for every $\alpha \in I$, then there exists a discrete family $(G_\alpha)_{\alpha \in I}$ of open subsets of X such that $B_\alpha \subset G_\alpha$ for every $\alpha \in I$.*

Proof. Let $U = \{x \in X: (H_\alpha)_{\alpha \in I}$ is discrete at $x\}$. Then U is open by 2.5. If $x \in \bigcup_{\alpha \in I} B_\alpha$, then $x \in B_\beta \subset H_\beta$ for some $\beta \in I$. Set $K = \{\beta\}$ and note that H_β is a neighborhood of x such that $H_\beta \cap H_\alpha = \varnothing$ for all $\alpha \notin K$. Thus $x \in U$ and $\bigcup_{\alpha \in I} B_\alpha \subset U$. Since X is normal and $\bigcup_{\alpha \in I} B_\alpha$ is closed, there exists an open set W such that $\bigcup_{\alpha \in I} B_\alpha \subset W \subset \operatorname{cl} W \subset U$. For each $\alpha \in I$, set $G_\alpha = H_\alpha \cap W$ and note that $B_\alpha \subset G_\alpha$. Consider any $x \in X$. If $x \in U$, then $(H_\alpha)_{\alpha \in I}$ is discrete at x and therefore $(G_\alpha)_{\alpha \in I}$ is discrete at x. And if $x \notin U$, then $X - (\operatorname{cl} W)$ is a neighborhood of x such that

$$X - \operatorname{cl} W \cap G_\alpha = \varnothing$$

for all $\alpha \in I$, so again $(G_\alpha)_{\alpha \in I}$ is discrete at x. Thus $(G_\alpha)_{\alpha \in I}$ is discrete and the proof is complete. □

2.7 COROLLARY *If $(F_\alpha)_{\alpha \in I}$ is a discrete family of closed subsets of the collectionwise normal space X, then there exists a discrete family $(G_\alpha)_{\alpha \in I}$ of open subsets of X such that $F_\alpha \subset G_\alpha$ for every $\alpha \in I$.*

We know that a topological space is a completely regular space if and only if the collection of all cozero-sets form a base for the open subsets. Consequently we will direct many questions toward the cozero-sets. What operations, then, on cozero-sets will again give us a cozero-set?

2.8 PROPOSITION *If X is a topological space and if $(G_n)_{n \in \mathbf{N}}$ is a countable family of cozero-sets then $G = \bigcup_{n \in \mathbf{N}} G_n$ is a cozero-set of X.*

Proof. For each $n \in \mathbf{N}$ let $f_n \in C(X)$ such that $X - G_n = Z(f_n)$. Then $f = \sum_{n \in \mathbf{N}} |f_n \wedge \mathbf{1}|/2^n$ is in $C(X)$ because the series converges uniformly and one easily shows that $X - G = Z(f)$. □

Note that the above also shows that a countable intersection of zero-sets is a zero-set.

2.9 PROPOSITION *Suppose that X is a topological space and that $(U_\alpha)_{\alpha \in I}$ is a locally finite family of cozero-sets of X. Then $U = \bigcup_{\alpha \in I} U_\alpha$ is a cozero-set of X.*

Proof. For each $\alpha \in I$ there exists an $f_\alpha \in C(X, E)$ such that $U_\alpha = X - Z(f_\alpha)$. Let $f = \sum_{\alpha \in I} f_\alpha$. Then one may easily verify that $f \in C(X)$ and $U = X - Z(f)$. □

Let us introduce another type of real-valued function that will be useful in later work (see Exercise 6).

2.10 DEFINITION Let f be a real-valued function on the topological space X. The function is *upper semicontinuous* if for each real number r in $\mathbf{R}$, $\{x \in X : f(x) < r\}$ is open in X. The function f is called *lower semicontinuous* if for each $r \in \mathbf{R}$,

$$\{x \in X : f(x) > r\}$$

is open in X.

Now we can investigate the relationships that exist between cozero-sets, continuous functions and pseudometrics. Intertwined in this is the role of an open cover and the concepts of normality and paracompactness.

2.11 DEFINITION *A pseudometric* on a non-empty set X is a function from $X \times X$ into $\mathbf{R}^+$ such that, for all $x, y, z \in X$,

(1) $d(x,y) = d(y,x)$;

(2) $x = y$ implies $d(x,y) = 0$;

(3) $d(x,y) \leqslant d(x,z) + d(z,y)$.

If '$d(x,y) = 0$ implies $x = y$' also holds, then d is called a *metric*.

If X is a topological space and if d is a pseudometric on X, then d is *continuous* in case d is continous relative to the product topology on $X \times X$.

Now suppose that X is a topological space and that d is a pseudometric on X. If $x \in X$ and $\epsilon > 0$, set $S(x; d, \epsilon)$ equal to $\{y \in X \colon d(x,y) < \epsilon\}$ and call $S(x; d, \epsilon)$ the *open ball with center x and radius* ϵ. When confusion is unlikely, $S(x; d, \epsilon)$ will be written instead as $S(x; \epsilon)$. Let $\mathscr{B}_d = \{S(x; d, \epsilon) \colon x \in X \text{ and } \epsilon > 0\}$. Then $\mathscr{B}_d$ is a base for a unique topology on X. This topology is denoted by $\mathscr{T}_d$ and $\mathscr{T}_d$ is called the *topology on X determined by d.*

It is now possible to state and prove several basic results about pseudometrics. First it is possible to give an obvious characterization of continuity for a pseudometric.

2.12 PROPOSITION *Suppose that* $(X, \mathscr{T})$ *is a topological space and that d is a pseudometric on X. Then d is continuous if and only if* $\mathscr{T}_d \subset \mathscr{T}$.

Proof. Suppose first that $d \colon X \times X \to \mathbf{R}^+$ is continuous. Let $x \in X$ and let $\epsilon > 0$. It must be shown that $S(x; d, \epsilon) = \{y \in X \colon d(x,y) < \epsilon\}$ is in $\mathscr{T}$. Define the map f_x from X into $X \times X$ by $f_x(y) = (x,y)$ for all y in X. One easily verifies that f_x is continuous. Now $[0,\epsilon)$ is open in $\mathbf{R}^+$, so $(d \circ f_x)^{-1}([0,\epsilon))$ is open in X. But

$$(d \circ f_x)^{-1}([0,\epsilon)) = S(x; d, \epsilon),$$

whence $S(x; d, \epsilon)$ is open in X. Therefore $\mathscr{T}_d \subset \mathscr{T}$.

Conversely, since d is continuous relative to $\mathscr{T}_d$ if $\mathscr{T}_d \subset \mathscr{T}$, then, d is continuous relative to $\mathscr{T}$. The proof is now complete. □

Let A and B be non-empty subsets of a pseudometric space (X, d). Let $d(x, A)$ be the $\inf\{d(x, a): a \in A\}$ and let $d(A, B)$ be the $\inf\{d(a, b): a \in A \text{ and } b \in B\}$. If $A = \varnothing$ then $d(A, B) = \infty$.

If S is a non-empty subset of a topological space X, and if d is a continuous pseudometric on X then define the function d^S from X into $\mathbf{R}^+$ by

$$d^S(x) = d(x, S) \quad (x \in X).$$

The definition shows us how to construct new functions from the pseudometric. In fact, if the pseudometric is continuous the next proposition shows that these new functions are also continuous. The proof is left as an exercise.

2.13 PROPOSITION *Suppose that S is a non-empty subset of a topological space X, and that d is a continuous pseudometric on X. Then the function d^S is continuous.*

If d is a continuous pseudometric on a space X, then the topology $\mathscr{T}_d$ generated by d is a subcollection of the original topology on the space. It makes sense then to speak of the d-closure of any subset S, written d-cl S, of the space. This proposition then says that

$$d\text{-cl}\, S = \{x \in X: d(x, S) = 0\}$$

is a zero-set of X. The reader should verify this statement as well as the following fact. *If (X, d) is a pseudometric space and if S is a non-empty subset of X, then $x \in \operatorname{cl} S$ if and only if $d(x, S) = 0$.*

The following results follow immediately.

2.14 PROPOSITION *Suppose that X and Y are topological spaces, that f is a continuous map from X into Y, and that d is a continuous pseudometric on Y. Then*

$$r = d \circ (f \times f)$$

is a continuous pseudometric on X.

2.15 PROPOSITION *If G is an open subset of the pseudometric space (X,d), then G is a cozero-set.*

2.16 COROLLARY *Let d be a continuous pseudometric on the topological space $(X,\mathscr{T})$. If G is an open subset of X relative to $\mathscr{T}_d$ then G is a cozero-set of X relative to $\mathscr{T}$.*

In the previous section the generalization of compactness to paracompactness was stated as pertinent to the problem of metrizability. In fact a topological space X is pseudometrizable if and only if it satisfies either one of the following equivalent conditions.

(1) The space X is a regular space and has a σ-locally finite base of open sets.

(2) The space X is a regular space and has a σ-discrete base of open sets.

Clearly statement (2) *implies* that of (1). Exercise 9 assists in showing that statement (1) *implies* that X is pseudometrizable. On the other hand, to show that every pseudometric space (X,d) satisfies statement (2), a stronger result may be shown which also yields the paracompactness of X. We show that every open cover of a pseudometric space has a σ-discrete refinement. Consequently, for each $n \in \mathrm{N}$, $\mathscr{B}_n = \{S(x; d, 1/n) : x \in X\}$ has a σ-discrete cozero-set refinement $\mathscr{R}_n$ and $\mathscr{R} = \bigcup_{n \in \mathbf{N}} \mathscr{R}_n$ is a σ-discrete cozero-set base.

2.17 THEOREM *If (X,d) is a pseudometric space, then every open cover of X has a σ-discrete cozero-set refinement.*

Proof. Suppose that $\mathscr{U} = (U_\alpha)_{\alpha \in I}$ is an open cover of (X,d). For each $n \in \mathrm{N}$ and $\alpha \in I$, let

$$U_n^\alpha = \{x \in X : x \in U_\alpha \quad \text{and} \quad d(x, X - U_\alpha) > 1/2^n\}.$$

From the triangle inequality it is clear that

$$d(U_n^\alpha, X - U_{n+1}^\alpha) > 1/2^{n+1}.$$

Let $<$ be a well ordering of I (see 1.6) and for each $n \in \mathrm{N}$ and $\alpha \in I$ let $V_n^\alpha = U_n^\alpha - \cup \{U_{n+1}^\beta : \beta < \alpha\}$. If $\beta < \alpha$ then $V_n^\alpha \subset X - U_{n+1}^\alpha$ while if $\alpha < \beta$ then $V_n^\alpha \subset X \backslash U_{n+1}^\beta$.

In either case, $d(V_n^\alpha, V_n^\beta) \geqslant 1/2^{n+1}$. Now set

$$W_n^\alpha = \{x \in X : d(x, V_n^\alpha) < 1/2^{n+3}\}.$$

Note that $d(W_n^\alpha, W_n^\beta) \geqslant 1/2^{n+1}$ if $\alpha \neq \beta$ and hence for each $n \in \mathbf{N}$, $\mathscr{W}_n = (W_n^\alpha)_{\alpha \in I}$ is discrete. If $x \in X$ then there exists a first $\alpha \in I$ such that $x \in U_\alpha$. But then $x \in W_n^\alpha$ for some $n \in \mathbf{N}$. Clearly $W_n^\alpha \subset U_\alpha$. Therefore $\mathscr{W} = \bigcup_{n \in \mathbf{N}} \mathscr{W}_n$ is a σ-discrete open refinement of $\mathscr{U}$. By 2.15 it follows that each element in $\mathscr{W}$ is a cozero-set. The proof is now complete. □

2.18 Corollary *Every pseudometric space is paracompact.*

The proof of this corollary depends on first showing Exercise 10.

2.19 Exercises

(1) Prove 2.14, 2.15, and 2.16.

(2) Suppose that X is a topological space, that $f \in C(X)$ and that $r \in \mathbf{R}$. Then the sets $\{x \in X : f(x) < r\}$ and $\{x \in X : f(x) > r\}$ are cozero-sets in X, whereas the sets

$$\{x \in X : f(x) \leqslant r\} \quad \text{and} \quad \{x \in X : f(x) \geqslant r\}$$

are zero-sets. Every zero-set in X is a G_δ-set in X and therefore every cozero-set is an F_σ-set.

(3) A topological space X is collectionwise normal if and only if for every discrete family of closed subsets $(F_\alpha)_{\alpha \in I}$ of X there is a discrete family $(G_\alpha)_{\alpha \in I}$ of cozero-sets of X such that $F_\alpha \subset G_\alpha$ for every $\alpha \in I$.

(4) Prove that a pseudocompact normal T_1-space is countably compact.

(5) If X is a completely regular space then X is pseudocompact if and only if every locally finite open cover of X has a finite subcover (see [203]). (If $\mathscr{U}$ is a locally finite open cover of X with no finite subcover then there exists a countable subfamily $(U_n)_{n \in \mathbf{N}}$ of $\mathscr{U}$ such that every finite subfamily does not cover X. Choose $a_n \in U_n$ and $f_n \in C(X)$ such that $f_n(a_n) = n$ and $f_n(x) = 0$ if $x \in X - U_n$. Then $f = \sum_{n \in \mathbf{N}} f_n$ is an unbounded continuous function

on X. Conversely, if $f \in C(X)$, $f \geqslant \mathbf{0}$, consider the sets

$$O_1 = \{x \in X: f(x) < 2\}$$

and

$$O_n = \{x \in X: n-1 < f(x) < n+1\} \quad \text{for} \quad (n = 2, 3, \ldots).)$$

(6) Suppose that X is a topological space and that f is a real-valued function.
(*a*) Prove that f is continuous if and only if f is both upper and lower semicontinuous.
(*b*) Prove that f is lower semicontinuous (respectively, upper semicontinuous) if and only if for all $r \in \mathbf{R}$, $\{x \in X: f(x) \leqslant r\}$ (respectively, $\{x \in X: f(x) \geqslant r\}$) is closed in X. What may be said about characteristic functions?

(7) Let F be a zero-set on the topological space X. Then there is a sequence $(A_n)_{n \in \mathbf{N}}$ of cozero-sets and a sequence $(B_n)_{n \in \mathbf{N}}$ of zero-sets such that

$$X - F = \cup\{A_n: n \in \mathbf{N}\} = \cup\{B_n: n \in \mathbf{N}\}$$

and $A_n \subset B_n \subset A_{n+1}$ for each $n \in \mathbf{N}$.

(8) Every compact subset of a Tychonoff space X or of a locally compact Hausdorff space X is C-embedded in X.

(9) If X is a regular topological space with a σ-locally finite base then X is a normal space. Moreover normal T_1-spaces with σ-locally finite bases are metrizable. Thus every regular Lindelöf space is normal.

(10) If every open cover of a space has an open σ-locally finite refinement, then every open cover has a locally finite refinement. (Can 'locally finite' be replaced by 'discrete' in either instance?) Thus every pseudometric space and every regular Lindelöf space are paracompact.

(11) Let X be a topological space and let W be a subset of $X \times X$. For $x \in X$, let $W(x) = \{y \in X: (x, y) \in W\}$. A cover $\mathscr{U}$ of X is said to be a *Lebesgue cover* if there is a neighborhood W of the diagonal of $X \times X$ such that the family $\mathscr{W} = \{W(x): x \in X\}$ refines $\mathscr{U}$.
(*a*) (Lebesgue's lemma) Every open cover of a compact pseudometric space is a Lebesgue cover. Consequently if W is a neighborhood of a compact subset K of the pseudometric space (X, d) then $d(K, X - \mathrm{W}) > 0$.

(*b*) More generally, if an open cover of a space has a closed locally finite refinement, then it is a Lebesgue cover. Thus each open cover of a compact regular space is a Lebesgue cover.
(*c*) If every open cover of a topological space is a Lebesgue cover, then every open cover of X has an open σ-discrete refinement. Consequently a regular space is paracompact if and only if every open cover is a Lebesgue cover.

With these results, we see how the concept of a Lebesgue cover is related to the concepts of compactness and paracompactness. Intuitively speaking, in a closed interval of the real line, the results say that open covers of the interval cover the interval 'uniformly'. This points to yet another aspect of the pseudometric which is definitely non-topological. We will say more about this in Section 13.

(12) A non-negative function p on a linear space L is a *pseudonorm* if $p(x+y) \leqslant p(x)+p(y)$ and if $p(ax) = |a|\,p(x)$ for x and y in L and for all scalars a. (The first condition says that p is *subadditive* whereas the second says that p is *absolutely homogeneous*.)
(*a*) If p is a pseudonorm on the linear space L then the set

$$N_p = \{x \in L \colon p(x) = 0\}$$

is a linear subspace. If $N_p = \{0\}$ then p is called a *norm*, that is a pseudonorm p is a norm if and only if $p(x) = 0$ implies x is the zero vector in L.
(*b*) If p is a pseudonorm for L then the function d_p defined by $d_p(x, y) = p(x-y)$ is a pseudometric for L and L with the pseudometric topology $\mathscr{T}_{d_p}$ is a linear topological space (not necessarily Hausdorff). In this case $\mathscr{T}_{d_p}$ is called the *pseudonorm topology* and d_p is called the *pseudometric defined by* p. The linear space L is said to be *pseudonormable* if and only if there is a pseudonorm whose topology is that of the space. If p is a norm for L then the function d defined above is a metric for L and then L with the topology $\mathscr{T}_d$ is a Hausdorff linear space. The linear space L is said to be *normable* if and only if there is a norm p whose topology $\mathscr{T}_{d_p}$ is that of the space. We will see later in Exercise 15 that these pseudometrics may be used to determine the original topology of L.

(13) A subset A of a linear space L is called *radial at a point* x in L if for each $y \neq x$ in L there is a $z \neq x$ in L such that the line segment joining x and z is contained in the intersection of A with the line segment joining x and y (geometrically speaking, A must contain a line segment through x in each direction). The *radial kernel* of a set A is the set of all points at which A is radial.

(*a*) In $\mathbf{R}^2$ there is a set that is radial at only one point.

(*b*) If U is a set that is radial at 0 then the *Minkowski functional* for U is defined to be the real-valued function p_U defined on L by

$$p_U(x) = \inf\{a\colon (1/a)\,x \in U \text{ and } a > 0\}.$$

If $a \geqslant 0$ then $p_U(ax) = a \cdot p_U(x)$, that is p_U is *non-negatively homogeneous*. If U is convex then p_U is subadditive. If U is circled then p_U is absolutely homogeneous. Thus if U is a convex circled set that is radial at 0 then the Minkowski functional p_U for U is a pseudonorm.

(*c*) Minkowski functionals and convex sets are very closely related, for certain convex sets can be described by means of these real-valued functions. In particular if U is a convex set that is radial at 0, $\{x\colon p_U(x) < 1\}$ is the radial kernel of U and

$$U \subset \{x\colon p_U(x) \leqslant 1\}.$$

Furthermore, if p is a non-negative, non-negatively homogeneous subadditive function on E and if $U = \{x\colon p(x) \leqslant 1\}$ then U is a convex set that is radial at 0 and p is the Minkowski functional for U. The set U is circled if and only if p is a pseudonorm.

(14) A subset A of L is *bounded* if for each neighborhood U of 0 there is a real number a such that $A \subset aU$. Thus if a linear space is pseudonormable then there is a convex bounded neighborhood of 0. On the other hand if a linear topological space L has a bounded convex neighborhood of 0 then there is a bounded circled convex neighborhood U (see 1.26(18*l*)) of 0 and the Minkowski functional p_U for U is a pseudonorm whose topology is that of the space. Thus a linear space is pseudonormable if and only if there is a bounded convex neighborhood of 0.

(15) A linear topological space L is called *locally convex* if the collection of convex neighborhoods of 0 is a local base. In such a case the topology of L is called a *locally convex topology*. Each

subspace of a locally convex space is locally convex. (The class of locally convex spaces has many important properties. Much of the theory of linear topological spaces is concerned with locally convex spaces. In fact, we will see below that pseudonorms (and therefore pseudometrics) play an important role in such spaces.)

If L is a locally convex linear topological space then there is a local base $\mathscr{U}$ such that

(*a*) the sets in $\mathscr{U}$ are convex circled sets each of which is radial at 0,
(*b*) the intersection of two sets in $\mathscr{U}$ contains a set in $\mathscr{U}$,
(*c*) for each U in $\mathscr{U}$ and each scalar $a \neq 0$, $aU \in \mathscr{U}$, and
(*d*) each set in $\mathscr{U}$ is closed.

Conversely, if L is a linear space, then any non-empty family of subsets that satisfies (*a*), (*b*), and (*c*) is a local base for a unique locally convex topology for L.

(16) Let $\mathscr{T}$ be a locally convex topology for a linear space L and let $\mathscr{P}$ be the collection of all continuous pseudonorms on L.
(*a*) The collection of p-unit spheres about 0 for $p \in \mathscr{P}$ is a local base for $\mathscr{T}$, that is, a set U is a $\mathscr{T}$-neighborhood of 0 if and only if there is a $p \in \mathscr{P}$ and a real number $r > 0$ such that

$$S(0; d, r) = \{x: p(x) < r\} \subset U,$$

where d is the pseudometric defined in terms of p.
(*b*) A linear map T of a linear topological space F into L is continuous if and only if the composition $p \circ T$ is continuous for each p in $\mathscr{P}$.
(*c*) A subset of L is bounded if and only if it is of finite p-diameter for each p in $\mathscr{P}$.

(17) Let d be a pseudometric on a space $(X, \mathscr{T})$. A family of subsets $(A_\alpha)_{\alpha \in I}$ of X is said to be a *d-discrete family of gauge δ*, for δ a fixed positive real number, if $d(A_\alpha, A_\beta) \geqslant \delta$ for $\alpha \neq \beta$. If the A_α are single points x_α, $\alpha \in I$, satisfying this definition, then $(x_\alpha)_{\alpha \in I}$ is called a *d-discrete set of points* (*of gauge δ*).
(*a*) If d is a continuous pseudometric on X then every d-closed subset of X is a zero-set in $(X, \mathscr{T})$.
(*b*) If d is a continuous pseudometric on X then the union of a d-discrete family of closed sets is closed. Moreover if each set is d-closed then their union is d-closed. Hence the union is a zero set. Also any d-discrete set in X is closed.

(c) If d is a continuous pseudometric on X then any d-discrete set S in X is C-embedded in X. (If $g \in C(S)$ consider

$$f(x) = g(s)(1 - 3d(x,s)/\delta)$$

for $d(x,A) \leqslant \frac{1}{3}\delta$, $s \in S$, δ the gauge, and $f(x) = 0$ otherwise.)

3 Uniform spaces

Any discussion of uniform spaces should be prefaced by some historical remarks. Two fundamental concepts grew out of the classic works of Fréchet (see [126]), Hausdorff (see [XIV]) and the early volumes of *Fundamenta Mathematicae*. These are the notions of a topological space and of a uniform space as abstractions of a metric space. However, several important properties of metric spaces (like the property of a sequence being Cauchy or of a space being complete) are not invariant under topological homeomorphism. These properties however are closely related to topological properties since it is often the case that topological properties can be deduced from or can imply these non-invariant or 'quasi-topological' properties.

Uniform spaces which were first introduced by André Weil (see [XXIX]) in 1937 are the natural mathematical structure in which to consider such quasi-topological properties. Weil's definition for a uniform space looked at a particular filter on $X \times X$ for which he had a certain base of sets generated by a family of pseudometrics. Because of the inconvenience of Weil's axioms, Tukey's approach to uniform spaces was through the notion of uniform covers (see [XXVIII]). An earlier use of this sort was made by Alexandroff and Urysohn (see [5]). Bourbaki (see (III]) gave a version similar to Weil's but with some adjustments. In Chapter 9, Bourbaki shows also (as Weil had done) that a uniformity can be defined by a family of pseudometrics. Our approach to uniformities will be through collections of subsets of $X \times X$. In Section 13 we will indicate the equivalence of this approach to those using uniform covers or collections of pseudometrics. First we need to introduce some terminology.

If X is a set and if $U, V \subset X \times X$, then $U \circ V$ is defined to be

$$\{(x, y) \in X \times X: \text{for some } z \in X, (x, z) \in V \quad \text{and} \quad (z, y) \in U\}.$$

We sometimes write UV for $U \circ V$. Set

$$U^n = (U^{n-1}) \circ U = U \circ (U^{n-1})$$

and write U for U^1. Set U^{-1} equal to $\{(x, y) \in X \times X: (y, x) \in U\}$. The subset U is *symmetric* in case $U = U^{-1}$. The set

$$\{(x, x) \in X \times X: x \in X\}$$

is called the *diagonal* of X and is denoted by Δ_X or when confusion is unlikely simply by Δ.

If A is a subset of X and if U is a subset of $X \times X$ we define $U(A) = \{y \in X: (x, y) \in U \text{ for some } x \in A\}$. In particular, if $A = \{x\}$ we write $U(x)$ for $U(\{x\})$.

3.1 Definition Let X be a set. By a *uniformity* (or *uniform structure*) *on* X we mean a non-empty collection $\mathscr{U}$ of subsets of $X \times X$ that satisfies the following conditions.

(U. 1) The diagonal of X is a subset of every element in $\mathscr{U}$.

(U. 2) If $U \in \mathscr{U}$ then $U^{-1} \in \mathscr{U}$.

(U. 3) If $U \in \mathscr{U}$ then there exists a $V \in \mathscr{U}$ such that $V \circ V \subset U$.

(U. 4) If U and V are in $\mathscr{U}$ then $U \cap V \in \mathscr{U}$.

(U. 5) If $U \in \mathscr{U}$ and if $U \subset V \subset X \times X$, then $V \in \mathscr{U}$.

A uniformity $\mathscr{U}$ is said to be *Hausdorff* if the following condition is also satisfied:

(U. 6) $\cap \mathscr{U} = \Delta_X$.

If $\mathscr{U}$ is a uniformity on X, then the pair $(X, \mathscr{U})$ is called a *uniform space*. The elements in $\mathscr{U}$ are called *entourages*.

Note the similarity between the conditions above and the metric axioms. Condition (U. 1) is derived from the metric condition $d(x, x) = 0$; condition (U. 2) generalizes symmetry,

$$d(x, y) = d(y, x);$$

and (U. 3) is a modification of the triangle inequality. Conditions (U. 4) and (U. 5) are added to ensure that we can define a topology in a reasonable way. They are merely the neighborhood axioms for a local base, that is, the collection $\mathscr{U}$ forms a filter so that $\mathscr{U}_x = \{U(x): x \in X\}$ is a local base for a unique topology.

3.2 Examples Let X be a non-empty set.

(1) The collection of all subsets of $X \times X$ containing Δ is a uniformity. It is the *finest* (largest) *uniformity* on X.

(2) The set $\{X \times X\}$ is a uniformity on X and is the *coarsest* (smallest) *uniformity* on X.

(3) If (X, d) is a pseudometric space, let

$$U(d, \epsilon) = \{(x, y) \in X \times X \colon d(x, y) < \epsilon\}.$$

Then $\{U(d, \epsilon) \colon \epsilon > 0\}$ is a uniformity on X.

As in topological spaces we now introduce the important and usually more workable concepts of base and subbase for a uniformity.

3.3 Definition Suppose that $\mathscr{U}$ is a uniformity on X. A non-empty collection $\mathscr{B}$ of subsets of $\mathscr{U}$ is called a *base* for $\mathscr{U}$ if for each $U \in \mathscr{U}$ there is a V in $\mathscr{B}$ such that $V \subset U$. The collection $\mathscr{B}$ is a *subbase* for $\mathscr{U}$ if for each $U \in \mathscr{U}$ there exists a finite collection $B_1, \ldots, B_n$ in $\mathscr{B}$ such that $B_1 \cap \ldots \cap B_n \subset U$.

3.4 Proposition *Suppose that $\mathscr{B}$ is a non-empty collection of subsets of $X \times X$. Then $\mathscr{B}$ is a base for a unique uniformity on X in case the following conditions hold.*

(B. 1) *The diagonal of X is a subset of every element in $\mathscr{B}$.*

(B. 2) *If $B \in \mathscr{B}$ then $B^{-1} \in \mathscr{B}$.*

(B. 3) *If $B \in \mathscr{B}$ there exists a $D \in \mathscr{B}$ such that $D \circ D \subset B$.*

(B. 4) *If B_1 and B_2 are in $\mathscr{B}$ then $B_1 \cap B_2 \in \mathscr{B}$.*

Proof. Let $\mathscr{U} = \{U \subset X \times X \colon U \supset B \text{ for some } B \in \mathscr{B}\}$. We first show that $\mathscr{U}$ is a uniformity. We prove (U. 4) and leave the others as an exercise. If $U, V \in \mathscr{U}$ then by the definition of $\mathscr{U}$ there exist $B_1, B_2 \in \mathscr{B}$ such that $B_1 \subset U$ and $B_2 \subset V$. By (B. 4), $B_1 \cap B_2 \in \mathscr{B}$ and since $B_1 \cap B_2 \subset U \cap V$, it follows that $U \cap V \in \mathscr{U}$.

To prove $\mathscr{U}$ is unique suppose that $\mathscr{B}$ is a base for another uniformity $\mathscr{U}'$. If $U \in \mathscr{U}'$ then by 3.3, there exists $B \in \mathscr{B}$ such that $B \subset U$ and therefore by the definition of $\mathscr{U}$, $U \in \mathscr{U}$. Whence $\mathscr{U}' \subset \mathscr{U}$. On the other hand it follows from (U. 5) that any uniformity containing $\mathscr{B}$ must contain $\mathscr{U}$ and therefore $\mathscr{U} \subset \mathscr{U}'$. □

3.5 **Proposition** *Suppose that $\mathscr{S}$ is a non-empty collection of subsets of $X \times X$. Then $\mathscr{S}$ is a subbase for a unique uniformity on X in case the following conditions hold.*

(S. 1) *The diagonal of X is a subset of every element in $\mathscr{S}$.*

(S. 2) *If $S \in \mathscr{S}$ then there exists $T \in \mathscr{S}$ such that $T \subset S^{-1}$.*

(S. 3) *If $S \in \mathscr{S}$ then there exists $T \in \mathscr{S}$ such that $T \circ T \subset S$.*

The proof follows from 3.4 and is left as an exercise.

It should be noted that in general a union of uniformities need not be a uniformity. However as a consequence of 3.5 it can be shown that the union of any collection of uniformities is a subbase for a unique uniformity.

We now show that a uniform space $(X, \mathscr{U})$ induces a topology on X in a natural way. To understand this concept, recall how the pseudometric topology was defined. Namely, if (X, d) is a pseudometric space, a set G is in $\mathscr{T}_d$ if and only if for each $x \in G$ there exists $\epsilon > 0$ such that $S(x; d, \epsilon) \subset G$. We are thus led to

3.6 **Definition** Suppose that $(X, \mathscr{U})$ is a uniform space and let $\mathscr{T}(\mathscr{U})$ be the collection of sets G in X defined as follows:

The subset $G \in \mathscr{T}(\mathscr{U})$ if and only if for all $x \in G$ there exists $U \in \mathscr{U}$ such that $U(x) \subset G$.

We show (see 3.7) that $\mathscr{T}(\mathscr{U})$ is indeed a topology on X and we call $\mathscr{T}(\mathscr{U})$ the *uniform topology* on X. When confusion is unlikely we write $\mathscr{T}$ instead of $\mathscr{T}(\mathscr{U})$.

3.7 **Proposition** *The collection $\mathscr{T}(\mathscr{U})$ is a topology on X and in fact, $\mathscr{B}_x = \{U(x) \colon x \in X\}$ is a local base at x for each point of X.*

Proof. We show that if G_1 and G_2 belong to $\mathscr{T}(\mathscr{U})$ then $G_1 \cap G_2$ belongs to $\mathscr{T}(\mathscr{U})$, leaving the rest as an exercise. Thus suppose that $x \in G_1 \cap G_2$. Then $x \in G_1$ implies that there exists $U \in \mathscr{U}$ such that $U(x) \subset G$. Similarly there exists V in $\mathscr{U}$ such that $V(x) \subset G$. By (U. 4), $W = U \cap V$ is in $\mathscr{U}$ and

$$W(x) \subset U(x) \cap V(x) \subset G_1 \cap G_2.$$

Therefore by definition $G_1 \cap G_2$ belongs to $\mathscr{T}$. □

Thus uniform spaces do have a very natural topology associated with them. What separation axioms does this topology have? Moreover what topological spaces, if any, will yield a uniformity in some natural way? Let us keep these two questions in mind as we proceed in the development at hand.

3.8 DEFINITION Suppose that $(X,\mathscr{T})$ is a topological space and that $\mathscr{U}$ is a uniformity for X. If $\mathscr{T}(\mathscr{U}) = \mathscr{T}$ then we say that *X admits the uniformity $\mathscr{U}$* or that $\mathscr{U}$ is an *admissible uniformity on X*. We say that $(X,\mathscr{T})$ is *uniformizable* if it has an admissible uniformity. We say that the uniformity $\mathscr{U}$ is a *continuous uniformity* if $\mathscr{T}(\mathscr{U}) \subset \mathscr{T}$. If $(X, \mathscr{U})$ is a uniform space then it will always be understood that X is equipped with the uniform topology.

Note the similarity between continuous pseudometrics and continuous uniformities. We will make this more precise in Section 13.

We now state several propositions to show the interrelationship between the various concepts introduced above.

3.9 PROPOSITION *If $(X, \mathscr{U})$ is a uniform space, if*

$$M \subset X \times X$$

and if $U \in \mathscr{U}$ then UMU is a neighborhood of M in $X \times X$ (where $X \times X$ has the product topology).

Proof. Suppose without loss of generality that U is symmetric (otherwise prove it for $W = U \cap U^{-1}$ which is in $\mathscr{U}$ and contained in U). If $(x,y) \in M$ then $U(x)$ is a neighborhood of x in X and $U(y)$ is a neighborhood of y in X so $U(x) \times U(y)$ is a neighborhood of (x,y) in $X \times X$. One can easily show that $U(x) \times U(y) \subset UMU$ and therefore UMU is a neighborhood of M. □

3.10 PROPOSITION *If $(X, \mathscr{U})$ is a uniform space and if $U \in \mathscr{U}$ then* int $U \in \mathscr{U}$.

Proof. It is easy to show that there exists $V \in \mathscr{U}$ such that $V^3 \subset U$. By 3.9, V^3 is a neighborhood of V so U is a neighborhood of V. Thus $V \subset \text{int}\, U$ and therefore int $U \in \mathscr{U}$. □

3.11 Proposition *If $(X, \mathscr{U})$ is a uniform space, then*

$$\mathscr{B} = \{U \in \mathscr{U}\colon U \text{ is open and } U = U^{-1}\}$$

is a base for $\mathscr{U}$.

This follows immediately from 3.10 and the definition of base for a uniformity.

Another immediate result of 3.10 is the following.

3.12 Proposition *Suppose that $\mathscr{B}$ is a base (respectively, subbase) for a uniformity $\mathscr{U}$ on a non-empty set X. Then the collection*

$$\{U(x)\colon x \in X \quad \text{and} \quad U \in \mathscr{B}\}$$

is a base (respectively, subbase) for $\mathscr{T}(\mathscr{U})$.

We can now observe that our definition of a Hausdorff uniformity makes sense.

3.13 Proposition *If $(X, \mathscr{U})$ is a uniform space, then $(X, \mathscr{U})$ is a Hausdorff uniform space if and only if $(X, \mathscr{T}(\mathscr{U}))$ is a Hausdorff topological space.*

Proof. If $(X, \mathscr{U})$ is Hausdorff then $\cap \mathscr{U} = \Delta$. Suppose that $x \neq y$ are two points in X so that $(x, y) \notin \Delta = \cap \mathscr{U}$. Then there exists $U \in \mathscr{U}$ with $(x, y) \notin U$. By (U. 3), there exists a symmetric V such that $V \circ V \subset U$ and by 3.7, $V(x)$ and $V(y)$ are neighborhoods of x and y respectively. If $z \in V(x) \cap V(y)$, then $(x, z) \in V$ and $(y, z) \in V$ and therefore $(x, y) \in U$. It follows that $V(x) \cap V(y) = \varnothing$ and therefore $(X, \mathscr{T}(\mathscr{U}))$ is Hausdorff. The converse should be obvious and is left as an exercise. □

Topological spaces are the natural generalization of metric spaces for a meaningful study of continuous functions. However on a compact subset of a metric space, continuous functions satisfy a much stronger condition – they are uniformly continuous. In a general topological space, however, this more stringent requirement is not compatible with the set theoretical structure, that is more set theoretical structure is needed. The concept of a uniform space (which yields a topological space) has just enough of this additional structure for a meaningful study of uniform continuity.

3.14 Definition Suppose that f is a function from a uniform space $(X, \mathscr{U})$ into a uniform space $(Y, \mathscr{V})$. The function f is *uniformly continuous* in case for every $V \in \mathscr{V}$ there exists $U \in \mathscr{U}$ such that $(f \times f)(U) \subset V$.

If X and Y are two uniform spaces and if f is a one-to-one map of X onto Y such that both f and f^{-1} are uniformly continuous, then f is called a *uniform isomorphism*. The spaces X and Y are then said to be *uniformly equivalent*.

In topological spaces, a homeomorphism between two such spaces, tells us, from the point of view of topology, that the two spaces are the same. Such an analogy exists in many areas of mathematics, for example, the isomorphisms of algebra, the conformal maps of analysis and now the uniform isomorphisms for uniform spaces.

It is clear that a map f: $(X, \mathscr{U}) \to (Y, \mathscr{V})$ is uniformly continuous if and only if $(f \times f)^{-1}(V) \in \mathscr{U}$ for all $V \in \mathscr{V}$. However, often a more valuable characterization of uniform continuity is the following result.

3.15 Proposition *Suppose that* $(X, \mathscr{U})$ *and* $(Y, \mathscr{V})$ *are uniform spaces, that* $\mathscr{S}$ *is a subbase for* $\mathscr{V}$ *and that* f *maps* $(X, \mathscr{U})$ *into* $(Y, \mathscr{V})$. *Then* f *is uniformly continuous if and only if for each* S *in* $\mathscr{S}$ *there exists* U *in* $\mathscr{U}$ *such that* $(f \times f)(U) \subset S$.

Proof. If f is uniformly continuous then the result is trivial. Conversely, assume the condition holds and that $V \in \mathscr{V}$. Then, by 3.3, there exist $S_1, \ldots, S_n$ in $\mathscr{S}$ such that $S_1 \cap \ldots \cap S_n \subset V$. For each $i = 1, \ldots, n$, by hypothesis, there exists $U_i \in \mathscr{U}$ such that $(f \times f)(U_i) \subset S_i$. Moreover, by (U.4), $U_1 \cap \ldots \cap U_n = U \in \mathscr{U}$ and one can easily show that $(f \times f)(U) \subset V$. □

Now that we have the concepts of uniform continuity and uniform isomorphism, it is natural to ask questions of them in a manner analogous to what we would ask of their topological counterparts. Consequently it is an easy exercise to check the following statements.

(1) The composition of two uniformly continuous functions is uniformly continuous.

(2) The composition of two uniform isomorphisms, the inverse of a uniform isomorphism, and the identity map of a space onto itself are all uniform isomorphisms.

(3) Each uniformly continuous function is continuous in the relative topology, hence each uniform isomorphism is a homeomorphism.

For example, to show (3), let $x \in X$ and let G be a neighborhood of $f(x)$. By definition of uniform topology there exists $V \in \mathscr{V}$ such that $V(f(x)) \subset G$ and by hypothesis there exists $U \in \mathscr{U}$ such that $(f \times f)(U) \subset V$. But then $U(x)$ is a neighborhood of x and one easily shows that $f(U(x)) \subset V(f(x))$.

For uniform spaces, we have mimicked much of the usual topological concepts. Let's carry it further to see what the uniform counterpart is for the Tychonoff product topology.

3.16 Definition Suppose that $((X_\alpha, \mathscr{U}_\alpha))_{\alpha \in I}$ is a family of uniform spaces. By the *product uniformity* on $X = \prod_{\alpha \in I} X_\alpha$ we mean the smallest uniformity on X such that the projection map pr_α of X into X_α is uniformly continuous for all $\alpha \in I$.

For each $\alpha \in I$ and each $U \in \mathscr{U}_\alpha$ if we let

$$W_{(U,\alpha)} = \{(x,y) \in X \times X : (x_\alpha, y_\alpha) \in U\}$$

then $\{W_{(U,\alpha)} : \alpha \in I \text{ and } U \in \mathscr{U}_\alpha\}$ is a subbase for the product uniformity on X. (Note that $W_{(U,\alpha)} = (\mathrm{pr}_\alpha \times \mathrm{pr}_\alpha)^{-1}(U)$.)

As is done with topologies we may generalize this construction as follows.

If X is a set, if $((Y_\alpha, \mathscr{U}_\alpha))_{\alpha \in I}$ is a family of uniform spaces and if for each $\alpha \in I$, f_α is a function from X into Y then we equip X with the unique smallest uniformity $\mathscr{U}$ such that each map f_α from $(X, \mathscr{U})$ into $(X_\alpha, \mathscr{U}_\alpha)$ is uniformly continuous for all $\alpha \in I$. We call $\mathscr{U}$ the *uniformity on X induced by the family* $(f_\alpha)_{\alpha \in I}$.

Thus the product uniformity on $X = \prod_{\alpha \in I} X_\alpha$ is the uniformity induced by the family $(\mathrm{pr}_\alpha)_{\alpha \in I}$.

3.17 Theorem *Suppose that $((X_\alpha, \mathscr{U}_\alpha))_{\alpha \in I}$ is a family of uniform spaces, that $X = \prod_{\alpha \in I} X_\alpha$ is the topological product of*

$(X_\alpha)_{\alpha \in I}$ *with the product uniformity* $\mathscr{U}$, *that* $(Y, \mathscr{V})$ *is a uniform space and that* f *maps* $(Y, \mathscr{V})$ *into* $(X, \mathscr{U})$ *is a function. Then* f *is uniformly continuous if and only if* $\mathrm{pr}_\alpha \circ f$ *is uniformly continuous for each* $\alpha \in I$.

Proof. By the definition of the product uniformity each projection pr_α from X into X_α is uniformly continuous for all $\alpha \in I$. Hence if f is uniformly continuous then $\mathrm{pr}_\alpha \circ f$ is uniformly continuous for all $\alpha \in I$.

To show the converse, by 3.15, we need only consider subbasic elements of the form $U = (\mathrm{pr}_\alpha \times \mathrm{pr}_\alpha)^{-1}(U_\alpha)$ where $U_\alpha \in \mathscr{U}$ for some $\alpha \in I$. Now, by hypothesis, $((\mathrm{pr}_\alpha \circ f) \times (\mathrm{pr}_\alpha \circ f))^{-1}(U_\alpha)$ is an element of $\mathscr{V}$. Since $(f \times f)^{-1}(U) = ((\mathrm{pr}_\alpha \circ f) \times (pr_\alpha \circ f))^{-1}(U_\alpha)$, it follows that f is uniformly continuous. □

If $(X, \mathscr{U})$ is a uniform space and if S is a subset of X then the uniformity on S induced by the inclusion map j from S into X is called the *relative uniformity on* S. We call S equipped with the relative uniformity a *uniform subspace of* X.

Observe that if $(X, \mathscr{U})$ is a uniform space and if $S \subset X$, then the relative uniformity on S is $\mathscr{U} | S \times S$.

3.18 Exercises

(1) Prove 3.5.

(2) Show that the union of any collection of uniformities is a subbase for a unique uniformity. Give an example to show that a union of uniformities need not be a uniformity. What is the situation for the intersection of uniformities?

(3) Every member of a uniformity is a neighborhood of the diagonal (in the product topology induced by the uniformity). The family of all closed symmetric members of a uniformity is a base for the uniformity.

(4) If a uniform space is compact in its uniform topology then the uniformity consists of all neighborhoods of the diagonal.

(5) Let f be a uniformly continuous function from one uniform space into another. Then f is continuous in the uniform topologies.

(6) If the uniform space $(X, \mathscr{U})$ is compact in its uniform topology then every continuous function from $(X, \mathscr{U})$ into another uniform space is uniformly continuous.

(7) A function f from one metric space (X, d) to a metric space (Y, e) is said to be uniformly continuous if for all $\epsilon > 0$ there is a $\delta > 0$ such that $e(f(x), f(y)) < \epsilon$ whenever $d(x, y) < \delta$. Show that a function continuous on a compact subset of a metric space is uniformly continuous.

4 $\mathscr{Z}$-filters and convergence

Our approach to the Stone–Čech compactification of a space will follow an approach similar to that in [XI]. It is interesting to remark that this approach can be put very easily into a more general setting which gives a considerable amount of insight into the nature of this particular compactification. Moreover, the general setting yields also many other compactifications of a space. Since so much can be gained with just a little more effort (to introduce the new concepts), our approach will be that of the more general scheme.

To begin we need to make some slight modifications in the usual concept of a filter. The concept of a filter on a non-empty set X is just a particular collection of subsets of X. It is not intrinsically related to the topology on X, if X has one, until one begins to discuss convergence. However here it is the intention to relate the concept to the topology of the space concerned. In fact, practically speaking, our '$\mathscr{Z}$-filters' will be just the usual concept of a filter intersected with a base $\mathscr{Z}$ for the closed subsets of X. To be able to have a meaningful theory of convergence, however, certain axioms must be placed on this base $\mathscr{Z}$. To this end the concept of a normal base is given. This concept was introduced by Orrin Frink in 1964. Similar concepts had been considered in [129], [44], and [336].

4.1 DEFINITION Let $\mathscr{Z}$ be a non-empty collection of subsets of a topological space X. Members Z of $\mathscr{Z}$ will be referred to as *Z-sets* whereas complements of members will be referred to as *co-Z sets*. The collection $\mathscr{Z}$ is called a *ring of sets* if for members Z_1 and Z_2 of $\mathscr{Z}$ it follows that $Z_1 \cup Z_2 \in \mathscr{Z}$ and $Z_1 \cap Z_2 \in \mathscr{Z}$. If finite (countable) unions or intersections of members of $\mathscr{Z}$ belong to $\mathscr{Z}$ then $\mathscr{Z}$ is said to be *closed under finite* (*countable*) *unions* or *intersections*. The collection $\mathscr{Z}$ is said to be *disjunctive* if for any closed subset F and for any point x not in F there is a set Z in $\mathscr{Z}$ such that $x \in Z$ and $Z \cap F = \varnothing$. The collection is *normal* if for any pair Z_1 and Z_2 of disjoint members of $\mathscr{Z}$ there is a pair of sets C_1 and C_2 in $\mathscr{Z}$ which cover X and such that $F_i \subset X - C_i$ for $i = 1, 2$. The collection $\mathscr{Z}$ is said to be a *normal base* for $(X, \mathscr{T})$ if $\mathscr{Z}$ is a disjunctive normal ring of sets which is also a base for the closed subsets of X.

Examples of normal bases are many. In a regular topological space the collection $\mathscr{Z}$ of all closed subsets of X is a disjunctive ring of sets. In a normal space such a collection $\mathscr{Z}$ is also a normal collection. For completely regular spaces the collection $\mathscr{Z}$ of all zero-sets of $(X, \mathscr{T})$ will be shown to be a normal base.

For a collection $\mathscr{Z}$ of subsets of X *closed under finite intersections* only (the other axioms for a normal base presently are not needed) it is possible to define the concept of filter with respect to $\mathscr{Z}$.

4.2 Definition Let $\mathscr{Z}$ be a collection of subsets of a non-empty set X closed under finite intersections and let $\mathscr{F}$ be a subcollection of $\mathscr{Z}$. The collection $\mathscr{F}$ is called a $\mathscr{Z}$*-filter* if the following conditions are satisfied.

(F. 1) If Z_1 and Z_2 are in $\mathscr{F}$ then $Z_1 \cap Z_2 \in \mathscr{F}$.

(F. 2) If $Z_1 \in \mathscr{F}$ and if $Z \in \mathscr{Z}$ with $Z \supset Z_1$ then $Z \in \mathscr{F}$.

(F. 3) The collection $\mathscr{F}$ is non-empty and $\varnothing \notin \mathscr{F}$.

If $\mathscr{Z}$ is the collection of all subsets of X, then $\mathscr{Z}$-filters are simply the familiar (*Bourbaki*) *filters* (see [III]).

The collection $Z(X)$ of all zero-sets on a Tychonoff space X is a normal base on X. In fact, it has even more properties as will be seen. Also, $Z(X)$ will serve as a good candidate for $\mathscr{Z}$ when discussing $\mathscr{Z}$-filters. The following lemma gives some of the properties of $Z(X)$.

4.3 Lemma *For the collection $Z(X)$ of all zero-sets on a topological space X, the following statements hold.*

(1) *The collection $Z(X)$ contains the intersection of any countable subcollection.*

(2) *Every zero-set may be represented as a countable intersection of cozero-sets.*

(3) *A topological space X is completely regular if and only if $Z(X)$ is a base for the closed subsets of X.*

(4) *If X is completely regular then every neighborhood of a point contains a zero-set neighborhood of the point.*

(5) *The collection $Z(X)$ contains the union of any finite subcollection.*

We leave the proof as an exercise. Note that (1) follows from 2.8.

Sometimes it is necessary to look at a subcollection $\mathscr{B}$ of the $\mathscr{Z}$-filter $\mathscr{F}$ such that $\mathscr{B}$ gives us sufficient information about $\mathscr{F}$. Thus the definition of a base for a $\mathscr{Z}$-filter.

4.4 DEFINITION A collection $\mathscr{B}$ of members of a $\mathscr{Z}$-filter $\mathscr{F}$ is called a *base* for $\mathscr{F}$ if for each $Z \in \mathscr{F}$ there is a $B \in \mathscr{B}$ such that $B \subset Z$.

When can we say that a particular subcollection of $\mathscr{Z}$ forms a base for some $\mathscr{Z}$-filter $\mathscr{F}$? This is answered in the next lemma, the proof of which is left as an exercise.

4.5 LEMMA *The subcollection $\mathscr{B}$ of $\mathscr{Z}$ is a base for a unique $\mathscr{Z}$-filter $\mathscr{F}$ if and only if both of the following conditions hold.*

(1) *The collection $\mathscr{B}$ is non-empty and $\varnothing \notin \mathscr{B}$.*

(2) *If B_1 and B_2 are in $\mathscr{B}$ then there is a $B_0 \in \mathscr{B}$ such that $B_0 \subset B_1 \cap B_2$.*

Assuming conditions (1) and (2) of the preceding lemma, one shows that $\mathscr{B}$ is a base for a $\mathscr{Z}$-filter $\mathscr{F}$ by constructing the $\mathscr{Z}$-filter $\mathscr{F}(\mathscr{B})$ defined by

$$\mathscr{F}(\mathscr{B}) = \{Z \in \mathscr{Z} \colon \text{for some } B \in \mathscr{B},\, B \subset Z\}.$$

The filter $\mathscr{F}(\mathscr{B})$ is called the *$\mathscr{Z}$-filter generated by the base $\mathscr{B}$.*

It is now easy to see that if $\mathscr{G}$ is a subcollection of $\mathscr{Z}$ with the finite intersection property then $\mathscr{G}$ is contained in a $\mathscr{Z}$-filter. Of course the converse is obviously true. Moreover, if $\mathscr{B}$ is the collection of all finite intersections of members of $\mathscr{G}$, then by 4.5, $\mathscr{B}$ is a base for a unique $\mathscr{Z}$-filter on X. Consequently such collections $\mathscr{G}$ are called a *system of generators* for this $\mathscr{Z}$-filter $\mathscr{F}$ or it is said that $\mathscr{G}$ *generates* $\mathscr{F}$.

When working with $\mathscr{Z}$-filters, it is often necessary to decide when a particular member $A \in \mathscr{Z}$ can be joined to a $\mathscr{Z}$-filter $\mathscr{F}$ so that both $\mathscr{F}$ and A are contained in a $\mathscr{Z}$-filter.

4.6 LEMMA *Let $A \in \mathscr{Z}$ and let $\mathscr{F}$ be a $\mathscr{Z}$-filter on X. The collection $\mathscr{F} \cup \{A\}$ is contained in a $\mathscr{Z}$-filter if and only if A meets every member of $\mathscr{F}$.*

Proof. The necessity of the condition is obvious. On the other hand for sufficiency it is clear that $\mathscr{F} \cup \{A\}$ is a base for a $\mathscr{Z}$-filter on X. □

Let Φ be the collection of all $\mathscr{Z}$-filters considered as a partially ordered set under set-inclusion. We are then particularly interested in the maximal elements.

4.7 DEFINITION A $\mathscr{Z}$-filter $\mathscr{F}$ is called a *$\mathscr{Z}$-ultrafilter* if $\mathscr{F}$ is a maximal $\mathscr{Z}$-filter in the collection Φ of all $\mathscr{Z}$-filters where Φ is partially ordered by set inclusion.

4.8 THEOREM *Every $\mathscr{Z}$-filter is contained in a $\mathscr{Z}$-ultrafilter.*

Proof. See Exercise 3. □

When is a particular $\mathscr{Z}$-filter under consideration a $\mathscr{Z}$-ultrafilter? Zorn's lemma tells us that these $\mathscr{Z}$-ultrafilters exist, but it does not tell us when a $\mathscr{Z}$-filter is a $\mathscr{Z}$-ultrafilter. The next two propositions will assist us in answering this question.

4.9 PROPOSITION *Let $\mathscr{G}$ be a system of generators for a $\mathscr{Z}$-filter on X satisfying the following condition.*

For each $A \in \mathscr{Z}$ either $A \in \mathscr{G}$ or there is $Z \in \mathscr{Z}$ such that

$$Z \subset X - A \quad \textit{and} \quad Z \in \mathscr{G}.$$

Then $\mathscr{G}$ generates a $\mathscr{Z}$-ultrafilter on X.

Proof. Let $\mathscr{F}'$ be a $\mathscr{Z}$-filter containing the $\mathscr{Z}$-filter $\mathscr{F}$ which is generated by $\mathscr{G}$. Let A be any member of $\mathscr{F}'$. By the condition, either $A \in \mathscr{G} \subset \mathscr{F}$ or there is a $Z \in \mathscr{G} \subset \mathscr{F} \subset \mathscr{F}'$ and $Z \subset X - A$. Since $A \cap Z$ must be in $\mathscr{F}'$ it follows that $\mathscr{F}' = \mathscr{F}$. □

Let us note here that if $\mathscr{Z}$ is the power set of X, then our condition reads: for each $A \in \mathscr{Z}$ either $A \in \mathscr{G}$ or $X - A \in \mathscr{G}$. This is the usual condition for a (Bourbaki) filter to be an ultrafilter.

Let us now characterize those $\mathscr{Z}$-filters which are $\mathscr{Z}$-ultrafilters. This will be most helpful as a working tool.

4.10 PROPOSITION *Let $\mathscr{F}$ be a $\mathscr{Z}$-filter on X. Then the following statements are equivalent.*

(1) *The $\mathscr{Z}$-filter $\mathscr{F}$ is a $\mathscr{Z}$-ultrafilter.*

(2) *For each A in $\mathscr{Z}$, if A meets every member of $\mathscr{F}$ then $A \in \mathscr{F}$.*

(3) *For each $A \in \mathscr{Z}$ either $A \in \mathscr{F}$ or there is a $Z \in \mathscr{Z}$ such that $Z \subset X - A$ and $Z \in \mathscr{F}$.*

Proof. The proof that (1) *implies* (2) is obvious by 4.6 and the fact that $\mathscr{F}$ is a $\mathscr{Z}$-ultrafilter. Let us now assume condition (2). If $A \notin \mathscr{F}$ ($A \in \mathscr{Z}$) then condition (2) implies the existence of a set $Z \in \mathscr{F}$ for which $Z \cap A = \varnothing$. But then $Z \subset X - A$. Consequently (2) *implies* (3). Finally 4.9 shows that condition (3) implies that $\mathscr{F}$ must be a $\mathscr{Z}$-ultrafilter. □

A $\mathscr{Z}$-filter $\mathscr{F}$ is a *prime $\mathscr{Z}$-filter* if $A \cup B \in \mathscr{F}$ implies either $A \in \mathscr{F}$ or $B \in \mathscr{F}$ where A and B are members of $\mathscr{Z}$.

Using 4.10 it is easy to see that a $\mathscr{Z}$-ultrafilter $\mathscr{F}$ is always a prime $\mathscr{Z}$-filter. For, if $A \cup B \in \mathscr{F}$ but $A \notin \mathscr{F}$ and $B \notin \mathscr{F}$ then 4.10 (2) implies there exists Z_1 and Z_2 in $\mathscr{F}$ such that

$$A \cap Z_1 = \varnothing = B \cap Z_2.$$

On the other hand, if $Z = Z_1 \cap Z_2$, then $(A \cup B) \cap Z \in \mathscr{F}$ since $\mathscr{F}$ is closed under finite intersections. But this is a contradiction since $\mathscr{F}$ does not contain the empty set.

By considering the natural numbers $\mathbf{N}$ and a finite collection $\mathscr{Z}$ of its subsets, it is possible to construct a prime $\mathscr{Z}$-filter that is not a $\mathscr{Z}$-ultrafilter. For (Bourbaki) filters the two concepts are the same. However as a corollary to 4.10 the reader may prove:

4.11 COROLLARY *Every $\mathscr{Z}$-ultrafilter is a prime $\mathscr{Z}$-filter. Moreover if $\mathscr{F}$ is a prime $\mathscr{Z}$-filter and if (A_n) is a finite subfamily of $\mathscr{Z}$ with $\cup A_n \in \mathscr{F}$ then $A_n \in \mathscr{F}$ for some $n \in \mathbf{N}$.*

To orient the direction of developing further our results concerning $\mathscr{Z}$-filters let us look first at an application. Suppose that our distinguished subcollection $\mathscr{Z}$ of 2^X is the collection of all closed subsets of X. Then a space is compact if and only if every $\mathscr{Z}$-filter $\mathscr{F}$ is such that $\cap \mathscr{F} \neq \varnothing$.

A family $\mathscr{F}$ of subsets of a non-empty set X is said to be *fixed* if $\cap \mathscr{F} \neq \varnothing$. It is said to be *free* if $\cap \mathscr{F} = \varnothing$.

If $\mathscr{Z}$ is the collection of all closed subsets of X, using the above terminology we see that X is compact if and only if every $\mathscr{Z}$-filter on X is fixed. Actually for arbitrary topological spaces the above formulation of the concept of compactness may be made in terms of a usually smaller subcollection $\mathscr{Z}$ as the following theorem shows. In particular for Tychonoff spaces $\mathscr{Z}$ may be taken to be the collection of all zero-sets of X.

4.12 THEOREM *Let $\mathscr{Z}$ be a collection of subsets of X closed under finite intersections that is also a base for the closed subsets of a topological space X. Then the following statements are equivalent.*

(1) *The space X is compact.*

(2) *Every $\mathscr{Z}$-filter is fixed.*

(3) *Every prime $\mathscr{Z}$-filter is fixed.*

(4) *Every $\mathscr{Z}$-ultrafilter is fixed.*

Proof. Since the $\mathscr{Z}$-sets are closed subsets, condition (1) *implies* (2). Obviously condition (2) *implies* (3) and condition (3) *implies* (4). Finally let $\mathscr{F}$ be any family of closed subsets with the finite intersection property. Since $\mathscr{Z}$ is a base for the closed sets, it follows that $\mathscr{F}$ is contained in a larger collection $\mathscr{G}$, having the finite intersection property and consisting of $\mathscr{Z}$ sets. Precisely $\mathscr{G}$ is the collection of all basic sets $Z_\alpha \in \mathscr{Z}$ where $F = \bigcap_{\alpha \in I} Z_\alpha$ for $F \in \mathscr{F}$. Making use of 4.5 as we have stated previously, $\mathscr{G}$ is a system of generators for a $\mathscr{Z}$-filter which in turn is contained in a $\mathscr{Z}$-ultrafilter $\mathscr{F}'$ (4.8). Under the assumption of (4), the $\mathscr{Z}$-ultrafilter $\mathscr{F}'$ is fixed and therefore so is $\mathscr{G}$. Since each $F \in \mathscr{F}$ is an intersection of sets from $\mathscr{G}$, it follows that $\mathscr{F}$ is also fixed. Consequently (1) is shown and our proof is complete. □

Since a base for the closed subsets of a Tychonoff space is the collection $\mathscr{Z} = Z(X)$ of all zero-sets on X (it is also a ring of sets), the theorem gives the compactness of a Tychonoff space in terms of fixed zero-set filters. Note that the proof used the fact that any $\mathscr{Z}$-filter contained in a fixed $\mathscr{Z}$-filter must also be fixed. On the other hand any $\mathscr{Z}$-filter containing a free $\mathscr{Z}$-filter must also be free.

A fixed $\mathscr{Z}$-filter that will be of particular interest later on, is the following. For a fixed point $x \in X$ let

$$\mathscr{F}_x = \{Z \in \mathscr{Z} : x \in Z\}.$$

It is an easy exercise to check that this is a fixed $\mathscr{Z}$-filter as long as there is at least one $Z \in \mathscr{Z}$ containing the point $x \in X$. Let us assume always this last condition when discussing the $\mathscr{Z}$-filters $\mathscr{F}_x$. We call $\mathscr{F}_x$ a *point $\mathscr{Z}$-filter*.

4.13 PROPOSITION *If $\mathscr{Z}$ is a disjunctive collection of subsets closed under finite intersections, then the point $\mathscr{Z}$-filter $\mathscr{F}_x$ is a fixed $\mathscr{Z}$-ultrafilter.*

Furthermore if X is a T_1-space then $\cap \mathscr{F}_x = \{x\}$.

Proof. If $\mathscr{Z}$ is disjunctive and if $Z \notin \mathscr{F}_x$ then there is a $Z_1 \in \mathscr{Z}$ such that $x \in Z_1$ and $Z_1 \cap Z = \varnothing$. Consequently 4.10(3) applies. If $y \in \cap \mathscr{F}_x$ and $y \neq x$ then the fact that $\mathscr{Z}$ is disjunctive on a T_1-space implies there is a $Z \in \mathscr{Z}$ such that $x \in Z \in \mathscr{F}_x$ and $y \notin Z$ which is a contradiction. □

If $\mathscr{Z}$ is closed under finite intersections and is a disjunctive collection of sets in a T_1-space then one knows precisely what the fixed $\mathscr{Z}$-ultrafilters look like. They are precisely the point $\mathscr{Z}$-filters.

4.14 PROPOSITION *Let $\mathscr{Z}$ be a disjunctive ring of subsets of a T_1-space X. Then a $\mathscr{Z}$-ultrafilter $\mathscr{F}$ is fixed if and only if $\mathscr{F}$ is a point $\mathscr{Z}$-filter $\mathscr{F}_x$ for some x in X.*

Proof. By 4.13 every point $\mathscr{Z}$-filter is fixed. Conversely if $x \in \cap \mathscr{F}$ then $\mathscr{F}_x$ is a fixed $\mathscr{Z}$-ultrafilter and $\cap \mathscr{F}_x = \{x\}$ as well as $\mathscr{F} \subset \mathscr{F}_x$. Consequently $\mathscr{F} = \mathscr{F}_x$. □

Let us now direct our attention to the theory of convergence. The concept of convergence of (Bourbaki) filters plays a dominant role when one studies the concepts of compactness, continuity and completeness. Since $\mathscr{Z}$-filter convergence will be used to obtain compactifications of a space, our study then will consider these concepts with respect to $\mathscr{Z}$-filters. For example it has already been shown how compactness is characterized in terms of $\mathscr{Z}$-filters.

Let us remark again that throughout our discussion $\mathscr{Z}$ will be *closed under finite intersections*. Let us define then the concept of convergence for $\mathscr{Z}$-filters.

4.15 DEFINITION Let $\mathscr{B}$ be a $\mathscr{Z}$-filter base on a topological space $(X, \mathscr{T})$. A point x in X is a *limit point* of $\mathscr{B}$ if every neighborhood of x contains a member of $\mathscr{B}$. If this occurs then $\mathscr{B}$ is said to *converge to* x. The point x is said to be a *cluster point* of $\mathscr{B}$ if each neighborhood of x intersects every member of $\mathscr{B}$.

Thus if a $\mathscr{Z}$-filter base $\mathscr{B}$ converges to a point x in X then x is a cluster point of $\mathscr{B}$. Since every $\mathscr{Z}$-filter is a $\mathscr{Z}$-filter base for itself, one observes that discussion of a $\mathscr{Z}$-filter converging is also meaningful. Let us formalize some remarks about $\mathscr{Z}$-filters and $\mathscr{Z}$-filter bases. The proofs are straight forward and are left as exercises.

4.16 PROPOSITION *Let $\mathscr{B}$ be a base for the $\mathscr{Z}$-filter $\mathscr{F}$ and let $\mathscr{F}$ be contained in the $\mathscr{Z}$-filter $\mathscr{G}$. Then the following statements are valid.*

(1) *A point $x \in X$ is a limit (respectively, cluster) point of $\mathscr{F}$ if and only if x is a limit (respectively, cluster) point of $\mathscr{B}$.*

(2) *If x is a limit point of $\mathscr{F}$ then x is a limit point of $\mathscr{G}$.*

(3) *If x is a cluster point of $\mathscr{G}$, then x is a cluster point of $\mathscr{F}$.*

(4) *Every fixed $\mathscr{Z}$-filter has a cluster point.*

(5) *If $\mathscr{Z}$ is a disjunctive collection of subsets then every fixed $\mathscr{Z}$-ultrafilter $\mathscr{F}_x$ is convergent (to the point x).*

Let us consider some examples to clarify further the notion of convergence with respect to $\mathscr{Z}$-filters.

Let $(\mathbf{R}, \mathscr{D})$ be the real numbers with the discrete topology. For x and y distinct points in $\mathbf{R}$, let $\mathscr{Z}$ be all subsets of $\mathbf{R}$ which contain both points x and y. Of course $\mathscr{Z}$ contains the intersection of any finite number of elements from $\mathscr{Z}$. If $\mathscr{F}$ is any $\mathscr{Z}$-filter converging to x, then there must be a $Z \in \mathscr{F}$ such that $Z \subset \{x\}$, for $\{x\}$ is an open set. Consequently by definition of $\mathscr{Z}$, $y \in \{x\}$, which is a contradiction.

This pathological example would not occur if $\mathscr{Z}$ had been chosen above to be a disjunctive collection of sets. However when the condition that $\mathscr{Z}$ be disjunctive is applied in (5) of the previous proposition one sees that the point $\mathscr{Z}$-filters converge. Furthermore for (Bourbaki) filters, the neighborhoods of a given point $x \in X$ is a filter converging to x. Consequently one always has *a filter* converging to any point $x \in X$. To have such a thing occur with $\mathscr{Z}$-filters and $\mathscr{Z}$-sets which are also neighborhoods the following concept, stronger than disjunctive, is needed.

4.17 Definition The collection $\mathscr{Z}$ on a non-empty topological space X, is called a *discriminating base* for X if for every neighborhood $N(x)$ of a point x in X, there is a $Z \in \mathscr{Z}$ such that $x \in \operatorname{int} Z \subset Z \subset N(x)$.

Notice that the power set of X is a discriminating base for X with respect to any topology on X. In 4.3, it was stated that in completely regular spaces the collection $Z(X)$ of all zero-sets on X is a discriminating base. For discriminating collections $\mathscr{Z}$ one has the following properties.

4.18 Proposition *Let $\mathscr{Z}$ be a discriminating base on the topological space X. Then the following statements hold.*

(1) *The collection $\mathscr{Z}$ is also disjunctive.*

(2) *If $\mathscr{V}(x) = \{Z \in \mathscr{Z}: Z$ is a neighborhood of $x\}$ then $\mathscr{V}(x)$ is a $\mathscr{Z}$-filter converging to x.*

(3) *The point x is a limit point of a $\mathscr{Z}$-filter $\mathscr{F}$ if and only if $\mathscr{V}(x) \subset \mathscr{F}$.*

(4) *The point x is a cluster point of a $\mathscr{Z}$-filter $\mathscr{F}$ if and only if there is a $\mathscr{Z}$-filter $\mathscr{F}'$ converging to x and satisfying $\mathscr{F} \subset \mathscr{F}'$.*

Proof. The first three statements are left as exercises. To show (4) let $\mathscr{B} = \{F \cap Z \colon F \in \mathscr{F},\ Z \in \mathscr{V}(x)\}$ where x is a cluster point of $\mathscr{F}$. Then $\mathscr{B}$ generates a $\mathscr{Z}$-filter $\mathscr{F}'$ on X and $\mathscr{F} \subset \mathscr{F}'$ as well as $\mathscr{V}(x) \subset \mathscr{F}'$. By (3) above x is a limit point of $\mathscr{F}'$ (4.15).

Conversely by 4.16 if x is a limit point of $\mathscr{F}'$ then x is a cluster point of any $\mathscr{L}$-filter $\mathscr{F} \subset \mathscr{F}'$. □

4.19 DEFINITION The $\mathscr{L}$-filter $\mathscr{V}(x)$ defined in 4.18 above is called the *neighborhood* $\mathscr{L}$-filter *associated with the point* x.

These $\mathscr{L}$-filters are important ones in the study of convergence. More will be said about them later on. But first as a corollary to 4.18, one has the following relationship between cluster points and limit points of $\mathscr{L}$-ultrafilters.

4.20 COROLLARY *Let $\mathscr{L}$ be a discriminating base for a topological space X and let $\mathscr{F}$ be a $\mathscr{L}$-ultrafilter. A point $x \in X$ is a cluster point of $\mathscr{F}$ if and only if x is a limit point of $\mathscr{F}$.*

For $\mathscr{L}$-filters $\mathscr{F}$ which are not $\mathscr{L}$-ultrafilters, a cluster point x of $\mathscr{F}$ means that $\mathscr{F}$ is contained in the point $\mathscr{L}$-filter $\mathscr{F}_x$ (as long as $\mathscr{L}$ satisfies the hypotheses of the next proposition).

4.21 PROPOSITION *Let $\mathscr{L}$ be a disjunctive collection of closed subsets of X. A point $x \in X$ is a cluster point of a $\mathscr{L}$-filter $\mathscr{F}$ if and only if $\mathscr{F} \subset \mathscr{F}_x$. Consequently, distinct $\mathscr{L}$-ultrafilters cannot have a common cluster point.*

Proof. Let x be in X and let $x \notin F \in \mathscr{F}$. If x is a cluster point of the $\mathscr{L}$-filter $\mathscr{F}$ then $X - F$ must meet every member of $\mathscr{F}$ which clearly is not the case. Consequently $x \in \cap \mathscr{F}$ and $\mathscr{F} \subset \mathscr{F}_x$. On the other hand since $\mathscr{F}_x$ converges to x, it follows that if $\mathscr{F} \subset \mathscr{F}_x$ then x is a cluster point of $\mathscr{F}$. □

4.22 COROLLARY *If $\mathscr{L}$ is a discriminating base of closed subsets of X, then $\mathscr{F}_x$ is the unique $\mathscr{L}$-ultrafilter converging to the point $x \in X$.*

This proposition shows its merit in the fact that distinct $\mathscr{L}$-ultrafilters cannot have a common cluster point. This is not the case when $\mathscr{L} = 2^X$. Consequently for our purposes a distinct advantage is demonstrated by asking that $\mathscr{L}$ be naturally bound to the topology on X. The following example helps to clarify the situation.

Let us consider the one-point compactification $X = N^*$ of the natural numbers $\mathbf{N}$. Let V be the even integers and let O be the odd integers. It is now possible to obtain two (Bourbaki) ultrafilters $\mathscr{U}_1$ and $\mathscr{U}_2$ which both converge to the point α adjoined to $\mathbf{N}$ to obtain this compactification. In particular let $\mathscr{Z} = 2^X$ and let

$$\mathscr{F}_1 = \{Z \in \mathscr{Z}: (N^* - Z) \cap V \text{ is finite}\},$$

$$\mathscr{F}_2 = \{Z \in \mathscr{Z}: (N^* - Z) \cap O \text{ is finite}\}.$$

Then $\mathscr{F}_1$ and $\mathscr{F}_2$ are contained in ultrafilters $\mathscr{U}_1$ and $\mathscr{U}_2$ respectively. Now $\mathscr{U}_1 \neq \mathscr{U}_2$ since $V \in \mathscr{F}_1$ and $O \in \mathscr{F}_2$. On the other hand the neighborhood filter $\mathscr{V}(\alpha)$ associated with α is contained in both $\mathscr{F}_1$ and $\mathscr{F}_2$ since open sets G containing α have finite complements. Consequently both $\mathscr{U}_1$ and $\mathscr{U}_2$ converge to α.

For our purposes this situation is unacceptable. Later in our construction of compactifications for Tychonoff spaces, each $\mathscr{Z}$-ultrafilter $\mathscr{F}$ will be associated with a point (either inside or outside of the space) to which $\mathscr{F}$ will 'converge'. The reason for the quotes should become clearer later. However it will be essential to identify one point to one and only one $\mathscr{Z}$-ultrafilter. As 4.21 shows, this is the case when $\mathscr{Z}$ is bound to the topology on X. In fact even in the example above $\mathscr{Z}$ can be so chosen so that α is associated in this 'convergence' sense with one and only one $\mathscr{Z}$-ultrafilter.

On the other hand, it would be well to know when $\mathscr{Z}$-ultrafilters have unique limit points (whenever one exists). For Hausdorff spaces limit points must be unique.

4.23 Proposition *If $\mathscr{Z}$ is a collection of subsets closed under finite intersections and if X is a Hausdorff space then $\mathscr{Z}$-filters which converge have unique limit points. Moreover if $\mathscr{Z}$ is a discriminating base and if convergent $\mathscr{Z}$-filters have unique limit points then X must be Hausdorff.*

Proof. The first part is left as an exercise. If X is not a Hausdorff space, then for distinct points x and y in X, the neighborhood $\mathscr{Z}$-filters $\mathscr{V}(x)$ and $\mathscr{V}(y)$ have all respective members $N(x)$ and $N(y)$ satisfying $N(x) \cap N(y) \neq \varnothing$. Consequently x is a cluster point of $\mathscr{V}(y)$ and hence $\mathscr{V}(y)$ must be contained in a $\mathscr{Z}$-filter $\mathscr{F}$ converging to x (4.18). But then $\mathscr{V}(x)$ must also be contained in $\mathscr{F}$ (again by 4.18). Hence the $\mathscr{Z}$-filter $\mathscr{F}$ has distinct limit points x and y. □

In 4.12 it was shown how prime $\mathscr{Z}$-filters can also be used to determine the compactness of a space. The merits of prime $\mathscr{Z}$-filters are established by the fact that although they need not be $\mathscr{Z}$-ultrafilters, they are a subcollection of a unique $\mathscr{Z}$-ultrafilter (for suitable hypothesis on $\mathscr{Z}$).

4.24 PROPOSITION *Let $\mathscr{Z}$ be a normal ring of subsets of X. Then every prime $\mathscr{Z}$-filter is contained in a unique $\mathscr{Z}$-ultrafilter.*

Proof. Any $\mathscr{Z}$-filter $\mathscr{F}$ is contained in a $\mathscr{Z}$-ultrafilter $\mathscr{U}$ by 4.8. If $\mathscr{F}$ is also a prime $\mathscr{Z}$-filter contained in the $\mathscr{Z}$-ultrafilter $\mathscr{G} \neq \mathscr{U}$ then there are sets $A \in \mathscr{G}$ and $B \in \mathscr{U}$ for which $A \cap B = \varnothing$. Since $\mathscr{Z}$ is normal, there are also sets C and D in $\mathscr{Z}$ for which $A \subset X - C$, $B \subset X - D$ and $C \cup D = X$. Since $\mathscr{Z}$ is a ring, $X \in \mathscr{Z}$ and therefore $X \in \mathscr{F}$. Since $\mathscr{F}$ is prime, without loss of generality, assume $C \in \mathscr{F} \subset \mathscr{G}$. But since A was already in $\mathscr{G}$, this is impossible for $A \cap C = \varnothing$. A similar argument shows that D also may not be in $\mathscr{F}$. □

From this it appears that convergence properties of $\mathscr{Z}$-ultrafilters could be reduced to prime $\mathscr{Z}$-filters. This is the case to some extent as the next proposition and example demonstrate.

4.25 PROPOSITION *Let $\mathscr{Z}$ be a ring of sets that is a discriminating base and that is also a base for the closed subsets of a T_1-space X. If $\mathscr{F}$ is a prime $\mathscr{Z}$-filter and if $x \in X$, then the following statements are equivalent.*

(1) *The point x is a cluster point of $\mathscr{F}$.*

(2) *The $\mathscr{Z}$-filter $\mathscr{F}$ converges to x.*

(3) $\cap \{F: F \in \mathscr{F}\} = \{x\}$.

Proof. If x is a cluster point of $\mathscr{F}$ then $\mathscr{F} \subset \mathscr{F}_x$ (4.21). Since $\mathscr{F}_x$ converges to x, the neighborhood $\mathscr{Z}$-filter $\mathscr{V}(x) \subset \mathscr{F}_x$ (4.18(3)). Let $Z \in \mathscr{V}(x)$, that is $x \in \operatorname{int} Z$. Now $x \notin F = X - \operatorname{int} Z$. Consequently there is a $Z_1 \in \mathscr{Z}$ for which $x \notin Z_1$ and $F \subset Z_1$. Now $Z_1 \cup Z = X \in \mathscr{Z}$ (and therefore $X \in \mathscr{F}$ for $\mathscr{Z}$ is a base for the closed subsets) and $x \in X - Z_1 \subset \operatorname{int} Z \subset Z$. Since $\mathscr{F}$ is a family of closed subsets, the cluster point x must be in $\cap \mathscr{F}$. Consequently $Z_1 \notin \mathscr{F}$. Primeness of $\mathscr{F}$ says then that Z must be in $\mathscr{F}$. Thus $\mathscr{V}(x) \subset \mathscr{F}$ and $\mathscr{F}$ must converge to x.

The implication that (3) *implies* (1) is immediate and the implication that (2) *implies* (3) is left as an exercise. □

It might appear from this proposition that convergence of a prime $\mathscr{Z}$-filter $\mathscr{F}$ to the point x implies that $\mathscr{F}$ is the $\mathscr{Z}$-ultrafilter $\mathscr{F}_x$. However this is not the case.

Let X be the real numbers (with the usual topology) and let $\mathscr{Z} = \{Z = \operatorname{cl} Z \subset 2^X : 0$ is not a limit point of Z and on the boundary of $Z\}$. It can be shown that $\mathscr{Z}$ is a ring of sets which is a discriminating base and a base for the closed subsets of X. The only closed intervals which do not belong to $\mathscr{Z}$ are those which have 0 as an end point. Now the collection

$$\mathscr{F} = \{Z \in \mathscr{Z} : 0 \in \operatorname{int} Z\}$$

is a prime $\mathscr{Z}$-filter converging to 0. If $A \cup B \in \mathscr{F}$, $A, B \in \mathscr{Z}$ then

$$0 \in \operatorname{int}(A \cup B).$$

Assume $0 \in A$. If $0 \in \operatorname{int} A$ then $A \in \mathscr{F}$. If $0 \notin \operatorname{int} A$, then 0 cannot be a limit point of A by definition of $\mathscr{Z}$. Thus there are open sets U and V where $0 \in U \cap V$ and for which $(U \cap A) - \{0\} = \varnothing$ and $V \subset \operatorname{int}(A \cup B)$. Thus $U \cap V \subset B$ and $0 \in \operatorname{int} B$. Therefore $B \in \mathscr{F}$ and $\mathscr{F}$ is prime.

Now $\mathscr{F} \cup \{0\}$ clearly generates a $\mathscr{Z}$-filter and it properly contains $\mathscr{F}$. Thus the prime $\mathscr{Z}$-filter $\mathscr{F}$ cannot be the point $\mathscr{Z}$-filter $\mathscr{F}_0$.

In the next section, we will be interested in having extensions of our concepts of cluster point and limit point. The situation there is that X will be a dense subset of a Tychonoff space Y. If $\mathscr{Z}$ is a ring of sets on X, then we shall define a point y in Y being a cluster point of a $\mathscr{Z}$-filter $\mathscr{F}$ on X in the following manner.

4.26 Definition Let X be a subset of Y. The point $y \in Y$ is a *cluster point of a* $\mathscr{Z}$-*filter* F *on* X if every neighborhood (in Y) of y meets every member of $\mathscr{F}$. The point y is a *limit point of the*

$\mathscr{Z}$-filter $\mathscr{F}$ on X (or *$\mathscr{F}$ converges to the limit y*) if every neighborhood (in Y) of y contains a member of $\mathscr{F}$.

Thus y is a cluster point of $\mathscr{F}$ if y belongs to $\bigcap \{\mathrm{cl}_Y Z \colon Z \in \mathscr{F}\}$. If Y is compact, then every $\mathscr{Z}$-filter on $\mathscr{F}$ must have a cluster point in Y. On the other hand if $\mathscr{Z}$ is a ring of sets on Y satisfying the conditions of 4.25 and if every $\mathscr{Z}|X$ filter on X has a cluster point in Y, then Y is compact. Again limits will be unique as was the case in 4.23. However two distinct $\mathscr{Z}$-filters (or $\mathscr{Z}$-ultrafilters) on X may have the same limit in Y as was demonstrated in the example after 4.22.

4.27 EXERCISES

(1) Prove 4.3, 4.5, 4.16, 4.18, and 4.23.

(2) Show that for Tychonoff spaces $Z(X)$ is a normal base.

(3) If Φ is a totally ordered collection of $\mathscr{Z}$-filters on a non-empty set X, then $\bigcup \Phi$ is a $\mathscr{Z}$-filter on X. Consequently, using Zorn's lemma, 4.8 follows.

(4) Suppose that X is a Tychonoff space and that $\mathscr{F}$ is a $Z(X)$-filter on X. Then $\mathscr{F}$ is prime if and only if for all Z_1 and Z_2 in $Z(X)$ with $Z_1 \cup Z_2 = X$ it is true that Z_1 is in $\mathscr{F}$ or Z_2 is in $\mathscr{F}$.

(5) In discrete spaces all filters are $Z(X)$-filters and every prime filter is an ultrafilter. Give an example of a space X and a prime $\mathscr{Z}$-filter that is not a $\mathscr{Z}$-ultrafilter. Every (Bourbaki) prime filter is an ultrafilter.

(6) Let $(X, \mathscr{T})$ be a Tychonoff space and let d be a continuous pseudometric on X. If $\epsilon > 0$ then there exist sets $Z_{n,x}$ ($n \in \mathbf{N}, x \in X$) with the following properties.

(*a*) The space $X = \bigcup \{Z_{n,x} \colon n \in \mathbf{N} \text{ and } x \in X\}$.

(*b*) Each set $Z_{n,x}$ is d-closed (and therefore a zero-set) and of d-diameter less than ϵ.

(*c*) For each $n \in \mathbf{N}$, the family $(Z_{n,x})_{x \in X}$ of subsets of X is d-discrete. Thus from (*b*) and exercise 2.19(17) the union of this family is a zero-set in $(X, \mathscr{T})$.

(The convention is that $d(\varnothing) = 0$ and $d(\varnothing, A) = \infty > r$ for $r \in \mathbf{R}$. If $<$ is a well ordering of X, for each fixed $n \in \mathbf{N}$ proceed by induction on X by letting $\delta = \frac{1}{2}\epsilon$ and by putting z in $Z_{n,x}$ if and only if $d(Z_{n,y}, z) \geqslant \delta/n$ for all $y < x$ and $d(x, z) \leqslant \delta - \delta/n$. Then (*b*) and (*c*) are satisfied. For any $z \in X$, there is a least element x in X for which $d(x, z) < \delta$. Let $n \in \mathrm{N}$ be such that $d(x, z) < \delta - \delta/n$. If $y < x$ then $d(Z_{n,y}, z) \geqslant \delta/n$ and $z \in Z_{n,x}$.)

Thus if $\mathscr{F}$ is a *real* (see 5.7) $Z(X)$-ultrafilter on a Tychonoff space X there exists a family of non-empty subsets $(Z_\alpha)_{\alpha \in I}$ of X for which

(*d*) The union $\bigcup_{\alpha \in I} Z_\alpha \in \mathscr{F}$.

(*e*) Each Z_α is of d-diameter less than ϵ.

(*f*) The family is d-discrete.

(*g*) The union of every subfamily is a zero-set.

5 Compactifications and realcompletions

Let us now assume that $\mathscr{Z}$ is a normal base for a T_1 topological space X. It has already been seen that the point $\mathscr{Z}$-filters $\mathscr{F}_x = \{Z \in \mathscr{Z}: x \in X\}$ are $\mathscr{Z}$-ultrafilters. Since the space X is a T_1-space, there is a one-to-one correspondence between the points x in X and the set of all point $\mathscr{Z}$-ultrafilters. Let $\omega(\mathscr{Z})$ be the collection of all $\mathscr{Z}$-ultrafilters. Since there is this one-to-one correspondence between X and a subset of $\omega(\mathscr{Z})$, let us define a topology on $\omega(\mathscr{Z})$ that will permit X to be homeomorphic to a dense subset of $\omega(\mathscr{Z})$ with this topology.

To this end for each $Z \in \mathscr{Z}$, let $Z^\omega = \{\mathscr{F} \in \omega(\mathscr{Z}): Z \in \mathscr{F}\}$, and let $\mathscr{Z}^\omega = \{Z^\omega: Z \in \mathscr{Z}\}$.

5.1 PROPOSITION *For any normal base $\mathscr{Z}$ on a space X, the following statements hold.*

(1) *For any A and B in $\mathscr{Z}$,*

$$A^\omega \cap B^\omega = (A \cap B)^\omega \quad \textit{and} \quad A^\omega \cup B^\omega = (A \cup B)^\omega.$$

(2) *The collection $\mathscr{Z}^\omega$ is a normal collection.*

(3) *If $Z \in \mathscr{Z}$ then* $\omega(\mathscr{Z}) - Z^\omega = \{\mathscr{F} \in \omega(\mathscr{Z})$: for some $A \in \mathscr{F}$, $A \subset X - Z\}$.

Motivated by (3) above, let us write $\omega(\mathscr{Z}) - Z^\omega$ as $(X - Z)^\omega$ whenever $Z \in \mathscr{Z}$.

Since $\mathscr{Z}$ is a base for the closed subsets of X, this proposition says that $\mathscr{Z}^\omega$ may be taken as a base for a topology for $\omega(\mathscr{Z})$. This topology will manifest itself as a compact Hausdorff topology.

In a Tychonoff space, the collection of all zero-sets on X forms a normal base. Consequently let us assume that $\mathscr{Z}$ is a normal base on a T_1-space X and let us consider the normal collection $\mathscr{Z}^\omega$ as a base for the closed subsets of $\omega(\mathscr{Z})$. In fact $\mathscr{Z}^\omega$ will be a normal base for $\omega(\mathscr{Z})$.

For any $\mathscr{Z}^\omega$-filter and for $\mathscr{F}^\omega$ on $\omega(\mathscr{Z})$, let

$$\mathscr{F} = \{Z \in \mathscr{Z} : Z^\omega \in \mathscr{F}^\omega\}.$$

By 5.1, $\mathscr{F}$ is a family of $\mathscr{Z}$-sets with the finite intersection property. Consequently $\mathscr{F}$ is contained in a $\mathscr{Z}$-ultrafilter $\mathscr{G}$. If Z is in $\mathscr{G}$ then $\mathscr{G}$ is an element of Z^ω for each $Z^\omega \in \mathscr{F}^\omega$. Consequently $\mathscr{F}^\omega$ is fixed and $\omega(\mathscr{Z})$ is a compact space with this topology. If $\mathscr{F}$ and $\mathscr{G}$ are two distinct $\mathscr{Z}^\omega$-ultrafilters on X then there is an $F \in \mathscr{F}$ and $G \in \mathscr{G}$ for which $F \cap G = \varnothing$. Now the normality of $\mathscr{Z}$ gives $\mathscr{Z}$-sets A and B such that

$$F \subset X - A \subset B \subset X - G \quad \text{and} \quad A \cup B = X.$$

Thus $(X-A)^\omega$ and $(X-B)^\omega$ are disjoint open sets containing the distinct points $\mathscr{F}$ and $\mathscr{G}$. Consequently $\omega(\mathscr{Z})$ is Hausdorff. Let ϕ be the one-to-one map of X onto $\phi(X)$ taking points x in X to the point $\mathscr{Z}$-ultrafilters $\mathscr{F}_x$. Since the injection ϕ maps the base $\mathscr{Z}$ onto the base $\phi(X) \cap \mathscr{Z}^\omega$ for the relative topology on $\phi(X)$, ϕ is a homeomorphism of X onto $\phi(X)$, a subset of a compact Hausdorff space. Since compact Hausdorff spaces are normal Hausdorff spaces and therefore Tychonoff spaces, it follows that X must be a Tychonoff space since this separation axiom is hereditary. However, there has been constructed much more. If $\phi(X)$ is a dense subset of $\omega(\mathscr{Z})$ then a compactification of X has been constructed. Thus let us show that $\phi(X)$ is dense in $\omega(\mathscr{Z})$. A non-empty basic open set in $\omega(\mathscr{Z})$ is of the form $\omega(\mathscr{Z}) - Z^\omega$ for $Z \in \mathscr{Z}$. Thus if $\mathscr{F} \in (X-Z)^\omega$ then there is an $A \in \mathscr{F}$ such that $A \cap Z = \varnothing$. Let $x \in A$. Then $\mathscr{F}_x \notin Z$. Consequently

$$\mathscr{F}_x \in (X-Z)^\omega \cap \phi(X).$$

Thus $\phi(X)$ is dense in $\omega(\mathscr{Z})$.

5.2 Definition The compactification $\omega(\mathscr{Z})$ constructed above is called the *Wallman–Frink compactification* of a Tychonoff space with normal base $\mathscr{Z}$.

In [379], Wallman first gave a construction similar to the above for T_1 spaces. In general his compactification was not Hausdorff.

In a paper by Banaschewski [44] and in a paper by Shanin [336] similar results were obtained. In [129], Orrin Frink introduced normal bases and raised the question as to whether every compactification of a Tychonoff space is obtained from some normal base on X. Consequently these compactifications are now referred to as Wallman–Frink compactifications.

By constructing the above compactification for a T_1-space X with a normal base $\mathscr{Z}$, the following theorem has just been demonstrated.

5.3 THEOREM (*Frink*) *A T_1-space X is a Tychonoff space if and only if it has a normal base.*

In a first year course in set theoretic topology, it is usually shown (see [XVIII]) that the Tychonoff spaces are just the subspaces of of products of closed real line intervals (these products are often referred to as *cubes*). That is to say that the Tychonoff spaces have enough set theoretical structure to map them homeomorphically into a Tychonoff product space P which is composed of a product of closed intervals. Such a product space, by the Tychonoff product theorem for compactness, of course is a compact Hausdorff space. Consequently the closure $\mathrm{cl}_P X$ of this homeomorphic copy of the original space X is a compact Hausdorff subspace of the space P.

If this product space P is obtained by taking one copy of the unit interval $E = [0, 1]$ for each $f \in C(X, E)$, then $\mathrm{cl}_P X$ is called the *Stone–Čech compactification* βX of X. It is then shown that X (let us now identify X with its homeomorphic copy in P) is C^*-embedded in βX and that if there is any other compactification Y of X in which X is C^*-embedded then there is a homeomorphism of βX onto Y which leaves points of X fixed (see [XVIII]). Consequently, in this sense, βX is a unique compactification of X.

It is not difficult to check that X is C^*-embedded in βX. For in this case, $P = \Pi\{E_f : f \in C(X, E)\}$ where $E_f = E$ for each $f \in C(X, E)$ (for all closed intervals are homeomorphic). Consequently the projection map pr_f from P into $E_f = E$, when restricted to βX, will serve as the extension of $f \in C(X, E)$ to βX. From here the reader should also be able to check the uniqueness of βX as mentioned above.

With this in mind, it may now be conveniently shown that if $\mathscr{Z}$ is the normal base $Z(X)$ of all zero-sets on a Tychonoff space X, then $\omega(Z(X))$ is the Stone–Čech compactification βX of X. This follows as a corollary to the next theorem which is of interest in its own right.

For any normal base $\mathscr{Z}$ on a Tychonoff space X, it is possible to give some interesting equivalences for X to be C^*-embedded in a Tychonoff space Y in which X is dense. One fact that this theorem will use is given in the exercises. The reader is asked to prove for a normal base $\mathscr{Z}$ on X, if Z_1 and Z_2 are two disjoint members of $\mathscr{Z}$ then there is a continuous function $f \in C(X, E)$ such that $f(x) = 0$ if $x \in Z_1$ and $f(x) = 1$ if $x \in Z_2$. This is reasonable to expect since in Tychonoff spaces, $Z(X)$ is a normal base for X and moreover Tychonoff spaces are weaker than normal Hausdorff spaces yet stronger than regular Hausdorff spaces. In a normal space this statement of course is true for the normal base of all closed subsets of X (see 6.4). Here one asks the existence of such functions $f \in C(X, E)$ only for the members of the base $\mathscr{Z}$.

Below, the characterizations are given for the particular normal base of all zero-sets on a Tychonoff space X. The reader is asked to provide the proofs of the analogues for an arbitrary normal base $\mathscr{Z}$ on X.

5.4 THEOREM *If X is a dense subset of a Tychonoff space Y then the following statements are equivalent.*

(1) *Every continuous function from X into any compact Hausdorff space has a continuous extension to Y.*

(2) *The space X is C^*-embedded in Y.*

(3) *If Z_1 and Z_2 are in $Z(X)$ and if $Z_1 \cap Z_2 = \varnothing$ then*

$$\mathrm{cl}_Y Z_1 \cap \mathrm{cl}_Y Z_2 = \varnothing.$$

(4) *If Z_1 and Z_2 are in $Z(X)$ then $\mathrm{cl}_Y Z_1 \cap \mathrm{cl}_Y Z_2 = \mathrm{cl}_Y(Z_1 \cap Z_2)$.*

(5) *Every point of Y is the limit of a unique $\mathscr{Z}$-ultrafilter on X.*

Proof. If $f \in C^*(X)$ then f maps X into a bounded interval of the real line $\mathbf{R}$. Consequently f maps X into a compact Hausdorff subspace of $\mathbf{R}$ and thus (1) *implies* (2).

Let Z_1 and Z_2 be two disjoint members of $\mathscr{Z}$. By Exercise 3, there exists an $f \in C(X, E)$ such that $f(x) = 0$ for $x \in Z_1$ and $f(x) = 1$ for $x \in Z_2$. By (2) there is an $f^* \in C(Y, E)$ such that $f^*|X = f$. Consequently $\mathrm{cl}_Y Z_1 \cap \mathrm{cl}_Y Z_2 = \varnothing$ and (2) *implies* (3).

Of course $\mathrm{cl}_Y (Z_1 \cap Z_2)$ is always contained in $\mathrm{cl}_Y Z_1 \cap \mathrm{cl}_Y Z_2$. On the other hand, for $x \in \mathrm{cl}_Y Z_1 \cap \mathrm{cl}_Y Z_2$, if A is a zero-set (in Y) neighborhood of x, then $x \in \mathrm{cl}_Y (Z_1 \cap A)$ and $x \in \mathrm{cl}_Y (Z_2 \cap A)$. Since the finite intersection of zero-sets is a zero-set, it follows by (3) that A meets $Z_1 \cap Z_2$ and that $x \in \mathrm{cl}_Y (Z_1 \cap Z_2)$. Thus (3) *implies* (4).

The implications that (4) *implies* (5) and (5) *implies* (1) are left as exercises. □

5.5 Corollary *If X is a Tychonoff space, then the Wallman–Frink compactification, $\omega(Z(X))$ of X is homeomorphic to the Stone–Čech compactification of X.*

Proof. This follows from the uniqueness aspect of βX, 5.4 and Exercise 10. □

Having seen the Wallman compactification of a Tychonoff space X and having shown that it is the Stone–Čech compactification of a space when one considers the normal base of all zero-sets on X, let us consider one of its most interesting subspaces. To do this it is necessary to strengthen the concept of a normal base.

5.6 Definition A collection $\mathscr{Z}$ of subsets of a non-empty set X is said to be a *delta collection* if whenever $(Z_n)_{n \in \mathbf{N}}$ is a countable subcollection of $\mathscr{Z}$ then $Z = \bigcap_{n \in \mathbf{N}} Z_n$ is also a member of $\mathscr{Z}$. The collection $\mathscr{Z}$ is said to be *complement generated* if for each $Z \in \mathscr{Z}$ there is a countable collection $(Z_n)_{n \in \mathbf{N}}$ of members of $\mathscr{Z}$ such that $Z = \bigcap_{n \in \mathbf{N}} (X - Z_n)$. A normal base $\mathscr{Z}$ is said to be a *strong delta normal base* if $\mathscr{Z}$ is also a delta collection and complement generated. A subcollection $\mathscr{F}$ of $\mathscr{Z}$ is said to have the *countable intersection property* if whenever $(Z_n)_{n \in \mathbf{N}}$ is a countable subcollection of $\mathscr{F}$, the $\bigcap_{n \in \mathbf{N}} Z_n \neq \varnothing$.

If $\mathscr{Z}$ is a strong delta normal base then it is a normal base and one can construct $\omega(\mathscr{Z})$. Under the map ϕ which takes each point $x \in X$ to the fixed point $\mathscr{Z}$-ultrafilter $\mathscr{F}_x$, one has an identification of X with a subspace of $\omega(\mathscr{Z})$. These point $\mathscr{Z}$-ultrafilters also have the countable intersection property. Consequently there are at least as many $\mathscr{Z}$-ultrafilters with the countable intersection property as there are points in X. Our interest will now be centered on that subset $\rho(\mathscr{Z})$ of $\omega(\mathscr{Z})$ which consists of all $\mathscr{Z}$-ultrafilters with the countable intersection property. The topology on $\rho(\mathscr{Z})$ will be the relative topology it inherits as a subset of $\omega(\mathscr{Z})$. Thus X is homeomorphic (via the map ϕ) to a dense subset of $\rho(\mathscr{Z})$.

To have a fruitful discussion of $\rho(\mathscr{Z})$ it is important that $\mathscr{Z}$ be a strong delta normal base. Since our considerations will be for zero-set ultrafilters with the countable intersection property, let us investigate this concept a little further. Since there are $\mathscr{Z}$-filters without this property, let us make the following definition.

5.7 Definition A $\mathscr{Z}$-filter is said to be *real* if it has the countable intersection property.

5.8 Proposition *Let $\mathscr{Z}$ be a delta ring of sets which is a base for the closed subsets of X. Then X is Lindelöf if and only if every real $\mathscr{Z}$-filter is fixed.*

Proof. A Lindelöf space is one in which every open cover has a countable subcover. Consequently the proof is immediate using complementation and the techniques in 4.12. □

Since real $\mathscr{Z}$-filters are pertinent to the Lindelöf property, one might ask if Lindelöf may also be characterized, as was done in 4.12, in terms of real prime $\mathscr{Z}$-filters and real $\mathscr{Z}$-ultrafilters. This is not the case. It will be shown that if a space has every real $\mathscr{Z}$-ultrafilter fixed, then a Tychonoff product of any number of such spaces also has every real $\mathscr{Z}$-ultrafilter fixed. In Section 9 the *Sorgenfry* line (the real line with half open intervals as a base for the open sets) and the *Sorgenfry* plane (the product of the Sorgenfry line with itself) will be discussed. It is easy to show that the Sorgenfry line is Lindelöf and therefore every real $\mathscr{Z}$-ultrafilter is fixed. The Sorgenfry plane also has every real $\mathscr{Z}$-ultrafilter fixed but it will be shown not to be Lindelöf.

5.9 DEFINITION The space X is said to be *realcomplete* if every real $Z(X)$-ultrafilter on X is fixed. *A realcompletion* of X is a realcomplete Hausdorff space that contains X densely (up to homeomorphism).

Thus by the very definition of $\rho(Z(X))$, if X is a realcomplete Tychonoff space then X is homeomorphic to $\rho(Z(X))$ (for now the map ϕ is one-to-one onto). In a short while it will become clear however that $\rho(\mathscr{Z})$ is always realcomplete for any strong delta normal base on X.

We immediately have that the real line is realcomplete since

5.10 PROPOSITION *Every Lindelöf space is realcomplete.*

Thus the Sorgenfry line and the Sorgenfry plane are realcomplete. (The latter follows readily once it is shown that the product of realcomplete spaces is realcomplete. Exercise 11 tells how to do it.) Since the Sorgenfry plane is not Lindelöf, there must be some free real $Z(X)$-filters. Consequently the example also shows that not every *real* $Z(X)$-filter is contained in a real $Z(X)$-ultrafilter which is not the case for $Z(X)$-filters and $Z(X)$-ultrafilters. However for real prime $\mathscr{Z}$-filters one has the following.

5.11 THEOREM *If $\mathscr{Z}$ is a delta ring of sets which is normal and complement generated then every real prime $\mathscr{Z}$-filter $\mathscr{F}$ is contained in a unique real $\mathscr{Z}$-ultrafilter $\mathscr{U}$. Moreover $\mathscr{F}$ is closed under countable intersections if and only if $\mathscr{F} = \mathscr{U}$.*

Proof. By 4.24, $\mathscr{F}$ is embeddable in a unique $\mathscr{Z}$-ultrafilter $\mathscr{U}$. It remains then to show that $\mathscr{U}$ has the countable intersection property. Let $\mathscr{B} = (B_n)_{n\in\mathbf{N}}$ be a countable subcollection of $\mathscr{U}$ such that the intersection is empty. To prove the result, a countable subcollection of $\mathscr{F}$ will be obtained with an empty intersection. Since $\mathscr{Z}$ is complement generated, for each B_n there is a countable subcollection $\mathscr{A}_n$ of $\mathscr{Z}$ such that $X - B_n = \bigcup \mathscr{A}_n$. Now the collection $\mathscr{A} = (\mathscr{A}_n)_{n\in\mathbf{N}}$ is a countable collection of sets and $X = \bigcup_{n\in\mathbf{N}} X - B_n = \bigcup \mathscr{A}$. For each $A_n \in \mathscr{A}$, there is a $B \in \mathscr{B}$

such that $A_n \subset X - B$. (Note that there may be more than one A_n associated in this way with a $B \in \mathscr{B}$.) Since $\mathscr{Z}$ is normal there are sets C_n and D_n in $\mathscr{Z}$ such that $A_n \subset X - C_n \subset D_n \subset X - B$ and $C_n \cup D_n = X$. Since $\mathscr{Z}$ is a ring, $X \in \mathscr{Z}$ and therefore $X \in \mathscr{F} \subset \mathscr{U}$. Now for each n, $D_n \notin \mathscr{F}$ for then $\varnothing = D_n \cap B \in \mathscr{U}$ which is impossible. Since $\mathscr{F}$ is prime, C_n must be in $\mathscr{F}$ for each $n \in \mathbf{N}$. However $\bigcap_{n \in \mathbf{N}} C_n \subset \bigcap_{n \in \mathbf{N}} X - A_n = \varnothing$, since $\mathscr{A}$ covers X. □

As a corollary to this theorem, it is possible to give an equivalent formulation for realcompleteness. The Sorgenfry plane showed that realcompleteness, in contrast to compactness, does not imply that every real $Z(X)$-filter is fixed. However it does imply, as does compactness, that every prime real $Z(X)$-filter is fixed.

5.12 COROLLARY *A Tychonoff space X is realcomplete if and only if every prime real $Z(X)$-filter is fixed.*

Proof. The sufficiency of the condition for realcompleteness is, of course, obvious. On the other hand, if $\mathscr{F}$ is a real prime $Z(X)$-filter, then the theorem permits embedding $\mathscr{F}$ into a real $Z(X)$-ultrafilter. Consequently both must be fixed if X is realcomplete. □

When discussing realcomplete spaces the following generalization of the closure concept is very useful (see [286]).

5.13 DEFINITION Let A be a non-empty subset of a topological space X. The *realclosure* (also called the *Q-closure*) of A is the set of all points $x \in X$ for which every G_δ-set G containing x has a non-empty intersection with A. Let us write $\operatorname{rcl} A$ for the realclosure of A. The subset A is *realclosed* if $\operatorname{rcl} A = A$.

In general the realclosure of a set need not be closed but it is realclosed. Indeed any open interval of the real line is realclosed. However any subset A is contained in its realclosure which in turn, is contained in the closure of A.

In the exercises the reader is asked to show that not only every closed subset of a realcomplete space is realcomplete but also every realclosed subset is realcomplete. From this follows the following characterization of realcompleteness.

5.14 THEOREM *A Tychonoff space X is realcomplete if and only if it is realclosed in its Stone–Čech compactification.*

Proof. Since any compact Hausdorff space is realcomplete, the sufficiency of the condition implies that X must be realcomplete by Exercise 11. Consequently let $\mathscr{F}$ be any point in βX (that is, $\mathscr{F}$ is a $Z(X)$-ultrafilter) which is not in the realcomplete space X (really the homeomorphic image of X). Thus $\mathscr{F}$ is not a real $Z(X)$-ultrafilter and therefore there is a sequence of zero-sets $(Z_n)_{n\in\mathbf{N}}$ in $\mathscr{F}$ (as in Exercise 14) such that $\bigcap_{n\in\mathbf{N}} Z_n = \varnothing$. Now each zero-set Z_n is a G_δ-set in X and moreover $\bigcap_{n\in\mathbf{N}} Z_n^\omega$ is a G_δ-set in $\beta X = \omega(Z(X))$. The $Z(X)$-filter $\mathscr{F}$ belongs to the G_δ-set

$$G = \bigcap_{n\in\mathbf{N}} Z_n^\omega \quad \text{and} \quad G \cap X = \bigcap_{n\in\mathbf{N}} Z_n = \varnothing.$$

Thus X is realclosed in βX. □

5.15 COROLLARY *A Tychonoff space X is realcomplete if and only if it is realclosed in some compactification of X.*

Proof. This follows from 5.14 and Exercise 11.

For any strong delta normal base $\mathscr{L}$ on X, $\rho(\mathscr{L})$ is just the realclosure in $\omega(\mathscr{L})$ of X. Hence $\rho(\mathscr{L})$ is realcomplete.

5.16 THEOREM *For any strong delta normal base $\mathscr{L}$ on X, the realclosure of X in $\omega(\mathscr{L})$ is $\rho(\mathscr{L})$.*

Proof. Let $\mathscr{F}$ be any element of rcl X without the countable intersection property. Then as in the proof of 5.14, a G_δ-set may be constructed containing $\mathscr{F}$ and missing X. Thus the realclosure of X is contained in $\rho(\mathscr{L})$. On the other hand, if $\mathscr{F} \in \rho(\mathscr{L})$, let $\mathscr{F} \in G = \bigcap_{n\in\mathbf{N}} A_n^\omega$ be a G_δ-set in $\omega(\mathscr{L})$ where each A_n^ω is a basic open set in $\omega(\mathscr{L})$ (every G_δ-set containing $\mathscr{F}$ must contain such a G_δ-set containing $\mathscr{F}$). For each $n \in \mathbf{N}$ there is a Z_n in $\mathscr{F}$ for which $Z_n \subset A_n$. But $\bigcap_{n\in\mathbf{N}} Z_n \neq \varnothing$ since $\mathscr{F} \in \rho(\mathscr{L})$. Thus $G \cap X \subset \bigcap_{n\in\mathbf{N}} Z_n$ and $\mathscr{F}$ is in rcl X. □

5.17 Corollary *For any strong delta normal base $\mathscr{Z}$, X is homeomorphic to a dense subspace of the realcomplete space $\rho(\mathscr{Z})$.*

When $\mathscr{Z}$ is the particular strong delta normal base $Z(X)$ of all zero-sets on X, then $\rho(Z(X))$ is specially designated just as $\omega(Z(X))$ was.

5.18 Definition The realcompletion obtained by $\rho(Z(X))$ is called the *Hewitt–Nachbin realcompletion* of X and is denoted by υX (read upsilon X).

Thus from our remarks after 5.9, it is now clear that every realcomplete Tychonoff space X is its own Hewitt–Nachbin realcompletion.

Let us now look a little closer at the spaces $\rho(\mathscr{Z})$ and $\rho(Z(X))$ just constructed. Since $\rho(\mathscr{Z})$ was constructed as a subspace of $\omega(\mathscr{Z})$ it follows that X is C^*-embedded in $\rho(Z(X)) = \upsilon X$. However X is more than just C^*-embedded in υX, it is C-embedded there.

In Section 4 we defined what was meant by a point $w \in W$ being a cluster point or a limit point of a $\mathscr{Z}_X$-filter on the dense subset X. Now we may consider the realcompletion $\upsilon X = \rho(Z(X))$ of X. Then the identity map i on X is a map into the realcomplete space υX which contains X as a dense subset. Let us assume for the time being that every continuous function on X into a realcomplete space extends to a continuous function on W and let i^* be this extension of i. Thus for each $w \in W$ there is a unique real $Z(X)$-ultrafilter $i^*(w)$ on X. This follows from the very construction of $\rho(Z(X))$ and it follows from this construction that for $\phi^{\#}(\mathscr{F}) \cap \phi(X)$ on $\phi(X)$ each $\{\mathscr{F}\} \in \rho(Z(X))$ the real $Z(X)$-ultrafilter converges to $\{\mathscr{F}\}$ for $\{\mathscr{F}\}$ belongs to each $Z^{\rho} = \mathrm{cl}_{\rho(Z(X))}\phi(Z)$ where $Z \in \mathscr{F}$ (see Exercises 10(d) and 11). If U^{ρ} is any basic neighborhood in $\rho(Z(X))$ of $\{\mathscr{F}\}$ (that is, $X - U \in Z(X)$) then there is a $Z \in \mathscr{F}$ such that $Z \subset U$ (5.1). Consequently $\phi^{\#}(\mathscr{F})$ converges to $\{\mathscr{F}\}$ and it is unique. It is easy now to show that w is the limit of a unique real $Z(X)$-ultrafilter.

On the other hand in Exercise 11 we have given explicit hints on how to show the converse of the above. That is, if every point in W is the limit point of a unique real $Z(X)$-ultrafilter on the dense subset X, then every continuous function on X into any realcompact space extends to a continuous function on W. However note from our discussion above that each point in $\rho(Z(X))$ is the limit of a unique real $Z(X)$-ultrafilter on X.

Consequently the Hewitt–Nachbin realcompletion of X is a realcomplete Hausdorff space in which X is a C-embedded dense subset (see 6.9(1)). As was the analogous case with βX, if there is any other realcomplete Hausdorff space Y which contains X as a dense C-embedded subset then there is a homeomorphism of X onto Y which leaves X pointwise fixed. This follows from the following general proposition.

5.19 PROPOSITION *If X is dense in the Hausdorff spaces W and Y and if the identity map on X has continuous extensions i_1 from W into Y and i_2 from Y into W, then i_1 is a homeomorphism and $i_1^{-1} = i_2$.*

Now if W and Y are realcompletions of X each containing X as a C-embedded subset, let i_W be the identity map into W and let i_Y be the identity map into Y. Then each of the identity maps extend to their respective realcompletions and consequently W must be homeomorphic to Y.

Let us now state as a theorem what we have shown about the Hewitt–Nachbin realcompletion of X.

5.20 THEOREM *Every Tychonoff space X is contained in a unique realcomplete Hausdorff space vX as a C-embedded dense subset. That is, if Y is any realcomplete Tychonoff space that contains X as a dense C-embedded subset then there is a homeomorphism of Y onto vX that leaves the points of X fixed.*

5.21 EXERCISES

(1) Complete the proof of 5.1 and 5.4.

(2) A Tychonoff space X is pseudocompact if and only if $vX = \beta X$. It is compact if and only if it is both realcomplete and pseudocompact.

(3) If $\mathscr{Z}$ is a normal base on X and if Z_1 and Z_2 are disjoint members of $\mathscr{Z}$ then there is a continuous function $f \in C(X, E)$ such that $f(x) = 0$ if $x \in Z_1$ and $f(x) = 1$ if $x \in Z_2$.

(4) Suppose that X is a locally compact Hausdorff space. Let $\mathscr{Z}_C$ be the collection of all zero-sets of those $f \in C(X)$ that are constant on the complement of some compact subset of X. Then $\mathscr{Z}_C$ is a normal base on X and $\omega(\mathscr{Z}_C)$ is the Alexandroff one-point compactification on X (see [60] and [14]). This example is due to Fan and Gottesman, see [114] and [129].

(5) Suppose that $\mathscr{Z}$ is a normal base on a Tychonoff space X. Then X is open in $\omega(\mathscr{Z})$ if and only if X is locally compact. Isolated points of X are isolated in $\omega(\mathscr{Z})$. The only isolated points of $\beta\mathbf{N}$ are the points in $\mathbf{N}$.

(6) Let f be a continuous map of a space Y into a space Z and let X be a dense subspace of Y. If f restricted to X is a homeomorphism then $f(Y-X) \subset Y-f(X)$. Every compactification of X is a continuous image of X. If f is a homeomorphism of X into a compact space Z, then its continuous extension f^{β} to βX maps $\beta X - X$ into $Y-f(X)$.

(7) Are $\beta\mathbf{N}$, $\beta\mathbf{Q}$ and $\beta\mathbf{R}$ totally disconnected? What conditions on a Tychonoff space X will imply that $\omega(\mathscr{Z})$ is totally disconnected for any normal base $\mathscr{Z}$ on X? That $\omega(\mathscr{Z})$ is zero-dimensional?

(8) Let S be a subset of a Tychonoff space X. Then the following statements hold.

(*a*) S is C^*-embedded in X if and only if S is C^*-embedded in βX.

(*b*) S is C^*-embedded in X if and only if S is C^*-embedded in $\mathrm{cl}_{\beta X} S$.

(*c*) S is C^*-embedded in X if and only if $\mathrm{cl}_{\beta X} S = \beta S$.

(*d*) If S is open and closed in X, then $\mathrm{cl}_{\beta X} S$ is open in βX. May βX be relaced by $\omega(\mathscr{Z})$ for any normal base $\mathscr{Z}$?

(9) Let X be a dense subset of the Tychonoff space Y. Then the following statements (6) and (7) are equivalent to the statements (1) through (5) of 5.4, namely

(6) $X \subset Y \subset \beta X$,

(7) $\beta X = \beta X$.

Show that if Y is any compactification of X that satisfies any of these conditions then there is a homeomorphism of βX onto Y that leaves points of X fixed.

(10) Let $\mathscr{Z}$ be any normal base on a Tychonoff space X and

let ϕ be the embedding of X into $\omega(\mathscr{Z})$ (that is, $\phi(x) = \mathscr{F}_x$). Then the following statements hold.

(*a*) For each $Z \in \mathscr{Z}$, $\phi(Z) = \phi(X) \cap Z^\omega$.

(*b*) For each $Z \in \mathscr{Z}$, $\mathrm{cl}_{\omega(\mathscr{Z})}\, \phi(Z) = Z^\omega$.

(*c*) For disjoint members Z_1 and Z_2 of $\mathscr{Z}$,

$$\mathrm{cl}_{\omega(\mathscr{Z})} \phi(Z_1) \cap \mathrm{cl}_{\omega(\mathscr{Z})} \phi(Z_2) = \varnothing .$$

If in addition $\mathscr{Z}$ is a strong delta normal base, then the following are also true.

(*d*) For any $Z \in \mathscr{Z}$, $\mathrm{cl}_{\rho(\mathscr{Z})} \phi(Z) = Z^\rho$ where $Z^\rho = \rho(\mathscr{Z}) \cap Z^\omega$.

(*e*) If $(Z_n)_{n\in\mathbf{N}}$ is any sequence in $\mathscr{Z}$ then

$$\Big(\bigcap_{n\in\mathbf{N}} Z_n\Big)^\rho = \bigcap_{n\in\mathbf{N}} Z_n^\rho.$$

Consequently, if $\bigcap_{n\in\mathbf{N}} Z_n = \varnothing$ then $\bigcap_{n\in\mathbf{N}} \mathrm{cl}_{\rho(\mathscr{Z})} \phi(Z_n) = \varnothing$. Thus the collection $\mathscr{Z}^\rho = \{Z^\rho : Z \in \mathscr{Z}\}$ is a strong delta normal base on $\rho(\mathscr{Z})$. In fact, for any strong delta normal base $\mathscr{Z}$ on a T_1-space X, $\mathscr{Z} \subset Z(X)$. Does a statement analogous to (*e*) hold for the union?

(11) Let $\mathscr{Z}_X$ and $\mathscr{Z}_Y$ be rings of sets on the topological spaces X and Y respectively and let f be a continuous function from X into Y such that $f^{-1}(Z) \in \mathscr{Z}_X$ for all $Z \in \mathscr{Z}_Y$. If $\mathscr{F}$ is a $\mathscr{Z}_X$-filter on X then

$$f^{\#}(\mathscr{F}) = \{Z \in \mathscr{Z}_Y : f^{-1}(Z) \in \mathscr{F}\}$$

is a $\mathscr{Z}_Y$-filter on Y.

(*a*) If $\mathscr{F}$ is a prime (respectively, real) $\mathscr{Z}_X$-filter then $f^{\#}(\mathscr{F})$ is a prime (respectively, real) $\mathscr{Z}_Y$-filter. However, it need not be a $\mathscr{Z}_Y$-ultrafilter if $\mathscr{F}$ is a $\mathscr{Z}_X$-ultrafilter.

(*b*) The Tychonoff product of realcomplete spaces is realcomplete. (Consider the projection maps.)

(*c*) A realclosed subset of a realcomplete space is realcomplete.

(*d*) If Y is a compact (respectively, realcomplete) Hausdorff space and if $\mathscr{F}$ is a prime (respectively, realprime) $\mathscr{Z}_X$-filter on X, then $f^{\#}(\mathscr{F})$ has a limit point y_f in Y and $\cap f^{\#}(\mathscr{F}) = \{y_f\}$. Thus if X is dense in another space W and if $\mathscr{F}$ is a unique $\mathscr{Z}_X$-ultrafilter on X converging to the point $w \in W$ then define $f^*(w) = y_f \in Y$. The map f^* so defined is a continuous extension of f to the space $X \cup \{w\}$. Consequently show that in 5.4, statement (5) *implies* (1). To show that (4) *implies* (5) consider for each $w \in W$ the $Z(W)$-filter on W of all zero-set neighborhoods (in W) of w and use the fact that X is dense in W.

(12) Let X be a dense subset of a Tychonoff space W.
(*a*) If X is C-embedded in W and if $Z(f)$ is a zero-set in X, then $\mathrm{cl}_W Z(f) = Z(f^*)$ where f^* is the continuous extension of $f \in C(X)$. Conversely if X is C^*-embedded in W and if every zero-set $Z(f)$ in X satisfies $\mathrm{cl}_W Z(f) = Z(f^*)$ then X is C-embedded in W.
(*b*) If W is realcomplete, then every real $Z(X)$-ultrafilter on X converges in W.
(*c*) The smallest realcomplete space between X and βX is υX (see (*b*)). Thus if X is realcomplete then $X = \upsilon X$.
(*d*) If X is realcomplete and if $X \neq W$ then X is not C-embedded in W (see 5.21). Thus X is realcomplete if and only if it is not C-embedded in any Tychonoff space $W \neq X$ such that X is dense in W. Thus a C-embedded realcomplete subset of a Tychonoff space is closed.

(13) Let S be a subset of a Tychonoff space X.
(*a*) S is C-embedded in X if and only if S is C-embedded in υX.
(*b*) S is C-embedded in X if and only if S is C-embedded in $\mathrm{cl}_{\upsilon X} S$.
(*c*) If S is C-embedded in X then $\mathrm{cl}_{\upsilon X} S = \upsilon S$. If there exists a normal space Y such that $X \subset Y \subset \upsilon X$ then $\mathrm{cl}_{\upsilon X} S = \upsilon S$ implies that S is C-embedded in X.

Compare these results with Exercise 8.

(14) Let $\mathscr{F}$ be a $Z(X)$-ultrafilter on the Tychonoff space X. Then the following statements are equivalent.
(*a*) The $Z(X)$-ultrafilter $\mathscr{F}$ is not real.
(*b*) There is an $f \in C(X)$ such that for each $n \in \mathbf{N}$, the zero-sets

$$Z_n = \{x \colon |f(x)| \geqslant n\}$$

belong to $\mathscr{F}$.
(*c*) There is an $f \in C(X)$ such that f is unbounded on every zero-set of $\mathscr{F}$.

(15) *A* $\{0, 1\}$*-valued measure* on a non-empty set X is a countably additive function defined on the power set of X and assuming only the values 0 and 1. A cardinal number m is *measurable* if a set X of cardinal m admits a $\{0, 1\}$-valued measure such that $\mu(X) = 1$ and $\mu(\{x\}) = 0$ for every $x \in X$ where μ represents the measure.

(*a*) A discrete topological space is realcomplete if and only if every ultrafilter with the countable intersection property is fixed.
(*b*) Let $\mathscr{F}$ be an ultrafilter on a set X and let the characteristic function, $\chi_{\mathscr{F}}$, of $\mathscr{F}$ be defined on the power set 2^X of X as follows. For

$$A \in 2^X, \quad \chi_{\mathscr{F}}(A) = \begin{cases} 1 & \text{if } A \in \mathscr{F}, \\ 0 & \text{if } A \notin \mathscr{F}. \end{cases}$$

Then $\chi_{\mathscr{F}}$ is a non-zero finitely additive function defined on 2^X. Conversely if μ is any non-zero finitely additive function from 2^X into $\{0, 1\}$, then the collection $\mathscr{U}$ of all subsets of X for which $\mu(A) = 1$, is an ultrafilter, and $\mu_{\mathscr{U}} = \mu$. Thus there is a one-to-one correspondence from the set of all ultrafilters $\mathscr{U}$ on X onto the set of all non-zero finitely additive, $\{0, 1\}$-valued set functions μ defined on X. Moreover, if $\mathscr{U}$ is real, then the correspondent $\mu_{\mathscr{U}}$ is a *measure* and conversely if μ is a measure then the correspondent $\mathscr{U}$ is real.
(*c*) The measure $\mu_{\mathscr{U}}$ is said to be *fixed* or *free* according as the corresponding ultrafilter $\mathscr{U}$ is fixed or free. Then X is realcomplete if and only if every measure $\mu_{\mathscr{U}}$ is fixed. Moreover, the cardinality of X is non-measurable if every measure μ is fixed. Thus a discrete topological space is realcomplete if and only if its cardinal is non-measurable.

(For more results on non-measurable cardinals see [XI]. The class of non-measurable cardinals is quite large for it contains all cardinals obtained from the standard processes for forming cardinals from given ones, namely, addition, multiplication, exponentiation, the passage from a given cardinal to its immediate successor or to any smaller cardinal, and the formation of suprema. (All finite cardinals and $\aleph_0$ are non-measurable cardinals.) A celebrated unsolved problem is whether every cardinal is non-measurable.)

(16) Show that the Hewitt–Nachbin realcompletion of X may be obtained in the following way. Let P be the Tychonoff product space obtained by taking one copy of the real line $\mathbf{R}$ for each $f \in C(X)$, where X is a Tychonoff space. Consider the set $\mathrm{cl}_P \phi(X)$ in P where ϕ is the natural embedding of X in P. Then $\upsilon X = \mathrm{cl}_P \phi(X)$. Peruse the discussion after 5.3.

II. *Normality and real-valued continuous functions*

6 Normal spaces and C- and C^*-embedding

In 2.1 the concepts of C-embedding and C^*-embedding were introduced. It is clear that every C-embedded subspace of a topological space X is also C^*-embedded. Common to both of these concepts is the concept of separating two subsets of a space by a non-constant real-valued function as will be defined in the first definition. From this concept it will be possible to determine when a C^*-embedded subspace is also C-embedded. In fact in a normal space, it will be shown that for closed subsets, the concepts are equivalent.

Now more surprising is the fact that the set theoretical axioms for a space to be normal, guarantee the existence of non-constant real-valued continuous functions if the space has at least two disjoint closed sets. This amazing result was first shown by Urysohn in [373] and it is given here as 6.4. It shows immediately the usefulness of the fourth separation axiom.

6.1 DEFINITION If A and B are subsets of X then A is said to be *completely separated from* B in case there exists an $f \in C(X, E)$ such that $f(x) = 0$ if $x \in A$ and $f(x) = 1$ if $x \in B$. In such a case it is said that A and B are *completely separated* in X or that f *completely separates* A and B.

The concept of two subsets being completely separated is equivalent to asking that the subsets be contained in disjoint zero-sets as the next proposition shows. This then is good motivation for the concept of z-embedding which will be studied in the next section.

6.2 PROPOSITION *Suppose that* X *is a topological space and that* A *and* B *are subsets of* X*. Then* A *is completely separated from*

B if and only if A and B are contained in disjoint zero-sets. Moreover if A is completely separated from B, then A and B are contained in disjoint zero-set neighborhoods.

Proof. If A and B are completely separated then there exists $f \in C(X, E)$ such that $f(x) = 0$ if $x \in A$ and $f(x) = 1$ if $x \in B$. Let $Z = \{x \in X : f(x) \leqslant \frac{1}{3}\}$ and $Z^* = \{x \in X : f(x) \geqslant \frac{2}{3}\}$. Then one easily verifies that Z and Z^* are disjoint zero-set neighborhoods of A and B respectively.

Conversely, if A and B are contained in disjoint zero-sets $Z(f)$ and $Z(g)$ respectively then $h = |f|\,(|f| + |g|)^{-1}$ is a well defined real-valued continuous function defined on X that completely separates A from B. □

The crux of Urysohn's work with characterizing normal spaces is contained in the next lemma. It gives an interesting condition (condition (*c*)) on a collection of subsets of X so that it may be used to define a continuous real-valued function.

6.3 Lemma *Suppose that X is a topological space, that R_0 is a dense subset of* **R**, *and that $(U_r)_{r \in R_0}$ is a family of open subsets of X satisfying.*

(*a*) $\bigcup_{r \in R_0} U_r = X$,

(*b*) $\bigcap_{r \in R_0} U_r = \varnothing$,

(*c*) $\operatorname{cl} U_r \subset U_s$ *if* $r, s \in \mathbf{R}$ *and* $r < s$.

Then the function f mapping X into **R** *defined by*

$$f(x) = \inf\{r \in R_0 : x \in U_r\} \quad (x \in X)$$

is continuous.

Proof. First note that if $s \in R_0$ and if $f(x) < s$ then $x \in U_s$. For if $f(x) < s$ then there exists $r \in R_0$ such that $r < s$ and $x \in U_r$. But $U_r \subset U_s$ by (*c*) and therefore $x \in U_s$.

If $x \in \operatorname{cl} U_r$ then $f(x) \leqslant r$. For if $f(x) > r$, then there exists $s \in R_0$ such that $f(x) > s > r$. But $\operatorname{cl} U_r \subset U_s$ so $x \in U_s$. Therefore $f(x) \leqslant s$ which is a contradiction.

Now if $x \in X$, then $x \in U_r$ for some $r \in R_0$. Therefore

$$f(x) \leqslant r < +\infty.$$

Also by (*b*), $x \notin U_s$ for some $s \in R_0$. Then $-\infty < s$ and therefore $-\infty < s \leqslant f(x)$. Thus f is a real-valued function on X.

To show f is continuous, let $x \in X$ and let W be a neighborhood of $f(x)$. Since R_0 is dense in $\mathbf{R}$, there exist r and s in R_0 such that $r < f(x) < s$ and the closed interval $[r, s]$ is a subset of W. Let $V = U_s - \operatorname{cl} U_r$. Since $r < f(x)$, the point $x \notin \operatorname{cl} U_r$ by (*b*). Since $f(x) < s$, the point $x \in U_s$ by (*a*). Thus $x \in V$. Therefore V is a neighborhood of x. If $y \in V$ then $y \in U_s$. So $f(y) < s$. Moreover $y \notin \operatorname{cl} U_r$ so $f(y) \geqslant r$. Therefore $f(V) \subset [r, s]$ and it follows that f is continuous at x. □

The normality separation axiom implies the existence of continuous non-constant real-valued functions. If the space is also T_1, then all that is needed for the existence of such functions is that the space have at least two points. In 6.8 it will be shown that the normality separation axiom not only implies the above but it also implies the extension to the whole space of a certain class of continuous real-valued functions.

6.4 THEOREM (*Urysohn*) *If X is a normal space, then any two disjoint closed subsets of X are completely separated.*

Proof. Let A and B be disjoint closed subsets of X and let $(r_n)_{n \in \mathbf{N}}$ be a well ordering of the rational numbers in the closed unit interval E such that $r_1 = 1$ and $r_2 = 0$. One defines a sequence $(V_n)_{n \in \mathbf{N}}$ of open sets in X such that for each $m > 1$ the finite sequence $(V_n)_{n=1}^m$ satisfies

$$T_m\colon \textit{If } i, j \in \{1, \ldots, m\} \textit{ with } r_i < r_j, \textit{ then } \operatorname{cl} V_i \subset V_j.$$

Set $V_1 = X - B$ so V_1 is a neighborhood of A. Since X is normal, there exists a closed neighborhood W of A such that $W \subset V_1$ (1.10). Set $V_2 = \operatorname{int} W$. Now let $m > 1$ and assume inductively that $V_1, \ldots, V_m$ have been chosen such that $(V_n)_{n=1}^m$ satisfy T_m. Since $r_2 < r_{m+1} < r_1$, there exists a largest element r_k in the set $\{r_1, \ldots, r_m\}$ such that $r_k < r_{m+1}$. Also there exists a smallest element r_l in the set $\{r_1, \ldots, r_m\}$ such that $r_{m+1} < r_l$. By 1.11, there exists an open set V_{m+1} such that $\operatorname{cl} U_k \subset V_{m+1} \subset \operatorname{cl} V_{m+1} \subset U_l$. Then $(V_n)_{n=1}^m$ satisfies T_{m+1}.

Let Q be all rational numbers in $\mathbf{R}$ and define $(U_r)_{r\in Q}$ as follows:

$$U_r = \begin{cases} \varnothing & \text{if} \quad r < 0, \\ X & \text{if} \quad r > 1, \\ V_{r_n} & \text{if} \quad r = r_n \quad \text{for} \quad r_n \in [0, 1]. \end{cases}$$

Then $\bigcup_{r\in Q} U_r = X$, $\bigcap_{r\in Q} U_r = \varnothing$, and $\operatorname{cl} U_r \subset U_s$ whenever r and s belong to Q and $r < s$. Therefore by 6.3 a function f in $C(X)$ may be defined by the equation

$$f(x) = \inf\{r \in Q \colon x \in U_r\} \quad (x \in X).$$

Finally if $x \in A$ then $x \in V_2 = U_0$. So $f(x) \leqslant 0$. If $x \in B$ then $x \notin V_1 = U_1$. Hence $f(x) \geqslant 1$. Therefore A and B are completely separated. □

6.5 COROLLARY *Every normal T_1-space is Tychonoff.*

As it has been mentioned the extension of continuous real-valued functions from a subspace to the entire space is closely related to the separating of subsets. As in the previous theorem, this is relating structural properties of continuous real-valued functions to the underlying set theoretical structure – by all means an interesting comparison.

6.6 THEOREM (*Urysohn's extension theorem*) *If S is a non-empty subset of a topological space X, then the subset S is C^*-embedded in X if and only if any two completely separated sets in S are completely separated in X.*

Proof. Let $f \in C^*(S)$ such that f completely separates in S two subsets of S. The subset S being C^*-embedded in X implies that the extension f^* of f to X completely separates the subsets in X. Conversely let $f \in C^*(S)$ and let m be a positive integer such that $|f(x)| \leqslant m$ for all x in S. Let $r_n = \frac{1}{2}m(\frac{2}{3})^n$ for each n in $\mathbf{N}$. An inductive definition is utilized to define two sequences $(f_n)_{n\in\mathbf{N}}$ and $(g_n)_{n\in\mathbf{N}}$ of continuous real-valued functions that satisfy the conditions

(*a*) $f_n \in C^*(S)$ and $|f_n| \leqslant 3\mathbf{r}_n$,
(*b*) $g_n \in C^*(X)$ and $|g_n| \leqslant \mathbf{r}_n$,
(*c*) $f_{n+1} = f_n - (g_n|S)$.

To begin, let $f_1 = f$. Then $|f_1| = |f| \leqslant \mathbf{m} = 3\mathbf{r}_1$ as in condition (a). Let $g_1 = \mathbf{r}_1$ and let $f_2 = f_1 - g_1|S$. Now for $n > 1$ assume inductively that $(f_1, \ldots, f_n)$ and $(g_1, \ldots, g_{n-1})$ have been constructed satisfying conditions (a), (b), and (c). The subsets $A = \{x \in S : f_n(x) \leqslant -r_n\}$ and $B_n = \{x \in S : f_n(x) \geqslant r_n\}$ are completely separated in S and therefore completely separated in X. Let $g_n \in C^*(X)$ be such that $g_n(x) = -r_n$ if x is in A_n and $g_n(x) = r_n$ if x is in B_n and $|g_n| \leqslant \mathbf{r}_n$. Again letting

$$f_{n+1} = f_n - g_n|S,$$

then (b) and (c) are satisfied and it remains to show that

$$|f_{n+1}| \leqslant 3\mathbf{r}_{n+1}.$$

In the first case if x is in A_n, the induction hypothesis yields $-3r_n \leqslant f_n(x) \leqslant -r_n$ and $g_n(x) = -r_n$. Consequently $|f_{n+1}(x)| \leqslant 2r_n$. If x is in B, then

$$g_n(x) = r_n \quad \text{and} \quad r_n \leqslant f_n(x) \leqslant 3r_n.$$

Consequently again $|f_{n+1}(x)| \leqslant 2r_n$. Finally if x is in neither A_n nor B_n then $-r_n < f_n(x) < r_n$, $-r_n < g_n(x) < r_n$, and therefore $|f_{n+1}(x)| \leqslant 2r_n$. Condition ($a$) is always satisfied since

$$|f_{n+1}| \leqslant 2\mathbf{r}_n = 3\mathbf{r}_{n+1}.$$

Now take the extension of f to be the function g defined by $g(x) = \sum_{n=1}^{\infty} g_n(x)$ for each x in X. Since $\sum_{n=1}^{\infty} \mathbf{r}_n$ converges uniformly and since $|g_n| \leqslant \mathbf{r}_n$ for all n in N, it follows that g is in $C^*(X)$. If $x \in S$ and if $n \in \mathrm{N}$, then

$$\begin{aligned}(g_1 + \ldots + g_n)(x) &= (f_1 - f_2)(x) + (f_2 - f_3)(x) + \ldots + (f_n - f_{n+1})(x) \\ &= f_1(x) - f_{n+1}(x)\end{aligned}$$

and

$$\lim_{n \to +\infty} (g_1 + \ldots + g_n)(x) = f_1(x) - \lim_{n \to \infty} f_{n+1}(x) = f_1(x) = f(x).$$

This completes the proof. □

Let us now consider how 6.6 is related to the concept of C-embedding.

6.7 THEOREM *If S is a non-empty subset of a topological space X then the subset S is C-embedded in X if and only if it is C^*-embedded in X and completely separated from every zero-set disjoint from it.*

Proof. It is clear that if S is C-embedded in X then S is C^*-embedded in X. Let $f \in C(X)$. If $Z(f)$ is disjoint from S, then the function $f_1 = 1/f$ is an element of $C(S)$ and hence has an extension $f_1^* \in C(X)$. The function $f_1^* \circ f$ then completely separates S and $Z(f)$. Conversely if $f \in C(S)$ then the function

$$f_1 = \arctan \circ f \in C^*(S)$$

has a continuous extension $f_1^* \in C^*(X)$. The zero-set

$$Z = \{x \in X : f_1^*(x) \geqslant \tfrac{1}{2}\pi\}$$

is disjoint from S. If $h \in C^*(X)$ such that $\mathbf{0} < h \leqslant \mathbf{1}$, $h(x) = 0$ for $x \in Z$ and $h(x) = 1$ for $x \in S$. Thus

$$hf_1^* \in C^*(X) \quad \text{and} \quad f^* = \tan \circ (hf_1^*)$$

is a continuous extension of f. The function h maintains f^* as a single-valued mapping since $h(x) = 0$ for all x such that $f_1^*(x) \geqslant \frac{1}{2}\pi$. □

In the next result some of the basic facts pertaining to normal spaces are recalled. The well known Urysohn theorem will be recognized in (1) and (2). The Tietze extension theorem is recognized as (1) and (3). These can be found in any first year graduate topology text, for example see [IX] or [X]. Notice however that the other characterizations follow from our theorems once (1) and (2) are accepted. For example (2) *implies* (4) by 6.6. Since any two completely separated subsets A and B of a closed subset F are themselves contained in disjoint closed subsets of F, which by (2), are completely separated in X. Hence F must be C^*-embedded in X. Using this and also 6.7, it can be shown quickly that (2) *implies* also (3). Of course (3) *implies* (4). The fact that (4) (without the additional condition as in (2) of 6.7) *implies* normality is easy to see. If F_1 and F_2 are disjoint closed subsets of X, then the function f which assumes the value 0 for the points of F_1 and the value 1 for the points of F_2 is a bounded continuous function on the closed subspace $S = F_1 \cup F_2$.

6.8 THEOREM *If X is a topological space, then the following statements are equivalent.*

(1) *X is normal.*

(2) *Any two disjoint closed sets are completely separated.*

(3) *Every closed set is C-embedded in X.*
(4) *Every closed set is C^*-embedded in X.*

From 6.8 it is seen that every normal T_1-space is a Tychonoff space. It is easy to show that every Tychonoff space is a regular T_1-space, every regular T_1-space is a Hausdorff space, and every Hausdorff space is a T_1-space. Note also that if X is a collectionwise normal space, then X is a normal space.

6.9 Exercises

(1) Show that X is perfectly normal if and only if for each pair of disjoint closed sets A and B in X there exists $f \in C(X, E)$ such that $A = f^{-1}(\{0\})$ and $B = f^{-1}(\{1\})$.

(2) Let X be a dense subset of a Tychonoff space W. Then the following statements are equivalent.

(*a*) Every continuous function from X into any realcomplete space has a continuous extension to W.
(*b*) The space X is C-embedded in W.
(*c*) If a countable collection of zero-sets in X has empty intersection, then their closures in W have empty intersection.
(*d*) For any countable family of zero-sets $(Z_n)_{n \in \mathbf{N}}$ in X,

$$\mathrm{cl}_W\Big(\bigcap_{n\in\mathbf{N}} Z_n\Big) = \bigcap_{n\in\mathbf{N}} \mathrm{cl}_W Z_n.$$

(*e*) Every point of W is the limit of a unique real $Z(X)$-ultrafilter on X.
(*f*) $X \subset W \subset \upsilon X$.
(*g*) $\upsilon X = \upsilon W$.

(To prove this the following pattern is suggested. (*a*) *implies* (*b*) *implies* (*d*) *implies* (*c*) *implies* (*b*); (*e*) *implies* (*f*) *implies* (*g*) *implies* (*e*) *implies* (*a*) *implies* (*e*). The first implication follows readily as in 5.4. To show (*b*) *implies* (*d*) see 5.21(12). The implication (*c*) implies (*b*) follows from 6.7, 5.4 and the observation that every zero-set is complement generated. For (*f*) *implies* (*g*) consider 5.21(12). For the *equivalence* of (*a*) and (*e*) consider the discussion after 5.18.)

Thus υX is the largest subspace of βX in which X is C-embedded.

(3) Every Tychonoff space X has a realcompletion υX with the following equivalent properties:
(*a*) The space X is C-embedded in υX.
(*b*) Every continuous function f from X into a realcomplete space has a continuous extension to υX.
(*c*) Every point in υX is the limit of a unique real $Z(X)$-ultrafilter on X.
Moreover, if W is any realcompletion of X which satisfies any one of the above properties, then there is a homeomorphism of W onto υX which leaves points in X fixed.

7 Normal spaces and *z*-embedding

The concepts of C- and C^*-embedding were utilized in the previous chapter to characterize externally (so to speak) the important separation axiom of normality. In 6.7 it was shown that for a C^*-embedded subset to be C-embedded it had also to be completely separated from every zero-set which was disjoint from it. Viewing the 'completely separated part' of 6.6, in relation to this, the question prevails as to what will occur if one requires only that zero-sets on the subset S are traces of zero-sets on the space X. This is the concept of z-embedding. This concept although weaker than C^*-embedding also characterizes normal spaces in a manner analogous to 6.7.

DEFINITION Suppose that S is a non-empty subset of a topological space X. The subset S is *z-embedded in* X if every zero-set in S is the intersection of S with a zero-set in X (that is if $Z \in Z(S)$ then there exists $Z^* \in Z(X)$ such that $Z^* \cap S = Z$). If A and B are two subsets of X, it is said that A and B are *S-separated in* X if there are zero-sets Z_1 and Z_2 of X such that

$$Z_1 \cap Z_2 \cap S = \varnothing \quad \text{and} \quad A \subset Z_1, B \subset Z_2.$$

Let us note that if Z is a zero-set of X then there exists $f \in C(X)$ such that $f(x) = 0$ if and only if $x \in Z$. Clearly if such an f exists then there exists $g \in C(X, E)$ such that $Z(g) = Z(f)$.

If S is z-embedded in X then clearly every cozero-set of S can be extended to a cozero-set of X. Also if S is C^*-embedded in X

then S is z-embedded in X since every zero-set of S is the zero-set of a bounded real-valued continuous function.

If X is perfectly normal (see 1.9) then every subset of X is z-embedded. For let S be a subset of X and let Z be a zero-set of S. Then Z is a closed subset of S and hence there exists a closed subset F of X such that $F \cap S = Z$. But every closed subset of X is a zero-set (see 6.9(1)). Therefore F is a zero-set of X whose intersection with S is Z and hence S is z-embedded in X. This then gives numerous examples of z-embedded subsets that are not C^*-embedded. For example, any non C^*-embedded subset of the real line is z-embedded. Consequently z-embedding is a considerable weakening of C^*-embedding.

This concept of z-embedding will be shown later to be appropriate in the theory of realcompleteness. It yields sufficient conditions for a space to be realcomplete and it does assist in the preservation of realcompleteness under closed continuous mappings. But first its relationship to normality is demonstrated. A characterization of the concept in a manner analogous to 6.6 has been given by Blair in [52].

7.2 THEOREM (*Blair*) *If S is a non-empty subset of a topological space X then the following statements are equivalent.*

(1) *The subset S is z-embedded in X.*

(2) *If A and B are completely separated in S then there exists $g \in C(X)$ such that $g(x) = 0$ if $x \in A$ and $g(x) \neq 0$ if $x \in B$.*

(3) *If A and B are completely separated in S then they are S-separated in X.*

Proof. The proof proceeds by showing (1) *implies* (2) *implies* (3) *implies* (1). Assuming (1) let us suppose A and B are completely separated in the z-embedded subset S of X. Then there exist a zero-set Z in $Z(S)$ such that $A \subset Z$ and $Z \cap B = \varnothing$. By (1) there must be a zero-set $F \in Z(X)$ such that $F \cap S = Z$. Let $F = Z(g)$ where g is an element of $C(X, E)$. Moreover $g(x) = 0$ if $x \in A$ and $g(x) \neq 0$ if $x \in B$. Therefore (2) holds.

If A and B are completely separated in S they are contained in disjoint zero-sets in S (6.2). So suppose without loss of generality that A and B are themselves disjoint zero-sets. By (2), there exists a $Z \in Z(X)$ such that $A \subset Z$ and $Z \cap B = \varnothing$. Then $S \cap Z$

and B are disjoint zero-sets in S. By (2), there exists $Z^* \in Z(X)$ such that $B \subset Z^*$ and $S \cap Z \cap Z^* = \varnothing$. Consequently (3) holds.

Finally suppose $Z = Z(f) \in Z(X)$ where $f \in C(X, E)$. For all $n \in \mathbf{N}$ let $B_n = \{x \in S: f(x) \geqslant 1/n\}$. Then A and B_n are completely separated sets in S. Assuming (3) there must be a zero-set Z_n in X which is disjoint from B_n and which contains Z. Let Z^* be the intersection of all such zero-sets Z_n. The zero-set $Z^* \cap S$ (see 2.8) clearly contains Z. On the other hand, if $x \in Z^* \cap S$ then $x \notin B_n$ for all $n \in \mathbf{N}$. Therefore $f(x) < 1/n$ for all n, whence $f(x) = 0$ and x must be in Z. Therefore $Z^* \cap S = Z$ and (1) holds. □

Our next results characterize C^*-embedding and C-embedding in terms of z-embedding. Note the similarity of 7.3 with 6.6.

7.3 THEOREM (*Blair*) *If S is a subset of a topological space X then the following statements are equivalent.*

(1) *The subset S is C^*-embedded in X.*

(2) *The subset S is z-embedded in X and for each $A \subset S$ and each $Z \in Z(X)$ if A and $S \cap Z$ are S-separated in X then A and $S \cap Z$ are completely separated in X.*

Proof. Clearly a C^*-embedded subset is z-embedded. Assuming (1) the second part of (2) is immediate from 6.6.

Assuming (2) suppose A_1 and A_2 are S-separated in X, that is there is a zero-set Z_1 in S which is disjoint from A_2 and contains A_1. But then A_2 and $S \cap Z_1$ are S-separated in S and therefore by (2), completely separated in X. Since $A_1 \subset S \cap Z_1$, A_1 and A_2 are completely separated in X. Therefore by 6.6 S is C^*-embedded in X. □

Here again let's note the similarity of 7.4 with 6.7.

7.4 THEOREM (*Blair*) *Suppose that S is a subset of a topological space. Then the following statements are equivalent.*

(1) *The subset S is C-embedded in X.*

(2) *The subset S is z-embedded in X and it is completely separated from every zero-set disjoint from it.*

Proof. That (1) *implies* (2) is immediate by 7.3 and 6.7. Conversely by 2.5 it is sufficient to show that S is C^*-embedded in X. Suppose A_1 and A_2 are S-separated in X. By 6.2 there are zero-sets Z_1 and $Z_2 \in Z(X)$ such that $A_i \subset Z_i$, $i = 1, 2$ and $S \cap Z_1 \cap Z_2 = \varnothing$. Then $Z_1 \cap Z_2$ is a zero-set in X that is disjoint from S. Therefore, by (2), there exists $Z \in Z(X)$ such that $S \subset Z$ and $Z \cap (Z_1 \cap Z_2) = \varnothing$. But then $A_1 \subset Z \cap Z_1$ and $A_2 \subset Z \cap Z_2$. Therefore A_1 and A_2 are completely separated in X. It follows that S is C^*-embedded in X. □

An answer is now at hand as to when these three concepts (C-, C^*-, and z-embedding) are equivalent. For normal spaces it will be shown that closed subsets satisfy all three conditions. However for any topological space the conditions are equivalent for subsets which are zero-sets.

7.5 COROLLARY (*Blair*) *If Z is a non-empty zero-set of the topological space X then the following statements are equivalent.*

(1) *The set Z is C-embedded in X.*

(2) *The set Z is C^*-embedded in X.*

(3) *The set Z is z-embedded in X.*

Proof. Clearly (1) *implies* (2) and (2) *implies* (3). To see that (3) *implies* (1) let Z^* be a zero-set of X that is disjoint from Z. Then Z and Z^* are completely separated (6.2) and therefore, by 7.4, S is C-embedded in X. □

One should ask if there are any other subsets of a topological space for which the three notions are equivalent. For example, is it true for an arbitrary closed subset?

Non C^*-embedded subsets which are z-embedded have been shown already to exist. However what is now needed are some more definite examples. To do this let us prove a small lemma about F_σ-subsets of a topological space.

7.6 DEFINITION The subset S of a topological space is called a *generalized F_σ-set* if for each open set G with $S \subset G \subset X$, there is an F_σ-set F such that $S \subset F \subset G$.

7.7 LEMMA *If X is a Lindelöf topological space and if S is an F_σ-subspace of X then S is Lindelöf.*

Proof. Suppose that $S = \bigcup_{n \in \mathbf{N}} F_n$, where each F_n is closed and

containing α. The immediate successor of β is an isolated point. Consequently W is not separable for the cardinality of Φ is $\aleph_1$.

The closed set $\{\omega_1\}$ in W^* is not an intersection of a countable collection of basic open sets $(\beta, \omega_1]$, that is, it is not a G_δ-set. Consequently W^* is not perfectly normal and therefore not metrizable. But it is paracompact.

If A and B are any two disjoint closed subsets of W, neither of which is bounded, then select an increasing sequence $C = (\alpha_n)_{n\in\mathbf{N}}$ in W for which $\alpha_n \in A$ if n is even and $\alpha_n \in B$ if n is odd. Then the $\sup C$ must be in both A and B which is a contradiction.

Let f be any function in $C(W)$. Then the family

$$\mathscr{F} = \{f(W - W(\sigma)) \colon \sigma \in W\}$$

has the finite intersection property. Since $W - W(\sigma)$ is homeomorphic to W for each $\sigma \in W$, $f(W - W(\sigma))$ is a countably compact, second countable (and thus Lindelöf) subspace that must then be compact. Thus $\mathscr{F}$ is a family of compact subsets of the compact space $f(W)$. Let r be any element in $\cap \mathscr{F}$ and let A be the unbounded closed subset $f^{-1}(r)$ of W. Since the set A is disjoint from each of the closed subsets

$$\mathscr{B}_n = \{\alpha \in W \colon |f(\alpha) - r| \geqslant 1/n\} \quad (n \in \mathbf{N}),$$

each subset $\mathscr{B}_n$ must be unbounded. Let β_n be an upper bound for $\mathscr{B}_n$, and let $\alpha = \beta + 1$ where β is the supremum of this sequence $\{\beta_n\}_{n\in\mathbf{N}}$. If $x \geqslant \alpha$, then $|f(x) - r| < 1/n$ for all $n \in \mathbf{N}$. Consequently f is constant on $W - W(\alpha)$ with value r. To extend f to W^* one needs only to define $f^*(\alpha) = f(\alpha)$ if $\alpha \in W$ and $f^*(\omega_1) = r$.

EXAMPLE 4 *The Tychonoff plank*

An example of a space that is not hereditarily normal is constructed. In fact it is a compact Hausdorff and therefore collectionwise normal space that is not hereditarily normal.

The space is obtained by taking the Tychonoff product of the one-point compactification $N^* = \mathbf{N} \cup \{\omega\}$ of the integers $\mathbf{N}$ with W^*. Let t be the corner point (ω_1, ω) in this product

$$T^* = W^* \times N^*.$$

The *Tychonoff plank* is defined to be the dense subspace

$$T = T^* - \{t\}$$

of T^*. Let us investigate the following properties.

4.1 *The space T^* is compact and it is the one-point compactification of the subspace T.*

4.2 *The subspace T is C-embedded in T^* and consequently T^* is both the Stone–Čech compactification as well as the Hewitt–Nachbin realcompletion of T.*

Clearly T^* is compact for it is just the product space obtained from compact Hausdorff spaces.

For any continuous real-valued function f on T and for each $n \in \mathbf{N}$, the space $W^* \times \{n\}$ is homeomorphic to W^*. But 9(3.9) then implies that there is an $\alpha_n \in W$ such that f is constant on the set $\{(\sigma, n) \in T: \alpha_n \leqslant \sigma \leqslant \omega_1\}$. Since the sequence $\{\alpha_n\}_{n \in \mathbf{N}}$ is countable, $\alpha = \sup\{\alpha_n: n \in \mathbf{N}\}$ is an ordinal in W. Thus for each $n \in \mathbf{N}$, f is constant on $\{(\sigma, n): \alpha \leqslant \sigma < \omega_1\}$. Now consider any two points (σ_1, ω) and (σ_2, ω) with $\alpha \leqslant \sigma_1 < \omega_1$ and $\alpha \leqslant \sigma_2 < \omega_1$. Since f is continuous $\lim f(\sigma_i, n) = f(\sigma_i \omega)$ for $i = 1, 2$. But for each $n \in \mathbf{N}$, $f(\sigma_1, n) = f(\sigma_2, n)$ and therefore $f(\sigma_1, \omega) = f(\sigma_2, \omega)$. Thus f is constant on $\{(\sigma, \omega): \alpha \leqslant \sigma \leqslant \omega_1\}$. The extension f^* of f to T^* is then defined to be precisely this constant value. Consequently T is C-embedded (and therefore C^*-embedded) in T^* and

$$T^* = vT = \beta T.$$

EXAMPLE 5 *The Space Λ*

In [220], Katětov constructed the following example that is obtained by taking the difference set $\beta\mathbf{R} - (\beta\mathbf{N} - \mathbf{N})$. Before giving some of its properties, let us make some remarks about the components which go into its construction.

Any map f of the integers $\mathbf{N}$ onto the rationals $\mathbf{Q}$ is a continuous map onto a dense subset of $\mathbf{R}$ and a continuous map into the compact Hausdorff space $\beta\mathbf{R}$. Consequently f has a continuous extension f^* from $\beta\mathbf{N}$ into $\beta\mathbf{R}$. The range of this extension must be all of $\beta\mathbf{R}$ for $f^*(\beta\mathbf{N})$ is a closed subset of $\beta\mathbf{R}$ containing the dense subset $\mathbf{Q}$. Thus the cardinality of $\beta\mathbf{R}$ is not greater than that of $\beta\mathbf{N}$. In fact, a slight adjustment above shows that the cardinality of $\beta\mathbf{Q}$ is not greater than the cardinality of $\beta\mathbf{N}$ also.

On the other hand $\mathbf{N}$ is C^*-embedded in $\mathbf{Q}$ and $\mathbf{Q}$ is C^*-embedded in $\mathbf{R}$. Therefore $\mathbf{N}$ is C^*-embedded in $\mathbf{R}$. Hence by 5.21(8c), it follows that

$$\mathrm{cl}_{\beta\mathbf{Q}} \mathbf{N} = \beta\mathbf{N},$$

and

$$\mathrm{cl}_{\beta\mathbf{R}} \mathbf{N} = \beta\mathbf{N}.$$

Consequently both $\beta\mathbf{Q}$ and $\beta\mathbf{R}$ have the same cardinality as $\beta\mathbf{N}$.

Having demonstrated above how both $\beta\mathbf{Q}$ and $\beta\mathbf{R}$ may be obtained as continuous images of $\beta\mathbf{N}$, it may be shown, analogously, that $\beta\mathbf{N}$ and $\beta\mathbf{R}$ are continuous images of $\beta\mathbf{Q}$. On the other hand, neither $\beta\mathbf{N}$ nor $\beta\mathbf{Q}$ is a continuous image of $\beta\mathbf{R}$. This will be clear from the following discussion on connectedness or lack of connectedness of the various spaces.

Both $\beta\mathbf{N}$ and $\beta\mathbf{Q}$ are totally disconnected spaces. If x and y are distinct points of $\beta\mathbf{N}$, then there is a zero-set Z on $\mathbf{N}$ such that $x \in Z^\beta$ and $y \notin Z^\beta$. But Z^β is a closed and open subset of $\beta\mathbf{N}$ (see 5.21(10)). Thus $\beta\mathbf{N}$ is totally disconnected. Let U and V be disjoint closed neighborhoods of the distinct points x and y of $\beta\mathbf{Q}$. Then $U \cap \mathbf{Q}$ and $V \cap \mathbf{Q}$ are disjoint closed subsets of the zero-dimensional normal space $\mathbf{Q}$. Consequently there is a closed open subset F (and therefore a zero-set) of $\mathbf{Q}$ for which

$$U \cap \mathbf{Q} \subset F \subset X - (V \cap \mathbf{Q}).$$

Thus F^β is a closed open subset of $\beta\mathbf{Q}$ and $\beta\mathbf{Q}$ must be totally disconnected. On the other hand the closure of any connected set must be connected. Thus $\beta\mathbf{R}$ is a connected space and is not the pre-image by a continuous map of either $\beta\mathbf{Q}$ or $\beta\mathbf{N}$.

Now let us consider a special subspace of $\beta\mathbf{R}$. Let us remove from $\beta\mathbf{R}$ the subset $\beta\mathbf{N} - \mathbf{N}$. The resulting space constructed by Katětov in [220] is referred to as Λ, that is, the set $\mathbf{N} \cup (\beta\mathbf{R} - \beta\mathbf{N})$. The following properties of Λ should be checked by the reader.

5.1 *The space* $\mathbf{R}$ *is a subspace of* Λ. *Since* Λ *itself is a subset of* $\beta\mathbf{R}$, *it follows that* $\beta\mathbf{R} = \beta\Lambda$.

5.2 *The closed neighborhoods of points in* $\beta\Lambda - \mathbf{R}$ *meet* $\mathbf{R} - \mathbf{N}$.

5·3 *The zero-set* $\Lambda - \mathbf{R}$ *is dense in* $\beta\Lambda - \mathbf{R}$ *and belongs to every free zero-set ultrafilter on* $\mathbf{N}$ (*that is,* $\Lambda - \mathbf{R}$ *belongs to every* $\mathscr{F} \in \beta\mathbf{N} - \mathbf{N}$).

5.4 *The space* Λ *is not countably compact but it is pseudo-compact. Thus it is neither normal nor realcompact.*

5.5 *The space* Λ *contains the closed subset* $\mathbf{N}$ *as a* C^**-embedded subset that is not* C*-embedded. In fact* $\mathbf{N}$ *is a closed* G_δ*-set that is not a zero-set in* Λ. *Consequently a closed, countable union of zero-sets in a space* X *need not be a zero-set in* X.

EXAMPLE 6 *Bing's space*

The following example was considered by Bing in [50] related to his work on the metrization problem. It was there that the concept of a collectionwise normal space was defined and the example was used to show that not every normal Hausdorff space is collectionwise normal. The Hausdorff condition is essential here since without it a very easy example may be constructed to show that a normal space need not be collectionwise normal (see [XVIII, Chapter 1, Problem D]).

If W is any uncountable set, let $D_S = \{0, 1\}$ for each S in the power set 2^W, where $\{0, 1\}$ has the discrete topology. Let X be the product set $\Pi\{D_S: S \in 2^W\}$ of two point discrete spaces. For each $w \in W$, let us define a point $f_w \in X$ by

$$f_w(S) = \begin{cases} 1 & \text{if} \quad w \in S, \\ 0 & \text{if} \quad w \notin S. \end{cases}$$

Thus we have a natural association established between W and the subset X_0 of X where $X_0 = \{f_w \in X: w \in W\}$. A topology on X is now defined which will make $X - X_0$ a collection of isolated points in X, and will make X_0 a discrete subset of X. For each $f_w \in X_0$ and for each non-empty finite subcollection $\mathscr{K} \subset 2^W$, let us define

$$B(f_w, \mathscr{K}) = \{f \in X: f(K) = f_w(K) \text{ for all } K \in \mathscr{K}\}.$$

The element f_w belongs to $B(f_w, \mathscr{K})$. So let us take as a base for a topology on X the following collection

$$\mathscr{B} = \{\{f\}: f \in X - X_0\} \cup \{B(f_w, \mathscr{K}): f_w \in X_0, \mathscr{K} \neq \varnothing,$$
$$\mathscr{K} \text{ a finite subcollection of } 2^W\}.$$

Then $\mathscr{B}$ satisfies the criterion for a topological base. Its topology is finer than the Tychonoff product topology. Consequently X must be Hausdorff.

Let A_1 and A_2 be disjoint closed subsets of X, let $X_i = X_0 \cap A_i$ and let $W_i = \{w \in W: f_w \in X_i\}$, $i = 1, 2$. If X_1 is empty then A_1 and $X - A_1$ are disjoint open sets containing A_1 and A_2. A similar argument holds if X_2 is empty. On the other hand if X_1 and X_2 are non-empty, for $i = 1, 2$ set

$$V_i = \{f \in X: f(W_i) = 1 \quad \text{and} \quad f(W_j) = 0, \quad j \neq i\}.$$

Now V_i is an open set containing X_i and $V_1 \cap V_2 = \varnothing$. Then

$$(V_1 - A_2) \cup (A_1 - X_1) \quad \text{and} \quad (V_2 - A_1) \cup (A_2 - X_2)$$

are disjoint open sets containing A_1 and A_2 respectively. Thus X must be normal.

To show that X is not collectionwise normal we must exhibit a discrete collection of closed subsets of X that is not contained in a collection of pairwise disjoint open sets. Now for each $w \in W$, $\{f_w\}$ is closed in the Hausdorff space X. If f is any element in $X - X_0$, then $\{f\}$ is an open neighborhood of f that misses X_0. On the other hand, for each $f_w \in X_0$, the basic set $B(f_w, \{\{w\}\})$ containing f_w meets X_0 only at the point f_w. Consequently X_0 is a discrete subset of X and the family $\mathscr{F} = \{\{f_w\}: f_w \in X_0\}$ is a discrete family of closed subsets of X.

It remains to show that no pairwise disjoint family $\{U_w: w \in W\}$ of open sets U_w has $\{f_w\} \subset U_w$ for all $w \in W$. To this end it is enough to assume that the U_w are pairwise disjoint basic open sets $B(f_w, \mathscr{K}_w)$. Since W is uncountable and since $\mathscr{K}_w$ is finite (and non-empty), there is an integer $n \geqslant 1$ and an uncountable subset V of W such that for each $w \in V$, the cardinality of $\mathscr{K}_w$ is n. For any two elements w, v in V, $\mathscr{K}_w \cap \mathscr{K}_v \neq \varnothing$. Let v be a fixed element in V. Then there is a set $K_1 \in \mathscr{K}_v$ and an uncountable subset V_1' of V such that $K_1 \in \mathscr{K}_w$ for all $w \in V_1'$. Moreover there is a $t_1 \in \{0, 1\}$ and an uncountable subset $V_1 \subset V_1'$ such that $f_w(K_1) = t_1$ for all $w \in V_1$. Choose $w_1 \in V_1$. Since $\{U_w: w \in V_1\}$ is pairwise disjoint, there is a $K_2 \in \mathscr{K}_{w_1} - \{K_1\}$ and an uncountable subset V_2' of V_1 such that $K_2 \in \mathscr{K}_w$ for all $w \in V_2'$. Again there is a $t_2 \in \{0, 1\}$ and an uncountable subset V_2 of V_2' such that $f_w(K_2) = t_2$ for all $w \in V_2$. By finite induction, one obtains sets K_i, V_i and points t_i, $i = 1, \ldots, n$. Let $\mathscr{K} = \{K_1, \ldots, K_n\}$ and let

$$U = \{f \in X: f(K_i) = t_i, i = 1, \ldots, n\}.$$

Then for all w in the uncountable set V_n, we have both $\mathscr{K}_n = \mathscr{K}$ and $U_w = U$ which contradicts the fact that our sets were pairwise disjoint.

EXAMPLE 7 *Σ-products*

Let $\{(X_\alpha, \mathscr{T}_\alpha): \alpha \in I\}$ be an uncountable family of topological spaces. A *Σ-product* of this family is a subspace Σ of the Tychonoff product space $X = \Pi\{X_\alpha : \alpha \in I\}$ for which there is a point $p = (p_\alpha)_{\alpha \in I}$ of X such that $q = (q_\alpha)_{\alpha \in I}$ is a point in Σ if and only if $q_\alpha \neq p_\alpha$ for at most a countable number of the α in I. The point p is called the *base point of this Σ-product.*

This subspace Σ is due to Corson in [83] and it has some interesting properties. Essentially it satisfies some strong types of normality and weak types of separability. More specifically we have

7.1 *A Σ-product of complete metric spaces is normal.*

7.2 *If Σ is a Σ-product of separable metric spaces then $X = \upsilon\Sigma$.*

7.3 *If Σ is a Σ-product of complete separable metric spaces each of which is not compact then Σ is a normal space whose realcompactification is not normal.*

7.4 *If Σ is a Σ-product of complete separable metric spaces then the collection of all neighborhoods of the diagonal in $\Sigma \times \Sigma$ is a uniformity for Σ (and thus Σ is collectionwise normal). If the product sets are not compact then Σ is not paracompact.*

To prove 7.1, let us suppose that $\{(M_\alpha, d_\alpha) : \alpha \in I\}$ is a family of complete metric spaces and that Σ is a Σ-product of the family with base point $p = (p_\alpha)_{\alpha \in I}$. For each $F \in [I]$, let $X_F = \prod_{\alpha \in F} M_\alpha$ and let X_F have the metric defined by

$$d_F(x, y) = \sup\{d_\alpha(x_\alpha, y_\alpha) : \alpha \in F\}.$$

To show that Σ is normal suppose that H and K are two subsets of Σ that cannot be separated in Σ. To obtain a contradiction, we construct by induction, two sequences, one in H and one in K that converge to the same point in Σ.

Let x_1 be any point in H. Then there is a countable set of indices C_1 such that $(x_1)_\alpha = p_\alpha$ if $\alpha \notin C_1$. Arrange C_1 in a simple sequence and let F_1 be a finite set of indices that includes the first element of C_1. Set $S_1 = \Sigma$. We thus have a quadruple (x_1, C_1, F_1, S_1).

Suppose that n quadruples (x_n, C_n, F_n, S_n) have been chosen with the following properties:

(i) $x_i \in H$ if i is odd and $x_i \in K$ if i is even.

(ii) Each C_i is a countable set of indices such that $(x_i)_\alpha = p_\alpha$ if $\alpha \notin C_i$.

(iii) Each C_i is arranged in a simple sequence.

(iv) F_i is a finite set of indices that contains $F_1 \cup \ldots \cup F_{i-1}$ as well as the first i elements of each C_j for $j \leqslant i$.

(v) $S_i \subset S_j$ for $2 \leqslant i \leqslant n$ and $j \leqslant i$.

(vi) For each $i \in \mathrm{N}$ there is a set T_i in $\mathrm{pr}_{i-1}(S_{i-1})$ such that $\operatorname{diam} T_i < 1/i$ and $S_i = \mathrm{pr}_{i-1}^{-1}(T_i)$ where pr_i mapping Σ into X_{F_i} is the natural projection.

(vii) The sets H and K cannot be separated in S_i and $x_i \in S_i$. We choose the $n+1$ quadruple as follows. Let $M = \mathrm{pr}_n(S_n)$ and let $\mathscr{U}$ be the cover of M consisting of all subsets of M of diameter less than $1/(n+1)$. Let $\Sigma' = \{q \in \prod_{\alpha \notin F_n} X_\alpha : q_\alpha = p_\alpha$ for all but at most a countable number of the $\alpha\}$. It must now be shown that S_n is homeomorphic to $M \times \Sigma'$ and that H and K cannot be separated in $M \times \Sigma'$ implies that H and K cannot be separated in $U^* \times M$ for some $U^* \in \mathscr{U}$. Let $T_{n+1} = U^*$ and observe that T_{n+1} is a subset of $M = \mathrm{pr}_n(S_n) \subset X_{F_n}$ of diameter less than $1/(n+1)$. By an argument similar to the above, S_{n+1} is homeomorphic to $T_{n+1} \times \Sigma'$ and H and K cannot be separated in S_{n+1}. Moreover $S_{n+1} \subset S_i$ for $i \leqslant n+1$. Thus in particular, $H \cap S_{n+1}$ and $K \cap S_{n+1}$ are non-empty. Thus choose $x_{n+1} \in H \cap S_{n+1}$ if $n+1$ is odd and $x_{n+1} \in K \cap S_{n+1}$ if $n+1$ is even. The sets C_{n+1} and F_{n+1} are chosen in the expected way and we have thus chosen $(x_{n+1}, C_{n+1}, F_{n+1}, S_{n+1})$. If the sequences $(h_n)_{n \in \mathbf{N}}$ and $(k_n)_{n \in \mathbf{N}}$ are chosen alternately from the sequence $(x_n)_{n \in \mathbf{N}}$ it can then be shown that they converge to the same point in Σ.

For 7.2 recall that a topological product of realcomplete spaces is realcomplete, so X is realcomplete. It is easy to show that Σ is dense in X. It thus remains to show that Σ is C-embedded in X. Let $f \in C(\Sigma)$ and let $\mathbf{Q}$ be the set of rational numbers. For each $r \in \mathbf{Q}$ set

$$U_r = \{p \in \Sigma : f(p) < r\}.$$

Similar to a theorem of Bockstein [55] we must show that for each $r \in \mathbf{Q}$, there exists a countable subset C of I and disjoint open subsets V_r and V_r' in X_{C_r} such that

$$\pi_{C_r}(U_r) \subset V_r \quad \text{and} \quad \pi_{C_r}(X - \operatorname{cl} U_r) \subset V_r'$$

where $\pi : \Sigma \to X_{C_r} = \prod_{\alpha \in C_y} X_\alpha$ is the canonical mapping. We then set $C = \bigcup_{r \in \mathbf{Q}} C_r$ and observe next that $(\pi_C(U_r))_{r \in \mathbf{Q}}$ satisfy the hypothesis of 6.3 and therefore the function $g: X_C \to \mathbf{R}$ defined by

$$g(x) = \inf\{r \in \mathbf{Q}: x \in \pi_C(U_r)\} \quad (x \in X_C)$$

is continuous. It can now be shown that $f = g \circ \pi_C$. Finally note that $f^* = g \circ \mathrm{pr}_C$ is a real-valued continuous function on X and that $f^*|\Sigma = f$ (pr_C is the projection of X onto the space

$$X_C = \prod_{\alpha \in C} X_C).$$

Now let us suppose that Σ is a Σ-product of complete separable metric non-compact spaces. By 7.1, Σ is a normal space and by 7.2, $X = \upsilon X$. To show that X is not normal we produce a closed subset F that is homeomorphic to the product of an uncountable number of copies of the integers. In [359], Stone has shown that for products of metric spaces, the normality of the Tychonoff product is equivalent to all but a countable number of the spaces being compact. The space F can be readily constructed since each X_α non-compact implies that there is a sequence $(x^i_\alpha)_{i \in \mathbf{N}}$ that has no convergent subsequence in X. Thus 7.3 follows.

Let Δ be the diagonal in $\Sigma \times \Sigma$ where Σ is the Σ-product of complete separable metric spaces and let U be an arbitrary neighborhood of Δ. It is sufficient to find a metric space (M, d) and a function $\phi: \Sigma \to M$ such that $\{(x, y): d(\phi(x), \phi(y)) < 1\} \subset U$. (See 8.10(1).) First let us note that $\Sigma \times \Sigma$ is normal. Hence there exists a continuous function $f: \Sigma \times \Sigma \to E$ such that $f(x) = 0$ if $x \in \Delta$ and $f(x) = 1$ if $x \in \Sigma \times \Sigma - U$. Next we observe that $\Sigma \times \Sigma$ is a Σ-product of $\prod_{\alpha \in I} (X_\alpha \times X_\alpha)$. It is easy to prove that $\prod_{\alpha \in J} (A_\alpha \times A_\alpha)$ is homeomorphic to $\prod_{\alpha \in J} A_\alpha \times \prod_{\alpha \in J} A_\alpha$. Then, similar to the proof of 7.2, we can find a countable subset C of I and a function $g \in C(X_C \times X_C)$ such that $f = g \circ \pi_C$ where $\pi_C: \Sigma \times \Sigma \to \Sigma_C \times X_C$ is the canonical mapping. Let Δ_C be the diagonal of $X \times X$ and observe that $\pi_C(\Delta) = \Delta_C$. Let $U_C = \{(x, y) \in X_C \times X_C: g(x, y) < 1\}$. Next note that $X_C \times X_C$ is a countable product of metric spaces hence it is metrizable and thus paracompact. In 8.10(2) it is asked to show that in a paracompact space all neighborhoods of the diagonal form a uniformity and therefore by 8.10(1) there exists a metric space (M,d) and a continuous function $\phi_C\colon X_C \to M$ such that

$$\{(x,y)\in X_C \times X_C\colon d(\phi_C(x), \phi_C(y)) < 1\} \subset U_C.$$

Define

$$\phi\colon \Sigma \to M \text{ by } \phi = \phi_C \circ k,$$

where $k\colon \Sigma \to X_C$ is the canonical mapping. It is now easy to show that

$$\{(x,y)\in \Sigma \times \Sigma\colon d(\phi(x), \phi(y)) < 1\} \subset U$$

and therefore the collection of all neighborhoods of the diagonal is a uniformity on Σ.

For the proof that Σ is not paracompact as well as for the details of the above results the reader is referred to [83].

Example 8 *Rudin's space*

The construction of a normal Hausdorff space which is not countably paracompact was for a long time one of the most difficult unsolved problems facing set theoretical topology. In the Appendix, we will say more about it and its relationship to other topological properties. However for now we present most of the details to the solution of this problem with enough hints so that the reader may complete the remaining on his own. The solution was announced by M. E. Rudin at the 1970 International Mathematics Congress in Nice. We give here the results as related in [333] and [413].

To simplify our notation here we make use of a convention which is to assume that an ordinal λ *is* the set of all ordinals less than λ. Then an ordinal γ is said to be *cofinal with* λ if there is a subset Γ of γ which is order isomorphic with λ and such that $\alpha < \gamma$ implies that there is a $\beta \in \Gamma$ for which $\alpha < \beta$. The smallest ordinal cofinal with λ is called the *cofinality* of λ and is written as, $\mathrm{cf}(\lambda)$. The following may readily be demonstrated.

(*a*) For all ordinals λ, $\mathrm{cf}(\lambda)$ is a cardinal.

(*b*) $\mathrm{cf}(\lambda) = 1$ if and only if λ is a non-zero, non-limit ordinal.

(*c*) $\mathrm{cf}(\lambda+\gamma) = \mathrm{cf}(\gamma)$ for all ordinals λ and γ.

An ordinal λ is said to be *regular* if $\mathrm{cf}(\lambda) = \lambda$. Otherwise λ is said to be *singular*. For such ordinals we have (*d*) every regular ordinal is a cardinal, but not conversely, and

(*e*) every non-limit ordinal λ, $\lambda > 1$ is singular.

We now construct the set X on which a normal Hausdorff, non countably paracompact topology will be placed. Let

$$F = \{f\colon \mathbf{N} \to \omega_\omega | f(n) \leqslant \omega_n \text{ for all } n \in \mathbf{N}\}.$$

Let $X = \{f \in F\colon$ there is $i \in \mathbf{N}$ such that $\omega_0 < \mathrm{cf}(f(n)) < \omega_i$ for all $n \in \mathbf{N}\}$. For f, g in F, define $f < g$ if and only if $f(n) < g(n)$ for all $n \in \mathbf{N}$; $f \leqslant g$ if and only if $f(n) \leqslant g(n)$ for all $n \in \mathbf{N}$; for $k \in \mathbf{N}, f <_k g$ if and only if $f(n) < g(n)$ for all $n \in \mathbf{N}$, $n > k$. Now it can be seen that

(*f*) the set F is just the countable Cartesian product of the sets of ordinals $\omega_n + 1$,

(*g*) the ordinal ω_ω is singular, $\omega_\omega = \sup_{n \in \mathbf{N}} \omega_n$, and

(*h*) for f, g in $F, f \leqslant g$ and $f \neq g$ need not imply $f < g$.

A topology on X is constructed by letting

$$U(f,g) = \{h \in X\colon f < h \leqslant g\}$$

where $f, g \in F$. Then a base for a topology on X is the collection $\mathscr{B} = \{U(f,g)\colon f, g \in F\}$. With this topology, X is a *normal Hausdorff, non-countably paracompact space.*

(*i*) If one places the box topology on F (see [XVIII]) then X with the prescribed topology is just a subspace of F.

(*j*) The topology on X is finer than the relative Tychonoff product topology on X. Thus X is a Hausdorff space.

(*k*) Verify that $\mathscr{B}$ is a base for a topology on X.

(i) The sequence $(x_n)_{n \in \mathbf{N}}$ for which $x_n = 0$ for each $n \in \mathbf{N}$ is in $F - X$.

(ii) The base is actually a base of open and closed subsets.

(*l*) With this topology X is a Tychonoff space.

(*m*) Every G_δ-set in X is open. Thus X is a P-space as defined in [XI]. If f is any point common to a countable collection of open sets U_n, $n \in \mathbf{N}$, then there is $a_n \in X$ such that

$$a_n < f \quad \text{and} \quad U(a_n, f) \subseteq U_n$$

or every $n \in \mathbf{N}$. Define an element $a \leqslant f$ by $a(k) = \sup_{n \in \mathbf{N}} a_n(k)$ for all $k \in \mathbf{N}$. Then $\operatorname{cf}(a(k)) \leqslant \omega_0 < \operatorname{cf}(a(k))$ for each $k \in \mathbf{N}$ and thus $a < f$. Consequently $f \in U(a,f) \subseteq \bigcup_{n \in \mathbf{N}} U_n$.

The space X is not countably paracompact. To prove this it is shown that there is a decreasing sequence $(D_i)_{i \in \mathbf{N}}$ of closed sets $D_i \subset X$, $i \in \mathbf{N}$ such that $\bigcap_{i \in \mathbf{N}} D_j = \varnothing$ and such that for every sequence $(U_i)_{i \in \mathbf{N}}$ of open sets for which $D_i \subset U_i$ for all i, then $\bigcap_{i \in \mathbf{N}} U_i \neq \varnothing$. (In the Appendix, the equivalence of this to countable paracompactness is given.)

(*n*) For each $n \in \mathbf{N}$, define

$$D_n = \{f \in X : \text{there is a } k \in \mathbf{N} \text{ such that } f(k) = \omega_k\}$$

and for all $n > 1$, define

$$C_n = \{f \in X : f(i) = \omega_i \text{ for all } i < n \text{ and } f(i) < \omega_i \text{ for all } i \geqslant n\}.$$

(i) For each $n \in \mathbf{N}$, $C_{n+1} \subseteq D_n$.

(ii) For each $n \in \mathbf{N}$, D_n is closed.

(iii) The sequence $(D_i)_{i \in \mathbf{N}}$ is decreasing.

(*o*) Let $(U_n : n \in \mathbf{N})$ be any sequence of open sets such that $D_n \subseteq U_n$ for all $n \in \mathbf{N}$ and such that $\bigcap_{n \in \mathbf{N}} U_n \neq \varnothing$. In fact it is shown that $\bigcap_{n \in \mathbf{N}} U_n \cap C_2 \neq \varnothing$.

(i) Suppose that $1 < n \in \mathbf{N}$, that $f \in C_{n+1}$ and that the open set $U \supset \{y \in C_{n+1} : f <_{n+1} y\}$. Then there is a $g \in C_n$ such that

$$\{y \in C_n : g <_n y\} \subset U.$$

To do this define $v \in C_n$ by

$$v(k) = \begin{cases} \omega_k & \text{for all } k < n \\ \omega_1 & \text{for } k = n \\ f(k) & \text{for all } k > n. \end{cases}$$

Let $V = \{g \in C_n - U : v <_n y\}$. If V is empty then take g to be v. If V is not empty, consider a maximal well ordered sequence $(v_\alpha)_{\alpha<\delta}$ such that $\beta < \alpha < \delta$ implies $v <_n v_\beta <_n v_\alpha$. Applying Zorn's lemma to the family of all such well ordered sequences yields a maximal such sequence $(v_\alpha)_{\alpha<\delta}$. Now $\delta \leqslant \omega_n$, however, actually $\delta \lneqq \omega_n$. To show this let y be defined by $y(k) = \sup_{\alpha<\delta} v_\alpha(k)$ for all $k \in \mathbf{N}$. Now $y(k) \leqslant \omega_k$ for all k and $y(k) = \omega_k$ for all $k < n$, $y \in C_{n+1}$

and $f <_{n+1} y$. Thus $y \in U$ and there is $\alpha \in F$, $\alpha < y$ such that the basic open set $U(\alpha, g) \subseteq U$. Let $\alpha = \sup\{\alpha_k : k \geqslant n\}$. Then $\alpha < v_\alpha \leqslant y$ and $v_\alpha \in U$ which is a contradiction. Consequently $\delta < \omega_n$. The function g may now be defined by letting

$$g(k) = \begin{cases} \omega_k & \text{for all } k < n, \\ y(k) + \omega_1 & \text{for all } k \geqslant n. \end{cases}$$

Then $\mathrm{cf}(g(k)) > \omega_0$ for all k, $g(k) = y(k) + \omega_1 < \omega_k$ for all $k \geqslant n$ and $g \in C_n$. By the maximality of $\{v_\alpha\}_{\alpha<\delta}$, it follows that

$$\{h \in C_n : g <_n h\} \subseteq U.$$

(ii) For each $n \in \mathbf{N}$, there is an $f_n \in C_2$ such that

$$\{y \in C_2 : f_n < y\} \subseteq U_n.$$

Let the g of (a) above be denoted by $g(f, U)$ since it depends only on f and U. For a fixed $n \in \mathbf{N}$ set

$$c_{n+1}(k) = \begin{cases} \omega_k & \text{for all } k \leqslant n, \\ \omega_1 & \text{for all } k > n. \end{cases}$$

Obviously, $c_{n+1} \in C_{n+1} \subseteq D_n \subseteq U_n$ and it follows that

$$\{y \in C_{n+1} : c_{n+1} <_{n+1} y\} \subseteq U_n.$$

By (i) above there is a $g(c_{n+1}, U_n)$ and so define by induction down $c_n \cdot g(c_{n+1}, U_n) \in C_n$. In general, for $1 \leqslant k \leqslant n$, define

$$c_k = g(c_{k+1}, U_n) \in C_k.$$

Thus set $f_n = c_2$.

(iii) Having (ii) above ((i) was needed only to prove (ii)), it is possible to show that $\bigcap_{n \in \mathbf{N}} U_n \cap C_2 \neq \varnothing$. Let $f_n \in C_2$ be as in (ii) above. Since for each $k \in \mathbf{N}$, ω_k is a regular ordinal then for each $n \in \mathbf{N}$, $1 < n$, one may choose ordinals $\alpha_{k\ n}$ for which

$$f_n(k) < \alpha_{k,n} < \omega_k$$

for all $k \in \mathbf{N}$. Set $\alpha'_k = \sup_n \alpha_{k,n}$ for each k. Then for $1 < n \in \mathbf{N}$, $f_n(k) < \alpha'_k < \omega_k$ for all k. Let $\alpha_k = \alpha'_k + \omega_1$. Then $\mathrm{cf}(\alpha_k) = \omega_1$, $f_n(k) < \alpha_k < \omega_k$ for all k, and $\mathrm{cf}(\alpha_k) = \omega_1$ for all k. Let $\alpha_1 = \omega_1$ and $x(k) = \alpha_k$ for all k. Then $x \in C_2 \cap U_n$ for all $n \in \mathbf{N}$.

Now comes the difficult task of showing that the space X *is collectionwise normal* and therefore normal. Let $\mathscr{H} = (H_i)_{i \in I}$ be a discrete collection of closed subsets of X. To show there is a collection $(U_i)_{i \in I}$ of pairwise disjoint open sets for which $H_i \subset U_i$ for all $i \in I$. Let $H = \bigcup_{i \in I} H_i$ and for $U \subset X$ define t_U by

$$t_U(n) = \sup\{f(n) : f \in U\} \quad \text{for all} \quad n \in \mathbf{N}.$$

Now $f \leqslant t_U$ for all $f \in U$ and t_U is the least element of X with respect to $\leqslant$ for which this is true. Consequently t_U is called the *least top* for U. For each countable ordinal α, a cover $\mathscr{G}_\alpha$ of $\mathscr{H}$ will be defined by induction with the following properties: (1) $\mathscr{G}_\alpha$ will be a collection of pairwise disjoint open sets. (2) If $\beta < \alpha < \omega_1$ and $V \in \mathscr{G}_\alpha$ then there is a $U \in \mathscr{G}_\beta$ for which $V \subseteq U$, $t_V \neq t_U$ whenever V intersects at least two members of $\mathscr{H}$, and $U = V$ whenever U intersects at most one member of $\mathscr{H}$.

(*p*) If such a family of covers exist, open sets $(U_i)_{i \in I}$ may be obtained as required. For $f \in H$, $\mathscr{G}_\alpha$ covers H with pairwise disjoint open sets. Thus for all $\alpha < \omega_1$ there is a unique $U_{f,\alpha} \in \mathscr{G}_\alpha$ for which $f \in U_{f,\alpha}$.

Consider $\mathscr{U} = \{U_{f,\alpha} : \alpha < \omega_1\}$. Then $U_{f,\alpha} \subset U_{f,\beta}$ for $\beta < \alpha < \omega_1$. There is a $U_{f,\alpha} \in \mathscr{U}$ such that $U_{f,\alpha}$ intersects at most one term of $\mathscr{H}$. This follows from the conditions above imposed on $\mathscr{G}_\alpha$ by assuming the contrary. Consequently $U_{f,\alpha} = U_{f,\beta}$. For all $j \in I$, let $U_j = \bigcup \{U_{f,\alpha} : f \in H_j\}$.

If f and g belong to different terms of $\mathscr{H}$, then there is a $\beta < \omega_1$ such that $\alpha_f < \beta$ and $\alpha_g < \beta$. Thus the terms of $\mathscr{G}_\beta$ to which f and g belong are $U_{f,\alpha}$ and $U_{g,\alpha}$ respectively. But the terms of $\mathscr{G}_\beta$ are pairwise disjoint, hence $U_j \cap U_k = \varnothing$ for $j \neq k$. Thus $(U_i)_{i \in I}$ are pairwise disjoint open sets and $H_i \subseteq U_i$.

(*r*) To show such a family of covers exists, let $\mathscr{G}_0 = \{X\}$ and assume $\mathscr{G}_\beta$ has been defined for all $0 \leqslant \beta < \alpha < \omega_1$. To define $\mathscr{G}_\alpha$, it is necessary to consider whether α is a limit ordinal or $\alpha = \beta + 1$.

If α is a limit ordinal, $\beta < \alpha$ and if $f \in H$, then let $U_{f,\beta}$ be the unique term of $\mathscr{G}_\beta$ to which f belongs and set $U_f = \bigcap_{\beta < \alpha} U_{f,\beta}$. Since every G_δ is open, U_f is an open subset of X. Define $\mathscr{G}_\alpha$ then by taking the family $(U_f)_{f \in H}$ of pairwise disjoint sets. The conditions on the cover $\mathscr{G}_\alpha$ remain to be checked.

If $\alpha = \beta+1$, then for each $U \in \mathscr{G}_\beta$, a collection $\mathscr{G}_U$ of disjoint open subsets of U which cover $U \cap H$ will be defined in such a way that every $V \in \mathscr{G}_U$ will satisfy the last two conditions requested above. Then $\mathscr{G}_\alpha = \bigcup \{\mathscr{G}_U: U \in \mathscr{G}_\beta\}$ will satisfy all of the conditions requested above. In defining $\mathscr{G}_U$ several cases arise dependent upon the sizes and positions of the elements of $\mathscr{G}_\beta$.

First, if $U \in \mathscr{G}_\beta$ intersects at most one term of $\mathscr{H}$ then set $\mathscr{G}_U = \{U\}$.

Secondly, assume U intersects at least two members of $\mathscr{H}$ and there is a $k \in \mathbf{N}$ such that $\mathrm{cf}(t(k)) \leqslant \aleph_0$. Now $\mathrm{cf}(t(k))$ is actually $\aleph_0$. Let $(\alpha_n)_{n\in\mathbf{N}} \subseteq t(k)$ be an increasing sequence with limit ordinal $t(k)$. Define $V_1 = \{f \in U: f(k) \leqslant \alpha_1\}$ and for $1 < n < \omega_0$ define $V_n = \{f \in U: \alpha_{n-1} < f(k) \leqslant \alpha_n\}$. Then $\mathscr{G}_U = (V_n)_{n\in\mathbf{N}}$ is a set of disjoint open subsets of U covering U no one of which has t as its top and $\mathscr{G}_U$ has the desired properties.

Thirdly, let us assume U intersects at least two members of $\mathscr{H}$ and that $\mathrm{cf}(t(n)) > \aleph_0$ for all $n \in \mathbf{N}$. For each $n \in \mathbf{N}$, set

$$U_n = \{f \in U: \text{ for all } k \in \mathbf{N},\ \mathrm{cf}(f(k)) \leqslant \mathrm{cf}(\omega_n)\}.$$

If $m < n$ then $U_m < U_n$ and also $U = \bigcup_{n\in\mathbf{N}} U_n$. For each fixed $n \in \mathbf{N}$, there is a $g \in F$ such that $g < t$ and $\{h \in U_n: g < h\}$ intersects at most one term of $\mathscr{H}$. If this statement is true, then for each $n \in \mathbf{N}$, let g_n be the element $g \in F$ guaranteed by the statement. Define f by letting $f(i) = \sup\{g_n(i): n \in \mathbf{N}\}$ for each $i \in \mathbf{N}$. Then $f < t$ by the assumption above. Now $\{h \in U: f < h\}$ intersects at most one term of $\mathscr{H}$, for suppose $f < h \in U \cap \mathscr{H}$ and $f < k \in U \cap \mathscr{H}$. There are i and j in $\mathbf{N}$ such that $h \in U_i$ and $k \in U_j$. Let $n = i+j$. Then $g_n < h$ and $g_n < k$ where h and k are in U_n. Thus h and k belong to the same term of $\mathscr{H}$. Define $\mathscr{G}_U$ then in the following way. If $M \subset \mathbf{N}$, define

$$V_M = \{h \in U: h(n) \leqslant f(n) \text{ for all } n \in M \text{ and } h(n) > f(n) \text{ for all } n \in \mathbf{N}-M\}.$$

If $t_{V_M} = t$ then M is empty. But $V_\varnothing$ intersects at most one term of $\mathscr{H}$. So if $\mathscr{G}_U = \{V_M: M \subset \mathbf{N}\}$ then $\mathscr{G}_U$ has the desired properties.

Thus it remains to show in the above statement the existence of a $g \in F$ with the desired properties. This requires considerable effort and to this end let us assume that it is false, that is for every $f \in F, f < t$, there are terms h and k of U_n such that $f < h, f < k$ while h and k belong to different terms of $\mathscr{H}$. For notational purposes define, for $i \leqslant n$,

$$M_i = \{j \in \mathbf{N}: \mathrm{cf}(t(j)) = \omega_i\}$$

and
$$M = \{j \in \mathbf{N}: \mathrm{cf}(t(j)) > \omega_n\}.$$

Then
$$\mathbf{N} = \bigcup_{i \leqslant n} M_i \cup M.$$

Let $R = \{r: (1, 2, \ldots, n) \to \omega_n: r(i) < \omega_i \text{ for all } i \in (1, \ldots, n)\}$. The cardinality of R is ω_n. If $\rho = \inf\{\sigma: \mathrm{cf}(\sigma) = \omega_n\}$, then ρ is a regular cardinal. Thus there is a well ordered sequence $(r_\beta)_{\beta < \rho}$ of terms of R such that whenever $\alpha < \rho$ and $r \in R$ then there is a $\alpha < \gamma < \rho$ such that $r_\gamma = r$.

For all $i \leqslant n$ and $j \in M_i$ choose an increasing ω_1 sequence $(s_{j,\sigma})_{\sigma < \omega_1}$ in $t(j)$ with limit $t(j)$. Now for each ordinal $\lambda < \omega_n$ select by induction a $f_\lambda \in F$, $h_\lambda \in U_n$, and $k_\lambda \in U_n$ as follows:

$$f_0(j) = \begin{cases} s_{j, r_0(i)} & j \in M_i, \\ 0 & j \in M. \end{cases}$$

Now choose an $h_0 \in U_n$ and a $k_0 \in U_n$ belonging to different terms of $\mathscr{H}$ such that $f_0 < h_0$ and $f_0 < k_0$. This choice is possible since we have assumed that precisely this holds for any $f \in F$ for which $f < t$ and f_0 is definitely less than t. If it is supposed that f_γ, h_γ, and k_γ have been chosen (h_λ and k_γ in U_n) for all $\gamma < \lambda < \omega_n$, we define

$$f_\lambda(j) = \begin{cases} s_{j, r_\lambda(i)} & j \in M_i, i \leqslant n, \\ \sup\{g(j): g \in \bigcup_{j < \lambda} \{h_\gamma, k_\gamma\}\} & j \in M. \end{cases}$$

If $f_\gamma < t$ then $f_\gamma \in F$. For $j \in M_i$, $i \leqslant n$, $f_\gamma(j) < t(j)$. For $j \in M$, let us note that if g is h_γ or k_γ for some $\gamma < \lambda$ then $g \in U_n$ by the induction hypothesis. Since $\lambda < \omega_n$ the definition of M gives $f_\lambda(j) < t(j)$ for all $j \in M$. Thus $f_\lambda \in F$ and $f_\lambda < t$ and the assumption that the statement above was false allows us once again to choose h_λ and $k_\lambda \in U_n$ such that $f_\lambda < h_\lambda$, $f_\lambda < k_\lambda$ and yet h_λ and k_λ belong to different terms of $\mathscr{H}$. Hence we have such elements for each ordinal $\lambda < \omega_n$.

Now we are in a position to define a function g such that when taken together with f, the basic set $U(f,g)$ will contain two members of the discrete collection $\mathscr{H}$.

Let g be defined by

$$g(j) = \begin{cases} t(j) & j \notin M, \\ \sup\{f_\lambda(j) \colon \lambda < \omega_n\} & j \in M. \end{cases}$$

Now for $j \notin M$, $\omega_0 < \mathrm{cf}(g(j)) \leqslant \omega_n$ and for $j \in M$, $\mathrm{cf}(g(j)) = \omega_n$. These follow from the assumptions made on the third case stated above and from the fact that for $j \in M$, the $f_\lambda(j)$ are increasing by definition.

Since $\mathscr{H}$ is a discrete collection of closed subsets, we may select a basic open set $U(f,g)$ with $f < g$ such that $U(f,g)$ intersects at most one term of $\mathscr{H}$. Since $f < g$ and $g \leqslant t$, it follows that $f < t$.

For $i \leqslant n$ and $j \in M_i$, $t(j)$ is cofinal with ω_i and $f(j) < g(j) = t(j)$. Thus we may choose an ordinal $\sigma_j < \omega_i$ such that $f(j) < s_{j,\sigma_j}$. If $\mu_i = \sup\{\sigma_j\}$, where the sup is taken over the countable set M_i, then $\mu_i < \omega_i$. Let us define $r \in R$ by

$$r(i) = \mu_i \quad i \leqslant n.$$

Now $f(j) < s_{j,r(i)}$ for all $j \in M_i$.

For each $j \in M$ there is an ordinal $\sigma_j \in \omega_n$ such that $f_{\sigma_j}(j) > f(j)$. Let σ be the sup of the σ_j for $j \in M$. Then $f(j) < f_\sigma(j)$ for all $j \in M$. Since the $f_\lambda(j)$ are increasing for all $j \in M$, we also have $f(j) < f_\delta(j)$ for all $\delta \geqslant \sigma$ and for all $j \in M$.

Now select γ with $\sigma < \gamma < \omega_n$ and $r_\gamma = r$. For this ordinal γ, consider f_γ, h_γ, and k_γ. For all $j \in M, f(j) < f_\gamma(j) < g(j)$. For $k \notin M$, $f_\gamma(k) = s_{k, r_\gamma(k)}$ by definition. But then $f_\gamma(k) = s_{k,\sigma_k} > f(k)$ for all $k \notin M$. Hence $f < f_\gamma < h_\gamma \leqslant g$ and $f < f_\gamma < k_\gamma \leqslant g$. Thus h_γ and k_γ are both in $U(f,g)$. But h_γ and k_γ are by construction in different terms of $\mathscr{H}$. This contradicts the fact that $U(f,g)$ intersects at most one term of $\mathscr{H}$.

III. *Normality and normal covers*

In this chapter the concept of normality will be investigated as it applies to the separation characteristics of the open sets of a space and as it also applies to covers of the space. The interpretation of a normal space in terms of a suitable cover will be given. Similarly a collectionwise normal space will be considered. In particular it will be shown that a topological space $(X,\mathscr{T})$ is normal if and only if for every point-finite open cover $(U_\alpha)_{\alpha\in I}$ of X there is an open cover $(V_\alpha)_{\alpha\in I}$ of X such that $\mathrm{cl}\, V_\alpha \subset U_\alpha$ for each α in I.

In addition uniform spaces and normal covers will be investigated. In Section 6, it was shown that among the uniformizable spaces, normal spaces are precisely that subcollection for which every continuous function on a closed subspace is the restriction of a continuous function on the whole space (Urysohn–Tietze extension theorem).

10 Definitions and elementary properties

In this section several known theorems and elementary propositions relating the interactions among various types of covers will be presented. It will be shown that normal covers are frequent. In 10.5 it is shown that locally finite open covers of a normal space are normal covers. This follows from 10.1 which shows that in normal spaces point-finite open covers may be strongly screened by an open cover. This is interesting for it shows that the notions appropriate to paracompactness are strongly linked to the separation permitted by a normal space and then also to normal covers. In fact a cover is normal if and only if it has a locally finite partition of unity subordinate to it (10.10).

For the sake of motivation, let us recall some elementary facts about normal spaces. The definition of a normal space is given by requiring that disjoint closed subsets may be separated by disjoint open sets. One then shows that this is equivalent to requiring that for each neighborhood U of a closed set F there is another neighborhood V of F whose closure is contained in U. That is, it may be said that each neighborhood of a closed set is strongly screened by an open set (see 1.7 and 1.8).

It is this characterization of a normal space which permits us to structurally analyze the concept of normality and then to utilize it for the construction of non-trivial real-valued continuous functions (this is the crux of Urysohn's lemma).

It is easy to see that if X is a normal space and if $\mathscr{U} = (U_1, U_2)$ is an open cover of X, then there exists a closed cover that screens $\mathscr{U}$. A stronger result than this is the following theorem. In the proof of this theorem Zorn's lemma is needed.

10.1 THEOREM *If X is a normal topological space, then every point-finite open cover $\mathscr{A}$ of X is strongly screened by an open cover.*

Proof. Let Φ be the collection of all pairs $((U_\alpha)_{\alpha\in I}, H)$ where $(U_\alpha)_{\alpha\in I}$ is a cover of X having the same set of indices as the point-finite open cover $\mathscr{A} = (A_\alpha)_{\alpha\in I}$ of X and where H is a subset of I possessing the following property:

$$\text{if} \quad \alpha\in H \quad \text{then} \quad \operatorname{cl} U_\alpha \subset A_\alpha,$$
$$\text{if} \quad \alpha\notin H \quad \text{then} \quad U_\alpha = A_\alpha \quad \text{and} \quad A_\alpha \neq \operatorname{cl} A_\alpha.$$

Note that $\Phi \neq \varnothing$ since the pair $((A_\alpha)_{\alpha\in I}, H_0)$ is an element of Φ where $H_0 = \{\alpha\in I: A_\alpha \text{ is both open and closed}\}$.

A partial order $\leqslant$ is defined on Φ as follows:

$$((U_\alpha)_{\alpha\in I}, H) \leqslant ((V_\alpha)_{\alpha\in I}, J)$$

if and only if

$$H \subset J \quad \text{and} \quad U_\alpha = V_\alpha \quad \text{for all} \quad \alpha\in H.$$

It is easy to verify that $(\Phi, \leqslant)$ is a partially ordered set. To utilize Zorn's lemma it is necessary to show that every chain $\psi \subset \Phi$ has a upper bound. Let $K = \{\alpha\in I: \alpha\in H \text{ for some } H \text{ such that } ((U_\alpha)_{\alpha\in I}, H)\in\psi\}$. For each $\alpha\in K$, let W_α be the element of index α of *any* cover $(U_\alpha)_{\alpha\in I}$ such that $\alpha\in H$. Observe that W_α

is truly independent of the choice of covers. For if $((U_\alpha)_{\alpha \in I}, H)$ and $((V_\alpha)_{\alpha \in I}, J)$ are members of Φ such that $\alpha \in H \cap J$ then, since Φ is a chain, either $((U_\alpha)_{\alpha \in I}, I) \leqslant ((U_\alpha)_{\alpha \in I}, H)$ or

$$((U_\alpha)_{\alpha \in I}, J) \leqslant ((U_\alpha)_{\alpha \in I}, H).$$

In the first instance, $J \subset H$ and $\alpha \in J$ implies $V_\alpha = U_\alpha = W_\alpha$. The other case is shown similarly. If $\alpha \notin K$ let $W_\alpha = A_\alpha$.

To show that $((W_\alpha)_{\alpha \in I}, K)$ is an upper bound of ψ in Φ we need only prove that $(W_\alpha)_{\alpha \in I}$ is indeed a cover of X. If $x \in X$ then there exists a finite subset F of I such that $x \in A_\alpha$ if and only if $\alpha \in F$. If there is an $\alpha \in F - K$, then $A_\alpha = W_\alpha$ and therefore $x \in W_\alpha$. On the other hand if $F \subset K$ then, since F is finite and ψ is a chain, there is an element $((V_\alpha)_{\alpha \in I}, H)$ in ψ such that $F \subset H$. Now $(V_\alpha)_{\alpha \in I}$ is a cover of X so $x \in V_{\alpha_0}$ for some $\alpha_0 \in I$. Since $V_{\alpha_0} \subset A_{\alpha_0}$, it follows that $\alpha_0 \in F$ and therefore in K. But $\alpha_0 \in K$ implies that $W_{\alpha_0} = V_{\alpha_0}$ so $x \in W_{\alpha_0}$.

Zorn's lemma (see 1.6) now gives Φ a maximal element $((M_\alpha)_{\alpha \in I}, L)$. To complete the proof we show that $L = I$. Suppose that there exists a $\beta \in I$ such that $\beta \notin L$. Thus $M_\beta = A_\beta \neq \operatorname{cl} A_\beta$. If $U = \cup \{M_\alpha \colon \alpha \in I \text{ and } \alpha \neq \beta\}$ then $X - M_\beta$ is a closed set contained in U and since X is normal there exists an open set V such that $X - M_\beta \subset V \subset \operatorname{cl} V \subset U$. Let $((P_\alpha)_{\alpha \in I}, Q)$ be defined as follows: let $P_\alpha = M_\alpha$ if $\alpha \neq \beta$, let $P_\beta = X - \operatorname{cl} V$, and let $Q = L \cup \{\beta\}$. The pair $((P_\alpha)_{\alpha \in I}, Q)$ belongs to Φ since $(P_\alpha)_{\alpha \in I}$ covers X and

$$\operatorname{cl} P_\beta = \operatorname{cl}((X - \operatorname{cl} V) \subset X - V \subset X - (X - M_\beta)) = M_\beta = A_\beta.$$

But $((P_\alpha)_{\alpha \in I}, Q) \geqslant ((M_\alpha)_{\alpha \in I}, L)$, is a contradiction to the maximality of $((M_\alpha)_{\alpha \in I}, L)$. The proof is now complete. □

The above theorem is one of the two most significant properties of a normal space. Later, in 11.7 (6), it will be shown that it actually characterizes normal spaces as those spaces among the uniformizable spaces that satisfy the stated condition. As a consequence of this fact, it is possible to state that a cover of a normal space is a normal cover if and only if the cover has an open locally finite refinement (10.5 and 10.6).

One of the basic tools for obtaining equivalences for normal spaces and collectionwise normal spaces is now at hand. Beyond these characterizations it is fundamentally important when working with covers for it states when a cover $\mathscr{U}$ which expands another cover $\mathscr{V}$ may be screened by a relatively large cover $\mathscr{W}$ which also expands $\mathscr{V}$.

10.2 THEOREM (*Morita* [270, 272]) *If a locally finite (respectively, star-finite) open cover $\mathscr{U}$ of a topological space X expands a closed cover $\mathscr{V}$ of X then there is a locally finite (respectively, star-finite) open cover $\mathscr{W}$ of X such that $\{\mathrm{st}(V, \mathscr{W}): V \in \mathscr{V}\}$ screens $\mathscr{U}$. Moreover if $\mathscr{U}$ is finite then also is $\mathscr{W}$.*

Proof. Let $\mathscr{U} = (U_\alpha)_{\alpha \in I}$, let $\mathscr{V} = (V_\alpha)_{\alpha \in I}$. Now for each $J \in [I]$, let

$$W_J = (\bigcap_{\alpha \in J} U_\alpha) \cap (\bigcap_{\alpha \notin J} X - V_\alpha)$$

and let $\mathscr{W} = (W_J)_{J \in [I]}$. If X is empty then the result is immediate, so suppose that $X \neq \varnothing$. If $x \in X$, let $J = \{\alpha \in I: x \in U_\alpha\}$, and note that $J \in [I]$ since $\mathscr{U}$ is point-finite. If $\alpha \notin J$, then $x \notin U_\alpha$. Therefore $x \in W_J$ and $\mathscr{W}$ is a cover of X. To see that $\mathscr{W}$ is locally finite, let $x \in X$. Then there exists a neighborhood G of X and a finite subset K of I such that $G \cap U_\alpha = \varnothing$ if and only if $\alpha \notin K$. If $W_J \cap G \neq \varnothing$ for some $J \in [I]$, then $\alpha \in J$ implies $W_J \cap G \neq \varnothing$ and therefore $U_\alpha \cap G \neq \varnothing$ so that $\alpha \in K$, whence $J \subset K$. If we set

$$\mathscr{L} = \{J \subset I: J \subset K\}$$

then $\mathscr{L}$ is a finite subset of $[I]$ and $W_J \cap G = \varnothing$ if $J \notin \mathscr{L}$. Thus $\mathscr{W}$ is locally finite.

To see that each W_J is open, choose $x \in W_J$ and let G be a neighborhood of x and K a finite subset of I such that

$$G \cap U_\alpha = \varnothing \quad \text{if} \quad \alpha \notin K.$$

Let
$$H = (\bigcap_{\alpha \in J} U_\alpha) \cap (\bigcap_{\alpha \in J - K} X - V_\alpha) \cap G.$$

We show that H is an open set containing x and contained in W_J. Since K and J are finite and since U_α, $X - V_\alpha$ and G are open sets it is clear that H is open. Now $x \in W_J$ so $x \in U_\alpha$ for all $\alpha \in J$. Also if $\alpha \notin K$ then $G \cap U_\alpha = \varnothing$ so $G \cap V_\alpha = \varnothing$ hence $G \subset X - V_\alpha$ and therefore $x \in X - V_\alpha$. It follows that $x \in H$. To show that $H \subset W_J$ suppose that $y \notin W_J$. If $y \notin U_\alpha$ for some $\alpha \in J$ then $y \notin H$. If $y \in U_\alpha$ for all

$\alpha \in J$ then $y \notin W_J$ implies $y \notin X - V_\alpha$ for some $\alpha \notin J$. If $\alpha \notin K$ then $G \subset V_\alpha$ so $y \notin G$ and therefore not in H. If $\alpha \in K$ then $\alpha \in K - J$ and $y \notin X - V_\alpha$ so $y \notin H$ by definition.

Finally, if $W_J \cap V_\alpha \neq \varnothing$ then $\alpha \in J$ and thus $W_J \subset U_\alpha$. Thus $\mathrm{st}(V_\alpha, \mathscr{W}) \subset U_\alpha$. Also if $\mathscr{U}$ has power at most γ then $\mathscr{W}$ has power at most $\aleph_0 \cdot \gamma = \gamma$.

The star-finite case is left as an exercise for the reader. □

10.3 Corollary (*Stone* [359]) *A locally finite open cover of a non-empty normal space has an open Δ-refinement that is also locally finite.*

This is an immediate consequence of 10.1 and 10.2.

The cardinality of the covers in the above corollary may be related as stated in the next corollary.

10.4 Corollary *If X is a non-empty normal topological space and if γ is an infinite cardinal number, then every locally finite open cover of X of power at most γ (respectively, power less than $\aleph_0$) has an open Δ-refinement that is locally finite and that has power at most γ (respectively, power less than $\aleph_0$).*

Again, this is an immediate consequence of 10.1 and 10.2.

Now from Stone's result the following interesting statement is obtained.

10.5 Corollary *A locally finite open cover of a normal space is a normal cover.*

Proof. If X is empty, then the result is immediate. If $X \neq \varnothing$, then a repeated application of 10.3 and an appeal to 1.18 yield the conclusion. □

Consequently normal open covers are quite frequent even on spaces which are not metrizable since every paracompact Hausdorff space is normal. On the other hand a normal open cover will generate a continuous pseudometric, from which much may be learned about the original topology.

Let us now take a closer look at 10.5. It would be helpful to know that for the normal sequence associated with the normal cover, each member of the sequence has the same cardinality as the cover.

10.6 COROLLARY *Suppose that X is a normal space, that γ is an infinite cardinal number, and that $\mathscr{U}$ is a locally finite open cover of X of power at most γ (respectively, power less than $\aleph_0$). Then there exists a normal sequence $(\mathscr{V}_i)_{i\in\mathbf{N}}$ of covers of X such that $\mathscr{V}_1 < \mathscr{U}$ and such that, for each $i \in \mathrm{N}$, $\mathscr{V}_i$ has power at most γ (respectively, power less than $\aleph_0$).*

Proof. If X is empty, then the result is immediate. If $X \neq \varnothing$, then a repeated application of 10.4 and an appeal to 1.18 yield the conclusion. □

Again let us look at another way of having covers 'agree'. Suppose two covers $\mathscr{U}$ and $\mathscr{V}$ of a space are intrinsically related, for example, as it has been seen, it is often required that $\mathscr{V}$ expands $\mathscr{U}$. It is useful to know that when $\mathscr{V}$ expands $\mathscr{U}$, collections of sets with the finite intersection property in $\mathscr{V}$ expand collections of sets in $\mathscr{U}$ with the finite intersection property. Thus the following definition is made.

10.7 DEFINITION Suppose that $(U_\alpha)_{\alpha\in I}$ and $(V_\alpha)_{\alpha\in I}$ are two families of subsets of X with the same set of indices. It is said that $(U_\alpha)_{\alpha\in I}$ *is similar to* $(V_\alpha)_{\alpha\in I}$ if, for every $J \in [I]$, $\bigcap_{\alpha\in J} U_\alpha = \varnothing$ if and only if $\bigcap_{\alpha\in J} V_\alpha = \varnothing$.

Note of course that '*is similar to*' is an equivalence relation among collections of subsets of a space.

With these ideas in mind consider again the hypothesis of 10.2. It is possible to obtain a collection $\mathscr{M}$ of open sets that strongly screens the locally finite collection $\mathscr{U}$ and $(\mathrm{cl}\, M)_{M\in\mathscr{M}}$ is similar to the closed collection $\mathscr{V}$. This result was first demonstrated by Morita [272, Theorem 1.3] and then proved independently by Katětov [222, Lemma 1.12].

10.8 THEOREM (*Morita–Katětov*) *Let $\mathscr{G}$ be a locally finite collection of open subsets of a normal space X that stretches a collection $\mathscr{F}$ of closed subsets of X. Then there exists a collection $\mathscr{M}$ of open sets such that* cl $\mathscr{M}$ *is similar to $\mathscr{F}$ and such that $\mathscr{M}$ strongly screens $\mathscr{G}$ and expands $\mathscr{F}$.*

Proof. Let $\mathscr{F} = (F_\alpha)_{\alpha\in I}$ and let $\mathscr{G} = (G_\alpha)_{\alpha\in I}$ be such that $F_\alpha \subset G_\alpha$ for each $\alpha \in I$. Let Φ be the collection of all pairs $((U_\alpha)_{\alpha\in I}, H)$ where $(U_\alpha)_{\alpha\in I}$ is an open family of subsets of X such that

$$F_\alpha \subset U_\alpha \subset \operatorname{cl} U_\alpha \subset G_\alpha,$$

and where H is a subset of I with the property that

$$(\operatorname{cl} U_\alpha)_{\alpha\in H} \cup (F_\alpha)_{\alpha\in I-H}$$

is similar to $(F_\alpha)_{\alpha\in I}$. The collection Φ is non-empty since the pair $((U_\alpha)_{\alpha\in I}, H_0)$ is an element of Φ where $H_0 = \varnothing$ and where $(U_\alpha)_{\alpha\in I}$ is chosen as in 10.1.

As in 10.1, a partial order $\leqslant$ is defined on Φ as follows: $((U_\alpha)_{\alpha\in I}, H) \leqslant ((V_\alpha)_{\alpha\in I}, J)$ if and only if $H \subset J$ and $U_\alpha = V_\alpha$ for all $\alpha \in H$. Again, as in 10.1, let ψ be a chain in Φ.

Let

$$K = \{\alpha\in I: \alpha \in H \text{ for some } H \text{ such that } ((U_\alpha)_{\alpha\in I}, H) \in \psi\}.$$

For each $\alpha \in K$, let W_α be the element of index α of any family $(U_\alpha)_{\alpha\in I}$ such that $\alpha \in H$. As in 10.1, W_α is independent of the choice of the family $(U_\alpha)_{\alpha\in I}$. If $\alpha \notin K$ then using the normality of X, choose W_α such that

$$F_\alpha \subset W_\alpha \subset \operatorname{cl} W_\alpha \subset G_\alpha.$$

The upper bound in Φ of ψ is $((W_\alpha)_{\alpha\in I}, K)$. To show this we need only prove that $(\operatorname{cl} W_\alpha)_{\alpha\in K} \cup (F_\alpha)_{\alpha\in I-K}$ is similar to $(F_\alpha)_{\alpha\in I}$. If $J_1 \in [K]$ and if $J_2 \in [I-K]$ then since J_1 is a finite subset of K and since ψ is a chain, there is an element $((U_\alpha)_{\alpha\in I}, H)$ in ψ for which $J_1 \subset H$. Now

$$(\operatorname{cl} U_\alpha)_{\alpha\in H} \cup (F_\alpha)_{\alpha\in I-H}$$

is similar to $(F_\alpha)_{\alpha\in I}$, so $(\bigcap_{\alpha\in J_1} \operatorname{cl} U_\alpha) \cap (\bigcap_{\alpha\in J_2} F_\alpha) = \varnothing$ if and only if $\bigcap_{\alpha\in J_1\cup J_2} F_\alpha = \varnothing$. However, by definition $W_\alpha = U_\alpha$ for every $\alpha \in H$

and since $J_1 \subset H$, it follows that $(\bigcap_{\alpha\in J_1} \mathrm{cl}\, W_\alpha) \cap (\bigcap_{\alpha\in J_2} F_\alpha) = \varnothing$ if and only if $\bigcap_{\alpha\in J_1\cup J_2} F_\alpha = \varnothing$.

By Zorn's lemma Φ has a maximal element $((M_\alpha)_{\alpha\in I}, L)$. To complete the proof we show that $L = I$. Suppose that β is in I but is not in L. Let $\mathscr{V} = \{H \subset X$: there is $A\in[L]$ and $B\in[I-L]$ with $H = (\bigcap_{\alpha\in A} \mathrm{cl}\, M_\alpha) \cap (\bigcap_{\alpha\in B} F_\alpha)$ and $(\bigcap_{\alpha\in A\cup B} F_\alpha) \cap F_\beta = \varnothing\}$. By the definition of Φ and the fact that $\mathscr{G}$ is locally finite it follows that the families $(\mathrm{cl}\, M_\alpha)_{\alpha\in I}$ and $(F_\alpha)_{\alpha\in I}$ are locally finite and consequently the collection $\mathscr{V}$ is closure preserving (1.4(2)). Thus the set $S = \cup\mathscr{V}$ is closed. In addition S and F_β are disjoint. To see this let $H\in\mathscr{V}$ so that there exist $A\in[L]$ and $B\in[I-L]$ with

$$H = (\bigcap_{\alpha\in A} \mathrm{cl}\, M_\alpha) \cap (\bigcap_{\alpha\in B} F_\alpha)$$

and

$$(\bigcap_{\alpha\in A\cup B} F_\alpha) \cap F_\beta = \varnothing.$$

Now $(\mathrm{cl}\, M_\alpha)_{\alpha\in L} \cup (F_\alpha)_{\alpha\in I-L}$ is similar to $(F_\alpha)_{\alpha\in I}$ implies that $(\bigcap_{\alpha\in A} \mathrm{cl}\, M_\alpha) \cap (\bigcap_{\alpha\in B} F_\alpha) \cap F_\beta = \varnothing$ whence $H \cap F_\beta = \varnothing$. Since X is normal there is an open set U for which

$$F_\beta \subset U \subset \mathrm{cl}\, U \subset G_\alpha \cap X - S. \tag{1.11}$$

Let the pair $((P_\alpha)_{\alpha\in I}, Q)$ be defined as follows: let $P_\alpha = M_\alpha$ for all $\alpha\in I, \alpha \neq \beta$, let $P_\beta = U$, and let $Q = M \cup \{\beta\}$. The pair $((P_\alpha)_{\alpha\in I}, Q)$ belongs to Φ. First of all $F_\alpha \subset P_\alpha \subset \mathrm{cl}\, P_\alpha \subset G_\alpha$ for all $\alpha\in I$. Secondly for the condition of similarity, let $T_1\in[Q]$ and let $T_2\in[I-Q]$. If $\beta\notin T_1$ then the result holds since $((M_\alpha)_{\alpha\in I}, L)$ is in Φ. So suppose that $\beta\in T_1$. If $(\bigcap_{\alpha\in T_1} \mathrm{cl}\, P_\alpha) \cap (\bigcap_{\alpha\in T_2} F_\alpha) = \varnothing$ then since $F_\alpha \subset \mathrm{cl}\, P_\alpha$ it follows that $(\bigcap_{\alpha\in T_1\cup T_2} F_\alpha) = \varnothing$. On the other hand suppose that $\bigcap_{\alpha\in T_1\cup T_2} F_\alpha = \varnothing$. Since $\beta\in T_1$ it follows that $(T_1-\{\beta\})\in[L]$ and hence

$$H = (\bigcap_{\alpha\in T_1-\{\beta\}} \mathrm{cl}\, M_\alpha) \cap (\bigcap_{\alpha\in T_2} F_\alpha)$$

is an element of $\mathscr{V}$. But then $H \subset S$ and $\mathrm{cl}\, U \subset X - S$ consequently $(\bigcap_{\alpha\in T_1} \mathrm{cl}\, P_\alpha) \cap (\bigcap_{\alpha\in T_2} F_\alpha) = \varnothing$ and it follows that $((P_\alpha)_{\alpha\in I}, \in Q)\,\Phi$. But this contradicts the maximality of $((M_\alpha)_{\alpha\in I}, L)$ and so it must be that $L = I$. Hence $(M_\alpha)_{\alpha\in I}$ is a family of open subsets of X such that

$$F_\alpha \subset M_\alpha \subset \operatorname{cl} M_\alpha \subset G_\alpha$$

and $(M_\alpha)_{\alpha\in I}$ is similar to $(F_\alpha)_{\alpha\in I}$. This completes the proof. □

Let us consider now a normal cover $\mathscr{U}$ and a continuous pseudometric d associated with it. It is possible to obtain a locally finite open cover which screens $\mathscr{U}$. But even more so this screening of $\mathscr{U}$ is a collection of sets open in $\mathscr{T}_d$.

10.9 Lemma *Suppose that $\mathscr{U}$ is an open cover of a topological space X, that $(\mathscr{U}_i)_{i\in\mathbf{N}}$ is a normal sequence of open covers of X such that $\mathscr{U}_1 < \mathscr{U}$, and that d is a continuous pseudometric associated with $(\mathscr{U}_i)_{i\in\mathbf{N}}$. Then there exists a locally finite open cover $\mathscr{W}$ of X relative to $\mathscr{T}_d$ such that $\mathscr{W}$ screens $\mathscr{U}$.*

Proof. If X is empty, then the result is immediate. Suppose that $X \neq \varnothing$ and let $\mathscr{V} = (S(x, 2^{-3}))_{x\in X}$. Then, by 8.6

$$\mathscr{V} < \mathscr{U} = (U_\alpha)_{\alpha\in I}.$$

For each $\alpha\in I$, let

$$A_\alpha = \bigcup \{S(x, 2^{-3}) : S(x, 2^{-3}) \subset U_\alpha\}.$$

Then $\mathscr{A} = (A_\alpha)_{\alpha\in I}$ is an open cover of X relative to $\mathscr{T}_d$ such that $A_\alpha \subset U_\alpha$ for each $\alpha\in I$. Now (X, d) is a pseudometric space, so it is paracompact. Therefore there exists a locally finite open cover of X relative to $\mathscr{T}_d$ such that $\mathscr{B} < \mathscr{A}$. Therefore, by 1.8 (letting $S = X$), there exists a locally finite open cover $\mathscr{W} = (W_\alpha)_{\alpha\in I}$ of X relative to $\mathscr{T}_d$ such that $W_\alpha \subset A_\alpha$ for each $\alpha\in I$. Thus $W_\alpha \subset U_\alpha$ for each $\alpha\in I$ and the proof is complete. □

Enough background has been established to give equivalent formulations for the concept of a normal open cover. Some clarifying remarks are appropriate. Note that (2) in the following theorem is stronger than requiring the existence of a locally finite cozero-set cover which strongly screens $\mathscr{U}$. Condition (4) permits one to work with collections of continuous real-valued functions. Condition (5) deserves more consideration, for the concept has been successfully utilized to define and to characterize classes of spaces (see Section 22). The equivalence of (1), (3), (4), (5), and (6) is stated by Morita [279, Theorem 1.2], who attributes it to Stone [359] and Michael [262]. Our proof of (4) implies (5) is taken from Morita's paper.

10.10 THEOREM *If $\mathscr{U} = (U_\alpha)_{\alpha\in I}$ is an open cover of a topological space X, then the following statements are equivalent.*

(1) *$\mathscr{U}$ is normal.*

(2) *There exists a locally finite cozero-set cover $\mathscr{V} = (V_\alpha)_{\alpha\in I}$ of X such that $\operatorname{cl} V_\alpha$ is completely separated from $X - U_\alpha$ for each $\alpha\in I$.*

(3) *The cover $\mathscr{U}$ has a locally finite cozero-set refinement.*

(4) *There exists a locally finite partition of unity subordinate to $\mathscr{U}$.*

(5) *There exists a metric space Y, a continuous map $f: X \to Y$, and an open cover $\mathscr{V}$ of Y such that $f^{-1}(\mathscr{V})$ is a refinement of $\mathscr{U}$.*

(6) *The cover $\mathscr{U}$ has a locally finite normal open refinement.*

Proof. (1) *implies* (2). By hypothesis, there exists a normal sequence $(\mathscr{U}_i)_{i\in\mathbf{N}}$ of open covers of X such that $\mathscr{U}_1 < \mathscr{U}$. By 8.4 there is an associated continuous pseudometric d on X. By 10.9 there exists a locally finite open cover $\mathscr{W} = (W_\alpha)_{\alpha\in I}$ of X relative to $\mathscr{T}_d$ such that $W_\alpha \subset U_\alpha$ for each $\alpha\in I$. Now (X,d) is a pseudometric space, so (X,d) is normal. Therefore, by 10.1, there exists an open cover $(V_\alpha)_{\alpha\in I}$ of X such that, for each $\alpha\in I$, $\operatorname{cl} V_\alpha \subset W_\alpha$ relative to $\mathscr{T}_d$. Thus for each $\alpha\in I$ there exists a $\mathscr{T}_d$-continuous map f_α mapping X to $\mathbf{R}$ such that $f_\alpha(x) = 1$ if $x\in\operatorname{cl} V_\alpha$ and $f_\alpha(x) = 0$ if $x\in X - W_\alpha$. From 2.12 and 2.15 it follows that each V_α is a cozero-set relative to the given topology $\mathscr{T}$ on X and that each f_α is a $\mathscr{T}$-continuous map. Thus $(V_\alpha)_{\alpha\in I}$ is a cozero-set cover of X such that $\operatorname{cl} V_\alpha$ is completely separated from $X - U_\alpha$ for each $\alpha\in I$.

(2) *implies* (3). This implication is trivial.

(3) *implies* (4). Let $\mathscr{V} = (V_\beta)_{\beta\in J}$ be a locally finite cozero-set refinement of $\mathscr{U}$. Thus, for each $\beta\in J$, there exists an $f_\beta\in C(X,E)$ such that $V_\beta = \{x\in X: f_\beta(x) > 0\}$. Define $f: X \to \mathbf{R}^+$ by the formula

$$f(x) = \sum_{\beta\in J} f_\beta(x) \quad (x\in X).$$

One easily verifies that f is a positive continuous function. For each $\beta\in J$, define $\phi_\beta = f_\beta - f$. Then $(\phi_\beta)_{\beta\in J}$ is a locally finite partition of unity subordinate to $\mathscr{U}$.

(4) *implies* (5). Let $\Phi = (\phi_\beta)_{\beta \in J}$ be a locally finite partition of unity subordinate to $\mathscr{U}$ and let $E_\beta = [0, 1]$ for each $\beta \in J$. Let $Y = \{(x_\beta)_{\beta \in J} : x_\beta \in E_\beta, \sum_{\beta \in J} x_\beta = 1$, and there exists a finite subset K of J such that $x_\beta = 0$ if $\beta \notin K\}$. Define $d: Y \times Y \to \mathbf{R}^+$ by

$$d(x, y) = \sum_{\beta \in J} |x_\beta - y_\beta| \quad (x, y \in Y).$$

One easily verifies that d is a pseudometric on Y.

Define $f: X \to Y$ by $f(x) = (\phi_\beta(x))_{\beta \in J}$ $(x \in X)$. Note that f is well defined since Φ is locally finite. To show that f is a continuous map, suppose that $x \in X$ and that $\epsilon > 0$. Since Φ is locally finite, there exist a neighborhood U' of x and a non-empty finite subset K of J such that $U' \cap H_\beta = \varnothing$ if $\beta \notin K$. For each $\beta \in K$ there exists a neighborhood V_β of x such that

$$|\phi_\beta(x) - \phi_\beta(y)| < \epsilon/|K|$$

if $y \in V_\beta$. Let $U = U' \cap (\bigcap_{\in K} V_\beta)$. If $y \in U$, then

$$d(f(x), f(y)) = \sum_{\beta \in J} |\phi_\beta(x) - \phi_\beta(y)| = \sum_{\beta \in K} |\phi_\beta(x) - \phi_\beta(y)| < \epsilon.$$

Therefore f is continuous. For each $\beta \in J$, let

$$V_\beta = \{(x_\beta)_{\beta \in J} : x_\beta > 0\}.$$

Then V_β is open in Y and $f(X) \subset \bigcup_{\beta \in J} V_\beta$. Finally,

$$\begin{aligned} f^{-1}(V_\beta) &= \{x \in X : f(x) \in V_\beta\} \\ &= \{x \in X : \phi_\beta(x) > 0\}. \end{aligned}$$

Hence (5) holds.

(5) *implies* (6). By hypothesis, there exists a metric space Y, a continuous map $f: X \to Y$, and an open cover $\mathscr{V}$ of Y such that $(f^{-1}(V))_{V \in \mathscr{V}} < \mathscr{U}$. Since Y is metric, it is paracompact, and therefore $\mathscr{V}$ has an open locally finite refinement $\mathscr{W}$. Since Y is normal, $\mathscr{W}$ is normal (10.5). Thus, by 1.21 $f^{-1}(\mathscr{W})$ is a locally finite normal open cover of X that refines $\mathscr{U}$.

(6) *implies* (1). This implication is trivial.

The proof is now complete. □

As an immediate corollary of 10.10 we have:

10.11 COROLLARY *If $\mathscr{U}$ is a normal open cover of a topological space X, then there exists a zero-set cover $\mathscr{V}$ of X such that $\mathscr{V}$ screens $\mathscr{U}$.*

In the final three results of this section some elementary information concerning covers will be discussed. These propositions are necessary for future reference.

Corollary 10.6 said that a locally finite open cover $\mathscr{U}$ of a normal space is normal and that the cardinality of each cover in the associated normal sequence is the same as that of $\mathscr{U}$. In view of the above this cardinality restriction can now be shown for any normal open cover of any topological space.

10.12 PROPOSITION *Suppose that $(X, \mathscr{T})$ is a topological space, that γ is an infinite cardinal number, and that $\mathscr{U} = (U_\alpha)_{\alpha \in I}$ is a normal open cover of X of power at most γ (respectively, power less than $\aleph_0$). Then there exists a normal sequence $(\mathscr{V}_i)_{i \in \mathbf{N}}$ of open covers of X such that $\mathscr{V}_1 < \mathscr{U}$ and such that, for each $i \in \mathrm{N}$, $\mathscr{V}_i$ has power at most γ (respectively, power less than $\aleph_0$).*

Proof. By hypothesis, there exists a normal sequence $(\mathscr{U}_i)_{i \in \mathbf{N}}$ of open covers of X such that $\mathscr{U}_1 < \mathscr{U}$. Let d be a continuous pseudometric associated with $(\mathscr{U}_i)_{i \in \mathbf{N}}$ (8.4). By 10.9 there exists a locally finite open cover $\mathscr{W}$ of X relative to $\mathscr{T}_d$ such that $\mathscr{W}$ screens $\mathscr{U}$. Therefore, by 10.6, there exists a normal sequence $(\mathscr{V}_i)_{i \in \mathbf{N}}$ of open covers of X, relative to $\mathscr{T}_d$, such that $\mathscr{V}_1 < \mathscr{U}$ and such that, for each $i \in \mathrm{N}$, $\mathscr{V}_i$ has power at most γ (respectively, power less than $\aleph_0$). Since $\mathscr{T}_d \subset \mathscr{T}$ and $\mathscr{W} < \mathscr{U}$, the result now follows. □

10.13 PROPOSITION *Suppose that X is a topological space, that $S \subset X$, that $\mathscr{U}$ is an open cover of S, and that $\mathscr{V}$ is a normal (respectively, locally finite normal, respectively, normal cozero-set, respectively, locally finite cozero-set) open cover of X such that $\mathscr{V}|S < \mathscr{U}$. Then there exists a normal (respectively, locally finite normal, respectively, normal cozero-set, respectively, locally finite cozero-set) open cover $\mathscr{W}$ of X such that $\mathscr{W}|S$ screens $\mathscr{U}$.*

Proof. Suppose that $\mathscr{V} = (V_\beta)_{\beta \in J}$ is a locally finite cozero-set cover of X. Since $\mathscr{V}|S < \mathscr{U} = (U_\alpha)_{\alpha \in I}$, there exists a function $\pi: J \to I$ such that $V_\beta \cap S \subset U_{\pi(\beta)}$ for each $\beta \in J$. For each $\alpha \in I$, let

$$W_\alpha = \bigcup_{\beta \in \pi^{-1}(\alpha)} V_\beta.$$

It can now be shown that $\mathscr{W} = (W_\alpha)_{\alpha \in I}$ is a locally finite cozero-set cover of X such that $\mathscr{W}|S$ screens $\mathscr{U}$ for each $\alpha \in I$.

The other cases follow from 10.10, 1.8, and the above. □

Let us comment here about paracompact spaces. Normal covers have just been characterized in 10.10 as those covers which have locally finite partitions of unity subordinate to it. Consequently not every open cover of a space is a normal open cover since T_1-spaces which satisfy this condition must be paracompact.

10.14 Corollary *A T_1 topological space is paracompact if and only if every open cover is a normal open cover.*

10.15 Exercise

Suppose that X is a paracompact space and that g and k are, respectively, lower and upper semicontinous real-valued functions on X such that $k(x) < g(x)$ for all $x \in X$. Then there exists $f \in C(X)$ such that $k(x) < f(x) < g(x)$ for all $x \in X$. (Let $\mathbf{Q}$ be the rational numbers and for each $r \in \mathbf{Q}$, let

$$U = \{x: k(x) < r\} \cap \{x: g(x) > r\}.$$

Then $\mathscr{U} = (U_r)_{r \in \mathbf{Q}}$ is an open cover of X. Now use 10.14 and 10.10 to obtain a locally finite partition of unity $(f_r)_{r \in \mathbf{Q}}$ subordinate to $\mathscr{U}$. Finally show that $f = \sum_{r \in \mathbf{Q}} r \cdot f_r$ is the desired function.)

(In [99], Dowker has shown that the converse implies X is countably paracompact and normal which is equivalent to the product $X \times E$ being normal.)

11 Normal spaces and open covers

In this section, after preliminary theorems on cozero-set covers, several necessary and sufficient conditions will be given for a topological space to be normal (11.7)

The cozero-sets of a completely regular space determine the topology of the space. In 11.1 the foundation is laid to show that the collection of cozero-sets possess an even nicer property for arbitrary topological spaces: *countable cozero-set covers are normal covers.* From this follows the existence of continuous pseudometrics. The idea for this theorem was first contained in the proof of Theorem 2 of [99] and later in the proof of Theorem 1 of [246].

11.1 THEOREM *Every countable cozero-set cover of a topological space X has a locally finite cozero-set cover that screens it.*

Proof. Let $(U_i)_{i\in\mathbf{N}}$ be a countable cozero-set cover of X. For each $i\in\mathrm{N}$, let $F_i = X - (\bigcup_{k<i} U_k)$. Then each F_i is a zero-set, $F_{i+1}\subset F_i$, and $\bigcap_{i\in\mathbf{N}} F_i = \varnothing$. Now, by 2.18(5), for each $j\in\mathrm{N}$, there exists a sequence $(A_{ji})_{i\in\mathbf{N}}$ of cozero-sets and a sequence $(B_{ji})_{i\in\mathbf{N}}$ of zero-sets such that $X-F_j = \bigcup_{i\in\mathbf{N}} A_{ji} = \bigcup_{i\in\mathbf{N}} B_{ji}$ and

$$A_{ji}\subset B_{ji}\subset A_{j,i+1}$$

for each $i\in\mathrm{N}$. Then $B_{ji}\subset X-F_j = \bigcup_{k\leqslant j} U_k$. For each $i\in\mathrm{N}$, let $V_i = U_i - \bigcup_{j<i} B_{ji}$. Then V_i is a cozero-set in X. If $j<i$, then $B_{ji}\subset \bigcup_{k\leqslant j} U_k \subset \bigcup_{k<i} U_k$. Hence, for each $i\in\mathrm{N}$,

$$\bigcup_{j<i} B_{ji}\subset \bigcup_{k<i} U_k.$$

Therefore $V_i\supset U_i - \bigcup_{k<i} U_k$. It is easy to show that $(V_i)_{i\in\mathbf{N}}$ is a cover of X, and hence is a refinement of $(U_i)_{i\in\mathbf{N}}$. It remains to show that $(V_i)_{i\in\mathbf{N}}$ is locally finite.

Let $x\in X$ and choose $j\in\mathrm{N}$ such that $x\notin F_j$. Hence, for some $k\in\mathrm{N}$, $x\in A_{jk}$. Now A_{jk} is a cozero-set, so A_{jk} is a neighborhood of x. Let $K=\{i\in\mathrm{N}\colon\ i\leqslant \max\ (j,k)\}$ and note that

K is finite. Suppose that $i \notin K$ so that $i > j$ and $i > k$. Then $A_{jk} \subset B_{ji}$. If $y \in A_{jk}$, then $y \in B_{ji}$. Thus $y \notin V_i = U_i - \bigcup_{m<i} B_{mi}$. Therefore $V_i \cap A_{jk} = \varnothing$ so $(V_i)_{i \in \mathbf{N}}$ is locally finite. The proof is now complete. □

In view of 11.1 and 10.10, what has been said is that countable cozero-set covers are normal.

11.2 Corollary *Every countable cozero-set cover of a topological space is normal.*

This corollary may now be used to strengthen 11.1. For open covers, star-finite implies locally finite. The proof of this is a modification of Morita's proof of [270, Theorem 3]. The corollary is Morita's original theorem.

11.3 Theorem *Every countable cozero-set cover of a topological space X has a star-finite normal cozero-set cover that screens it.*

Proof. Suppose that $\mathscr{A}$ is a countable cozero-set cover of X. By 11.1, there exists a locally finite cozero-set cover $\mathscr{G} = (G_i)_{i \in \mathbf{N}}$ of X such that $\mathscr{G} < \mathscr{A}$. Then, by 11.2 and 10.11, there exists a zero-set cover $\mathscr{F} = (F_i)_{i \in \mathbf{N}}$ of X such that $F_i \subset G_i$ for each $i \in \mathbf{N}$. Since F_i and $X - G_i$ are disjoint zero-sets, they are completely separated, so, for each $i \in \mathbf{N}$, there exists $f_i \in C(X, E)$ such that $f_i(x) = 0$ if $x \in F_i$ and $f_i(x) = 1$ if $x \in X - G_i$.

For each n, $i \in \mathbf{N}$ such that $i \leqslant n$, let

$$G_i^n = \{x \in X : f_i(x) < 1 - (n+1)^{-1}\}$$

and

$$H_i^n = \{x \in X : f_i(x) \leqslant 1 - (n+1)^{-1}\}.$$

Then G_i^n is a cozero-set, H_i^n is a zero-set, and $G_i^n \subset G_i$. Now for each $n \in \mathbf{N}$, let $U_n = \bigcup_{i=1}^{n} G_i^n$ and $V_n = \bigcup_{i=1}^{n} H_i^n$. Then one easily verifies that $X = \bigcup_{n \in \mathbf{N}} U_n$ and that $V_n \subset U_{n+1}$ for each $n \in \mathbf{N}$. Next, for each $n \in \mathbf{N}$, let $Q_n = U_n - V_{n-3}$ and $R_n = V_n - U_{n-1}$, where $U_0 = V_0 = V_{-1} = V_{-2} = \varnothing$. Then Q_n is a cozero-set and R_n is a zero-set for each $n \in \mathbf{N}$. Clearly $G_i^n \subset H_i^n$, so $U_n \subset V_n$ for all $n \in \mathbf{N}$. Thus $X = \bigcup_{n \in \mathbf{N}} V_n$. Now let $x \in X$. Then $x \in V_n$ for some $n \in \mathbf{N}$, and let us assume that n is the smallest positive integer with this property. If $n = 1$, then $x \in V_1 = R_1$. If $n > 1$, then $x \notin V_{n-1}$, whence $x \notin U_{n-1}$. Thus

$$x \in V_n - U_{n-1} = R_n$$

which implies that $X = \bigcup_{n \in \mathbf{N}} R_n$. To show that

$$(*) \qquad Q_m \cap Q_n = \varnothing \quad \text{if } m, n \in \mathbf{N} \quad \text{such that} \quad |m-n| > 3$$

it is supposed, without loss of generality, that $m, n \in \mathbf{N}$ such that $m-n > 3$ and $x \in Q_m$. Since $x \notin V_{m-3}$ and since $U_n \subset V_{m-3}$, it follows that $x \notin U_n$, whence $x \notin Q_n$.

Now let

$$M = \{(n, i) \in \mathbf{N} \times \mathbf{N} : i \leqslant n\}$$

and for each $(n, i) \in M$, let

$$B_{ni} = Q_n \cap G_i.$$

The assertion now is that $\mathscr{B} = (B_{ni})_{(n, i) \in M}$ is a countable star-finite normal cozero-set cover of X such that $\mathscr{B} < \mathscr{A}$. Let $x \in X$. Then $x \in R_m$ for some $m \in \mathbf{N}$ and $R_m \subset Q_{m+1}$. Let $n = m+1$, so $x \in Q_n$. Since $Q_n = \bigcup_{i=1}^{n} (Q_n \cap G_i)$, $x \in Q_n \cap G_i$ for some $n, i \in \mathbf{N}$ such that $i \leqslant n$. Moreover, since Q_n and G_i are cozero-sets for all n, $i \in \mathbf{N}$, and since $\mathscr{B} < \mathscr{G} < \mathscr{A}$, it follows that $\mathscr{B}$ is a countable cozero-set refinement of $\mathscr{A}$. Finally, let $j, m \in \mathbf{N}$ such that $j \leqslant m$ and set

$$N = \{(n, i) \in M : m-3 \leqslant n \leqslant m+3 \text{ and } i \leqslant n\}.$$

Then N is a finite subset of M. Moreover, if $(n, i) \notin N$, then $|m-n| > 3$, so, by (*), $Q_m \cap Q_n = \varnothing$ and it follows that

$$(Q_m \cap G_j) \cap (Q_n \cap G_i) = \varnothing .$$

Therefore $\mathscr{B}$ is star-finite and, since $\mathscr{B}$ is normal (11.2), the proof is complete. □

11.4 Corollary (*Morita* [270]) *Let $\mathscr{G} = (G_i)_{i \in \mathbf{N}}$ be a countable open cover of a non-empty topological space X which expands the closed cover $\mathscr{F} = (F_i)_{i \in \mathbf{N}}$ of X. If, for each $i \in \mathbf{N}$ there* *exists* $f_i \in C(X)$ *such that* $f_i(x) = 0$ *if* $x \in F_i$ *and* $f_i(x) = 1$ *if* $x \in X - G_i$ *then* $\mathscr{G}$ *has a countable star-finite open* Δ*-refinement.*

Proof. For each $i \in \mathbf{N}$, let $H_i = \{x \in X : f_i(x) < \frac{1}{2}\}$. Then $\mathscr{H} = (H_i)_{i \in \mathbf{N}}$ is a countable cozero-set cover of X. By 11.3, there exists a countable star-finite normal cozero-set cover $\mathscr{U}$ of X such that $\mathscr{U} < \mathscr{H}$. By 10.11, there exists a zero-set cover $\mathscr{V}$ of X such that $\mathscr{V}$ screens $\mathscr{U}$. Therefore, by 10.2, there exists a star-finite open Δ-refinement $\mathscr{W}$ of $\mathscr{G}$ and the proof of 10.2 shows that $\mathscr{W}$ is countable. □

11.5 COROLLARY *If* $\mathscr{G}$ *is a countable open cover of a non-empty normal topological space* X *and if* $\mathscr{F}$ *is a closed cover of* X *such that* $\mathscr{F}$ *screens* $\mathscr{G}$ *then* $\mathscr{G}$ *has a countable star-finite open star-refinement.*

Proof. The result follows from 11.4, 10.1, and 1.19. □

The conclusion of 11.3 can be strengthened even further. In fact the refinement that is obtained can be replaced by a star-refinement.

11.6 THEOREM *Every countable cozero-set cover of a topological space* X *has a countable star-finite normal cozero-set star-refinement.*

Proof. Suppose that $\mathscr{U}$ is a countable cozero-set cover of X. By 11.2, $\mathscr{U}$ is normal and hence, by 10.12, there exists a normal sequence $(\mathscr{V}_i)_{i \in \mathbf{N}}$ of covers of X such that $\mathscr{V}_1 < \mathscr{U}$ and such that for each $i \in \mathbf{N}$, the cardinality of $\mathscr{V}_i$ is countable. Thus,s ince $\mathscr{V}_2 <^* \mathscr{V}_1$, $\mathscr{V}_2$ is a star-refinement of $\mathscr{U}$. By '(1) *implies* (2)' of 10.10, there exists a countable locally finite cozero-set cover $\mathscr{W}$ of X such that $\mathscr{W} < \mathscr{V}_2$. By 11.3, $\mathscr{W}$ has a countable star-finite normal cozero-set refinement $\mathscr{A}$ and since

$$\mathscr{A} < \mathscr{W} < \mathscr{V}_2 <^* \mathscr{V}_1 < \mathscr{U},$$

it follows that $\mathscr{A}$ is a star-refinement of $\mathscr{U}$. □

It is now possible to give many equivalent formulations of a normal space. The importance of formulation (6) below has already been discussed. More pertinent now is the fact that every star-finite open cover is a locally finite open cover which in turn is a point-finite open cover. The cardinality of the cover may be two or it may be infinite with an appropriate modification by a localized finiteness condition. Since every discrete cover is a star-finite cover, the star-finite part of condition (6) brings one close to collectionwise normality with respect to its formulation in terms of covers. Also notice that it has not been asserted (as one might have expected) that normality is equivalent to every point-finite open cover being normal. This will be explained in 11.8.

Most of the following characterizations of a normal space in terms of various types of covers are known or implicitly proved in the literature. For this reason, there is hesitation to assert that any of these statements is completely new. However, it is believed that the equivalences of (1) through (5) and (12) and (13) were stated in [337] for the first time.

11.7 THEOREM *If X is a topological space, then the following statements are equivalent.*

(1) *The space X is normal.*

(2) *If $\mathscr{U}$ is a point-finite (respectively, locally finite, respectively, star-finite) open cover of X, then there exists a cozero-set cover $\mathscr{V}$ of X that screens $\mathscr{U}$.*

(3) *Every binary open cover of X has a binary cozero-set refinement.*

(4) *Every binary open cover of X has a countable cozero-set refinement.*

(5) *Every countable point-finite (respectively, locally finite, respectively, star-finite) open cover of X has a countable cozero-set refinement.*

(6) *If $\mathscr{U}$ is a point-finite (respectively, locally finite, respectively, star-finite, respectively, binary) open cover of X, then there exists an open cover $\mathscr{V}$ of X that strongly screens $\mathscr{U}$.*

(7) *If $\mathscr{U}$ is a countable point-finite (respectively, locally finite, respectively, star-finite) open cover of X, then there exists an open cover $\mathscr{V}$ of X that strongly screens $\mathscr{U}$.*

(8) *Every locally finite (respectively, star-finite, respectively, binary) open cover of X is normal.*

(9) *Every countable point-finite (respectively, locally finite, respectively, star-finite) open cover of X is normal.*

(10) *Every countable point-finite open cover of X has a locally finite cozero-set refinement.*

(11) *Every countable star-finite open cover of X has a locally finite cozero-set refinement.*

(12) *Every countable point-finite open cover of X has a countable star-finite normal cozero-set star-refinement.*

(13) *Every countable locally finite open cover of X has a countable star-finite normal cozero-set star-refinement.*

(14) *Every countable locally finite open cover of X has a countable locally finite open star-refinement.*

Proof. The order of proof will be (1) *implies* (2) *implies* (3) *implies* (4) *implies* (1), (2) *implies* (5) *implies* (4), (1) *is equivalent to* (6), (6) *is equivalent to* (7), (1) *is equivalent to* (8), (1) *implies* (9) *implies* (8), (5) *implies* (10) *implies* (11) *implies* (9), (5) *implies* (12) *implies* (13) *implies* (14) *implies* (8).

(1) *implies* (2). Let $\mathscr{U} = (U_\alpha)_{\alpha \in I}$ be a point-finite open cover of X. By 10.1 there exists an open cover $\mathscr{A} = (A_\alpha)_{\alpha \in I}$ of X that strongly screens $\mathscr{U}$. Since X is normal, for each $\alpha \in I$ there exists $f_\alpha \in C(X, E)$ such that $f_\alpha(x) = 0$ if $x \in \operatorname{cl} A$ and $f_\alpha(x) = 1$ if $x \in U_\alpha$. Let $V_\alpha = f_\alpha^{-1}([0, \frac{1}{2}])$. Then the family $\mathscr{V} = (V_\alpha)_{\alpha \in I}$ is a cozero-set cover of X that screens $\mathscr{U}$. Since a locally finite collection is point-finite and a star-finite collection is locally finite we have the results in the parentheses.

(2) *implies* (3), (3) *implies* (4). These implications are immediate.

(4) *implies* (1). Let F_1 and F_2 be disjoint closed subsets of X and let $U_i = X - F_i$ for $i = 1, 2$. Then $\mathscr{U} = (U_1, U_2)$ is a binary open cover of X. Hence, by (4), there exists a countable cozero-set refinement $\mathscr{W}$ of $\mathscr{U}$. Let $V_i = \bigcup \{W \in \mathscr{W} : W \subset U_i\}$ $(i = 1, 2)$. Then $\mathscr{V} = (V_1, V_2)$ is a cozero-set cover of X such that $\mathscr{V}$ screens $\mathscr{U}$. Now $X - V_1$ and $X - V_2$ are disjoint zero-sets and are therefore completely separated. Hence there exists $f \in C(X, E)$ such that $f(x) = 0$ if $x \notin V_1$ and $f(x) = 1$ if $x \notin V_2$. Let

$$W_1 = \{x \in X : f(x) < \tfrac{1}{3}\} \quad \text{and} \quad W_2 = \{x \in X : f(x) > \tfrac{2}{3}\}.$$

Then W_1 and W_2 are disjoint open sets such that

$$X - V_i \subset W_i \quad (i = 1, 2).$$

Since $V_i \subset U_i$, then $X - U_i \subset W_i$, and hence $F_i \subset W_i$ $(i = 1, 2)$. It follows that X is normal.

The implication (2) *implies* (5) is immediate. Since a binary open cover is a star-finite open cover, (5) *implies* (4). Moreover, (1) *implies* (6) follows from 10.1.

(6) *implies* (1). Let F_1 and F_2 be disjoint closed subsets of X. Then $(X - F_1, X - F_2)$ is a binary open cover of X. Hence, by (6), there exists a refinement (V_1, V_2) such that $\operatorname{cl} V_i \subset X - F_i$ $(i = 1, 2)$. Then $F_i \subset X - \operatorname{cl} V_i$ $(i = 1, 2)$, so $X - \operatorname{cl} V_1$ and $X - \operatorname{cl} V_2$ are disjoint open sets. Therefore X is normal.

The *equivalence* of (6) and (7) is evident and the implication (1) *implies* (8) follows from 10.5.

(8) *implies* (1). Let F_1 and F_2 be disjoint closed subsets of X. Then $\mathscr{U} = (X - F_1, X - F_2)$ is a binary open cover of X. By (8), there exists an open cover $\mathscr{V}$ of X such that $\mathscr{V} <^* \mathscr{U}$. One easily verifies that $\operatorname{st}(F_1, \mathscr{V})$ and $\operatorname{st}(F_2, \mathscr{V})$ are disjoint open sets such that $F_i \subset \operatorname{st}(F_i, \mathscr{V})$ $(i = 1, 2)$. Therefore X is normal.

(1) *implies* (9). If X is empty, then the assertion is clear. If $X \neq \varnothing$, then the result follows from 10.1, 11.4, and 10.5.

Clearly (9) *implies* (8). By 11.1 it follows that (7) *implies* (10) and the implication (10) *implies* (11) is immediate. Moreover, (11) *implies* (9) follows from 10.10 and (5) *implies* (12) from 11.6. The implications (12) *implies* (13) and (13) *implies* (14) are immediate.

(13) *implies* (8). If $\mathscr{U}$ is a locally finite open cover then a repeated application of (13) shows that $\mathscr{U}$ is normal.

The proof is now complete. □

11.8 REMARK Let X be a topological space and let us consider the statement:

(*) *Every point-finite open cover of X is normal.*

Clearly, if (*) holds, then X is normal, but the converse is not valid. To see this latter assertion, let us consider the normal non-collectionwise normal space G of [265, Example 2]. Michael shows that every open cover of G has a point-finite open refinement. If G normal implies that (*) holds, then G is paracompact and therefore collectionwise normal, a contradiction.

Thus the word 'countable' in the locally finite and star-finite parts of (9) of 11.7 can be omitted to obtain the corresponding parts of (8); but the word 'countable' cannot be omitted in the point-finite part of (9).

In 10.10 it was shown that a cover $\mathscr{U}$ of a space X is a normal cover if and only if there is a locally finite partition of unity subordinate to $\mathscr{U}$. Thus a T_1-space is paracompact if every open cover is a normal cover. Let us recall that every paracompact Hausdorff space is a normal Hausdorff space. There are T_1-spaces which are paracompact and not normal as well as normal Hausdorff spaces which are not paracompact. Theorems 10.10 and 11.7 now tell us what the difference between normality and paracompactness is in terms of coverings of the space.

11.9 Corollary *If X is a topological space then the following are equivalent.*

(1) *The space X is normal.*

(2) *For every locally finite open cover $\mathscr{U}$ of X there is a locally finite partition of unity subordinate to $\mathscr{U}$.*

(3) *For every star-finite open cover $\mathscr{U}$ of X there is a locally finite partition of unity subordinate to $\mathscr{U}$.*

(4) *For every binary open cover $\mathscr{U}$ of X there is a finite partition of unity subordinate to $\mathscr{U}$.*

(5) *For every countable point-finite open cover $\mathscr{U}$ of X there is a locally finite partition of unity subordinate to $\mathscr{U}$.*

(6) *For every countable star-finite open cover $\mathscr{U}$ of X there is a locally finite partition of unity subordinate to $\mathscr{U}$.*

Thus for a space to be paracompact every open cover must have a locally finite partition of unity subordinate to it, whereas for a space to be normal only the locally finite open covers, the star-finite open covers, the binary open covers, the countably point-finite open covers or the countably star-finite open covers must have the property.

12 Collectionwise normal spaces and open covers

In this section collectionwise normal spaces will be characterized in terms of covers. These characterizations are due to M. Katětov. For completeness and due to the fact that Katětov's original paper was in Russian, the proof of 12.4 is included which, for the most part, follows that of [225].

Recall that the condition of a cover being locally finite (and its related concepts such as point finiteness) was pertinent in localizing the finiteness condition of compactness (every open cover has a finite subcover) to yield the important generalization of paracompactness. The association was made between points of the space and members of a particular collection of subsets $\mathscr{U} = (U_\alpha)_{\alpha \in I}$ (for every point x in X there is a finite subset $K \subset I$ such that x (or a neighborhood of x) does not meet any of the U_α for $\alpha \notin K$). This concept is now taken with points replaced by another collection of subsets to obtain a stronger condition than locally finite.

12.1 Definition Suppose that X is a topological space and that $(F_\alpha)_{\alpha \in I}$ and $(G_\beta)_{\beta \in J}$ are two families of subsets of X. It is said that $(F_\alpha)_{\alpha \in I}$ is *finite with respect* to $(G_\beta)_{\beta \in J}$ in case for each $\beta \in J$ there exists a finite subset K of I such that $F_\alpha \cap G_\beta = \varnothing$ if $\alpha \notin K$. The family $(F_\alpha)_{\alpha \in I}$ is *uniformly locally finite* (*respectively, discrete*) *in* X in case there exists a locally finite (respectively, discrete) open cover $(U_\beta)_{\beta \in J}$ of X such that $(F_\alpha)_{\alpha \in I}$ is finite with respect to $(U_\beta)_{\beta \in J}$. If moreover we can take J of power at most γ, we say that $(F_\alpha)_{\alpha \in I}$ is γ-*uniformly locally finite* (*respectively, discrete*) *in* X.

Katětov originally defined 'uniformly locally finite' in [225]. It is evident that a uniformly locally finite family is locally finite. However, using 14.2 and Example 6 of Section 9, it can be shown that the converse does not hold (see also [225, Theorem 5.4]).

The following theorem is a somewhat more constructive formulation of uniform local finiteness. It says that uniformly locally finite families are precisely those families for which the closure of their members may be stretched to locally finite families of open sets.

12.2 THEOREM (*Katětov* [225]) *Suppose that $\mathscr{F}$ is a family of subsets of a topological space X. Then the following statements are equivalent.*

(1) *The collection $\mathscr{F}$ is uniformly locally finite.*

(2) *There exists an open locally finite family $\mathscr{G}$ of X such that $\mathscr{F}$ strongly screens $\mathscr{G}$.*

Proof. (1) *implies* (2). Since $\mathscr{F} = (F_\alpha)_{\alpha \in I}$ is uniformly locally finite, there exists a locally finite open cover $\mathscr{U} = (U_\beta)_{\beta \in J}$ of X such that $\mathscr{F}$ is finite with respect to $\mathscr{U}$. For each $\alpha \in I$, let $J_\alpha = \{\beta \in J : U_\beta \cap F_\alpha \neq \varnothing\}$ and let $G_\alpha = \bigcup_{\beta \in J_\alpha} U_\beta$. The family $\mathscr{G} = (G_\alpha)_{\alpha \in I}$ is a locally finite family of open subsets of X that strongly screens $\mathscr{G}$. Clearly $\mathscr{G}$ is a family of open subsets of X. Let $x \in X$. Since $\mathscr{U}$ is locally finite there exist a neighborhood H of x and a finite subset L of J such that $H \cap U_\beta = \varnothing$ if $\beta \notin L$. Moreover, for each $\beta \in L$, there exists a finite subset K_β of I such that $F_\alpha \cap U_\beta = \varnothing$ if $\alpha \notin K_\beta$. Let $K = \bigcup_{\beta \in L} K_\beta$ and note that K is finite. Suppose that $\alpha \in I$ and that $H \cap G_\alpha \neq \varnothing$. Then $H \cap U_\beta \neq \varnothing$ for some $\beta \in J$ such that $F_\alpha \cap U_\beta \neq \varnothing$. But then $\beta \in L$ and $\alpha \in K_\beta$, whence $\alpha \in K$. It follows that $\mathscr{G}$ is locally finite. Now let $\alpha \in I$ and suppose that $y \in \operatorname{cl} F_\alpha$. We have $y \in U_\beta$ for some $\beta \in J$, and since U_β is open, it follows that $U_\beta \cap F_\alpha \neq \varnothing$. Then $U_\beta \subset G_\alpha$, so $y \in G_\alpha$. Thus $\operatorname{cl} F_\alpha \subset G_\alpha$ and we have established (2).

(2) *implies* (1). Suppose that there exists a locally finite family $\mathscr{G} = (G_\alpha)_{\alpha \in I}$ of open subsets of X such that $\mathscr{F} = (F_\alpha)_{\alpha \in I}$ strongly screens $\mathscr{G}$. For each $J \in [I]$, let

$$U_J = \bigcap_{\alpha \in J} G_\alpha - \bigcup_{\alpha \notin J} \operatorname{cl} F_\alpha.$$

Let us now show that $\mathscr{U} = (U_J)_{J \in [I]}$ is a locally finite open cover of X and that $\mathscr{F}$ is finite with respect to $\mathscr{U}$. Suppose $x \in X$ and let $K = \{\alpha \in I : x \in \operatorname{cl} F_\alpha\}$. Since $\operatorname{cl} \mathscr{F}$ is locally finite (1.4) $K \in [I]$. Moreover, $x \in U_k$, so $\mathscr{U}$ is a cover of X. Since $\mathscr{G}$ is locally finite, it follows from 1.4(4) that $\mathscr{G}^F$ is locally finite. But, for each $J \in [I]$, $U_J \subset \bigcap_{\alpha \in J} G_\alpha$, so $\mathscr{U}$ is locally finite. Moreover, since $\mathscr{F}$ is closure preserving and $\mathscr{G}$ is open, it follows that $\mathscr{U}$ is open. Suppose finally that $J \in [I]$. If $\beta \in J$, then by the definition of U_J it is clear that $F_\beta \cap U_J = \varnothing$. Thus $\mathscr{F}$ is finite with respect to $\mathscr{U}$, so the proof is complete. □

For a uniformly locally finite family the open family to which it can be stretched may be taken to be also uniformly locally finite as the corollary shows.

12.3 Corollary (*Katětov* [225]) *Suppose that X is a normal topological space and that $\mathscr{F}$ is a non-empty family of subsets of X. If $\mathscr{F}$ is uniformly locally finite in X, then there exists a uniformly locally finite open cover $\mathscr{H}$ of X such that $\mathscr{F}$ strongly screens $\mathscr{H}$.*

Proof. If $\mathscr{F} = (F_\alpha)_{\alpha\in I}$, by 12.2 there exists a locally finite family $\mathscr{G} = (G_\alpha)_{\alpha\in I}$ of open subsets of X such that $\mathscr{F}$ strongly screens $\mathscr{G}$. Moreover, by 10.9 there is a family $\mathscr{U} = (U_\alpha)_{\alpha\in I}$ of open sets such that $\mathscr{F}$ strongly screens $\mathscr{U}$ and $\mathscr{U}$ strongly screens $\mathscr{G}$. Since $\mathscr{F}$ is locally finite, $X - (\bigcup_{\alpha\in I} \operatorname{cl} F_\alpha)$ is open. Let $\beta\in I$ be arbitrary and define $\mathscr{H} = (H_\alpha)_{\alpha\in I}$ and $\mathscr{V} = (V_\alpha)_{\alpha\in I}$ as follows: set $H_\beta = U_\beta \cup X - (\bigcup_{\alpha\in I} \operatorname{cl} F_\alpha)$ and $V_\beta = X$ and if $\alpha \neq \beta$, set $H_\alpha = U_\alpha$ and $V_\alpha = G_\alpha$. Clearly $\mathscr{V}$ is a locally finite family of open subsets of X such that $\mathscr{H}$ strongly screens $\mathscr{V}$. Therefore, by 12.2, $\mathscr{H}$ is uniformly locally finite. Since $\mathscr{H}$ is an open cover of X and $\mathscr{F}$ strongly screens $\mathscr{H}$, the proof is now complete. □

Uniform local finiteness of a cover is the proper strengthening of local finiteness to yield characterizations of normal spaces which are also collectionwise normal.

12.4 Theorem (*Katětov* [225]) *If X is a normal topological space, then the following statements are equivalent.*

(1) *X is collectionwise normal.*

(2) *If $\mathscr{H}$ is a locally finite collection of closed subsets of X with finite order, then there exists a locally finite family $\mathscr{G}$ of open subsets of X that stretches $\mathscr{H}$.*

(3) *For every closed $S \subset X$, if $\mathscr{F}$ (respectively $\mathscr{H}$) is a locally finite family of closed (respectively open) subsets of S such that $\mathscr{F}$ screens $\mathscr{H}$ then there exists a locally finite family $\mathscr{G}$ of open subsets of X such that $\mathscr{F}$ screens $\mathscr{G}|S$ and $\mathscr{G}|S$ screens $\mathscr{H}$.*

(4) *For every closed $S \subset X$, if $(H_\alpha)_{\alpha \in I}$ is a uniformly locally finite open cover of S, then there exists a uniformly locally finite open cover $(G_\alpha)_{\alpha \in I}$ of X such that $G_\alpha \cap S = H_\alpha$ for each $\alpha \in I$.*

(5) *For every closed $S \subset X$, if $\mathscr{F}$ is a uniformly locally finite family of subsets of S, then $\mathscr{F}$ is uniformly locally finite in X.*

Proof. (1) *implies* (2). Suppose that $\mathscr{H}$ is a locally finite family of closed subsets of X. To prove (2), we use induction on the order n of $\mathscr{H}$. If $n = 1$, then the members of $\mathscr{H} = (H_\alpha)_{\alpha \in I}$ are pairwise disjoint. By 1.4(3), it follows that $\mathscr{H}$ is discrete. By 2.7, there exists a discrete family $\mathscr{G}$ of open subsets of X such that $\mathscr{G}$ stretches $\mathscr{H}$.

Now suppose that (2) holds for $n \leqslant m$ and that $\mathscr{H} = (H_\alpha)_{\alpha \in I}$ has order $m+1$. Let $I^* = \{J \in [I]: |J| = m+1\}$. To show that $\mathscr{H}^* = (\bigcap_{\alpha \in J} H_\alpha)_{J \in I^*}$ is a discrete family of closed subsets of X, an appeal is made to 1.4(3). For this it is necessary to show that the members of the family are pairwise disjoint and closure preserving. One easily verifies that the members of $\mathscr{H}^*$ are pairwise disjoint closed subsets of X. In order to show that $\mathscr{H}^*$ is closure preserving, let $J^* \subset I^*$ and suppose that

$$x \notin \bigcup_{J \in J^*} \operatorname{cl}\,(\bigcap_{\alpha \in J} H_\alpha).$$

Since $\mathscr{H}$ is locally finite there exist a neighborhood W_0 of x and a finite subset K of I such that $W_0 \cap H = \varnothing$ if $\alpha \notin K$. Let

$$L^* = \{J \subset K: J \in J^*\}$$

and note that L^* is finite. Moreover, if $J \in L^*$, then $x \notin \operatorname{cl}\,(\bigcap_{\alpha \in J} H_\alpha)$, so there exists a neighborhood W_J of x such that

$$W_J \cap (\bigcap_{\alpha \in J} H_\alpha) = \varnothing.$$

Let
$$W = W_0 \cap (\bigcap_{J \in L^*} W_J)$$

and note that $W \cap (\bigcap_{\alpha \in J} H_\alpha) = \varnothing$ for all $J \in J^*$. From this it follows that $x \notin \operatorname{cl}\,(\bigcup_{J \in J^*} (\bigcap_{\alpha \in J} H_\alpha))$. Therefore

$$\operatorname{cl}\,(\bigcup_{J \in J^*} (\bigcap_{\alpha \in J} H_\alpha)) \subset \bigcup_{J \in J^*} \operatorname{cl}\,(\bigcap_{\alpha \in J} H_\alpha).$$

Since the other inclusion always holds, equality is obtained. Thus $\mathscr{H}^*$ is closure preserving and therefore, by 1.4(3), $\mathscr{H}^*$ is discrete.

Since X is collectionwise normal, there exists a locally finite family $(U'_J)_{J \in I^*}$ of open subsets of X such that $\bigcap_{\alpha \in J} H_\alpha \subset U'_J$ for each $J \in I^*$. For each $J \in I^*$, let

$$H_J = \cup \{H_\alpha : \alpha \in I \text{ and } \alpha \notin J\}.$$

Since $\mathscr{H}$ is a locally finite family of closed subsets of X of order $m+1$, it follows that H_J is closed and that $H_J \cap (\bigcap_{\alpha \in J} H_\alpha) = \varnothing$. For each $J \in I^*$, let

$$U_J = U'_J \cap X \backslash H_J$$

and note that $U_J \cap H_\alpha = \varnothing$ if $\alpha \notin J$. Next let $U = \bigcup_{J \in I^*} U_J$ and note that $(H_\alpha - U)_{\alpha \in I}$ has order at most m. Therefore, by the induction hypothesis, there exists a locally finite family $\mathscr{V} = (V_\alpha)_{\alpha \in I}$ of open subsets of X such that $(H_\alpha - U) \subset V_\alpha$ for each $\alpha \in I$.

For each $\alpha \in I$, let

$$J^*_\alpha = \{J \in I^* : \alpha \in J\}$$

and set

$$G_\alpha = V_\alpha \cup (\bigcup_{J \in J^*_\alpha} U_J).$$

The family $\mathscr{G} = (G_\alpha)_{\alpha \in I}$ is a locally finite family of open subsets of X expanding $\mathscr{H}$. Clearly $\mathscr{G}$ is an open collection. Next let $y \in H_\alpha$. If $y \notin U$, then $y \in V_\alpha \subset G_\alpha$. On the other hand, if $y \in U$, then there exists $J \in I^*$ such that $y \in U_J$ and $U_J \cap H_\alpha \neq \varnothing$. Therefore $\alpha \in J$ and $J \in J^*$. So $y \in G$. We conclude that $H_\alpha \subset G_\alpha$. Finally, let $x \in X$. Since $\mathscr{V}$ is locally finite, there exist a neighborhood W_1 of x and a finite subset K_1 of I such that $W_1 \cap V_\alpha = \varnothing$ if $\alpha \notin K_1$. Moreover, $\mathscr{U}^* = (U_J)_{J \in I^*}$ is locally finite, so there exist a neighborhood W_2 of x and a finite subset K^*_2 of I^* such that $W_2 \cap U_J = \varnothing$ if $J \notin K^*_2$. Let $K_3 = \cup \{J : J \in K^*_2\}$, let $K = K_1 \cup K_3$, and let $W = W_1 \cap W_2$. Then W is a neighborhood of x and K is a finite subset of I. Now suppose that $\alpha \in I - K_1$ and that $W \cap G_\alpha \neq \varnothing$. Then $W \cap V_\alpha = \varnothing$ and therefore $W_2 \cap U_J \neq \varnothing$ for some $J \in J^*_\alpha$. Then $\alpha \in J$ and $J \in K^*_2$, so $\alpha \in K_3$. It follows that $\mathscr{G}$ is locally finite, so (2) holds.

(2) *implies* (3). If S is empty, then the implication is immediate. So suppose that S is a non-empty closed subset of X and that $\mathscr{F} = (F_\alpha)_{\alpha \in I}$ (respectively $\mathscr{H} = (H_\alpha)_{\alpha \in I}$) is a locally finite family

of closed (respectively, open) subsets of S such that $\mathscr{F}$ screens $\mathscr{H}$. Let $\mathscr{I}_n$ be all subsets J of I such that $|J| < n$. For each $n \in \mathbf{N}$, let $S_n = \{x \in S$: for some $J \in \mathscr{I}_n$, $x \in F_\alpha$ if and only if $\alpha \in J\}$ and let $T_n = \{x \in S$: for some $J \in \mathscr{I}_n$, $x \in H_\alpha$ if and only if $\alpha \in J\}$. It is needed to show that S_n is open in S, that T_n is closed in S, that $S = \bigcup_{n \in \mathbf{N}} T_n$ and that $T_n \subset S_n$ for each $n \in \mathbf{N}$. Let $n \in \mathbf{N}$ and let $x \in S_n$. Then there exists a subset J of I such that $|J| < n$ and $x \in F_\alpha$ if and only if $\alpha \in J$. By hypothesis $\mathscr{F}$ is locally finite in S, so there exist a neighborhood G_0 of x in S and a finite subset K of I such that $G_0 \cap F = \varnothing$ if and only if $\alpha \notin K$. Note that $J \subset K$ and that $S - F_\alpha$ is a neighborhood of x that does not meet F_α if $\alpha \notin J$. Let $H = G_0 \cap (\bigcap_{\alpha \in K-J} S - F_\alpha)$. Then H is a neighborhood of x and one easily verifies that $H \subset S_n$, whence S_n is open. Now let $n \in \mathbf{N}$ and let $x \in S - T_n$. By hypothesis $\mathscr{H}$ is locally finite in S, so there exists a finite subset K of I such that $x \notin H_\alpha$ if and only if $\alpha \notin K$. Since $x \notin T_n$, $|K| \geqslant n$. Let $H = \bigcap_{\alpha \in K} H_\alpha$ and note that H is a neighborhood of x. If $y \notin S - T_n$, then there exists a subset J of I such that $|J| < n$ and $y \notin H_\alpha$ if and only if $\alpha \notin J$. If $y \in H$, then $y \in H_\alpha$ for all $\alpha \in K$ and therefore $K \subset J$, a contradiction. Thus $H \subset S - T_n$ and it follows that T_n is closed in S. The routine verification that $S = \bigcup_{n \in \mathbf{N}} T_n$ and that $T_n \subset S_n$ for each $n \in \mathbf{N}$ is omitted.

Since S is closed in X, S is normal. Hence, for each $n \in \mathbf{N}$, since T_n and $S - S_n$ are disjoint closed sets in S, there exists $f_n \in C(S, E)$ such that $f_n(x) = 0$ if $x \in S_n$ and $f_n(x) = 1$ if $x \in S - S_n$.

Since S is C-embedded in X (6.8) for each $i \in \mathbf{N}$, there exists $\bar{f}_n \in C(X, E)$ such that $\bar{f}_n | S = f_n$. Let

$$V'_n = \{x \in X : \bar{f}_n(x) < \tfrac{1}{2}\}$$

and let $V = \bigcup_{n \in \mathbf{N}} V'_n$. Then S and $X - V$ are disjoint closed sets in the normal space X so there exists a cozero-set W such that $X - V \subset W$ and $W \cap S = \varnothing$. Let $V_1 = V'_1 \cup W$ and for each $n \in \mathbf{N}$ with $n > 1$, let $V_n = V'_n$. Note that $\mathscr{V} = (V_n)_{n \in \mathbf{N}}$ is a countable cozero-set cover of X and hence, by 11.1, there exists a countable locally finite open cover $\mathscr{U} = (U_k)_{k \in \mathbf{N}}$ of X such that $\mathscr{U} < \mathscr{V}$. Observe that $\mathscr{U} | S < (S_n)_{n \in \mathbf{N}}$. By 10.11, there exists a closed cover

$\mathscr{B} = (B'_k)_{k\in\mathbf{N}}$ of X screening $\mathscr{U}$. Let $B_k = B'_k \cap S$. Then

$$\mathscr{B} = (B_k)_{k\in\mathbf{N}}$$

is a countable locally finite closed cover of S.

For each $k \in \mathbf{N}$, the family $(F_\alpha \cap B_k)_{\alpha\in I}$ is locally finite in X. If $k \in \mathbf{N}$, then there exists $n \in \mathbf{N}$ such that $B_k \subset U_k \cap S \subset S_n$. Let $J \subset I$ such that $|J| \geqslant n$ and suppose that $x \in \bigcap_{\alpha\in J}(F_\alpha \cap B_k)$. Then $x \notin S_n$ and therefore $x \notin B_k$, a contradiction. Thus

$$\bigcap_{\alpha\in J}(F_\alpha \cap B_k) = \varnothing$$

if $|J| \geqslant n$ and we conclude that $(F_\alpha \cap B_k)_{\alpha\in I}$ has order at most n. Therefore, by (2), there exists a locally finite family $(R_{\alpha k})_{\alpha\in I}$ of open subsets of X such that $F_\alpha \cap B_k \subset R_{\alpha k}$ for each $\alpha \in I$.

For each $\alpha \in I$, let W_α be an open subset of X such that $W_\alpha \cap S = H_\alpha$ and let

$$G_\alpha = W_\alpha \cap (\bigcup_{k\in\mathbf{N}}(R_{\alpha k} \cap U_k)).$$

The family $\mathscr{G} = (G_\alpha)_{\alpha\in I}$ is a locally finite family of open subsets of X such that $F_\alpha \subset G_\alpha \cap S \subset H_\alpha$ for each $\alpha \in I$. Clearly G_α is open in X for each $\alpha \in I$. Let $x \in X$. Since $(U_k)_{k\in\mathbf{N}}$ is locally finite, there exist a neighborhood H_1 of x and a finite subset K_1 of $\mathbf{N}$ such that $H_1 \cap U_k = \varnothing$ if $k \notin K_1$. For each $k \in K_1$, there exists a neighborhood H_k of x and a finite subset J_k of I such that $H_k \cap R_{\alpha k} = \varnothing$ if $\alpha \notin J_k$. Let $H = H_1 \cap (\bigcap_{k\in K_1} H_k)$ and let $K = \bigcup_{k\in K_1} J_k$. Then H is a neighborhood of x and K is a finite subset of I. Suppose that $\alpha \in I$ and that $H \cap G_\alpha \neq \varnothing$. Then $H \cap R_{\alpha k} \cap U_k \neq \varnothing$ for some $k \in \mathbf{N}$. This implies that $H_1 \cap U_k \neq \varnothing$ and therefore $k \in K_1$. But then $H_k \cap R_{\alpha k} \neq \varnothing$, so $\alpha \in J_k$. Thus $\alpha \in K$ and we conclude that $(G_\alpha)_{\alpha\in I}$ is locally finite. If $y \in F_\alpha$, then $y \in W_\alpha$. Moreover, since $(B_k)_{k\in\mathbf{N}}$ is a cover of S, there exists $k \in \mathbf{N}$ such that $y \in B_k$. Therefore $y \in F_\alpha \cap B_k \subset R_{\alpha k}$. Since $B_k \subset U_k$, it follows that $y \in G_\alpha \cap S$, and therefore $F_\alpha \subset G_\alpha \cap S$. Clearly $G_\alpha \cap S \subset H_\alpha$ and therefore (3) holds.

(3) *implies* (4). Since $\mathscr{H} = (H_\alpha)_{\alpha\in I}$ is a uniformly locally finite open cover of S there exists, by 12.3, a uniformly locally finite open cover $\mathscr{A}$ of S such that $\mathscr{H}$ strongly screens $\mathscr{A}$. By (3), there exists a locally finite family $\mathscr{U} = (U_\alpha)_{\alpha\in I}$ of open subsets of X such that $\mathscr{H}$ strongly screens $\mathscr{U}|S$ which screens $\mathscr{A}$. Since S is closed and $H_\alpha \subset S$ for each $\alpha \in I$ it follows that $\mathrm{cl}_S H_\alpha = \mathrm{cl}_X H_\alpha$. Moreover,

X is normal, so, by 1.11, for each $\alpha \in I$ there exists an open set V_α of X such that $\mathrm{cl}_X H_\alpha \subset V_\alpha \subset \mathrm{cl}_X V_\alpha \subset U_\alpha$. For each $\alpha \in I$, let W_α be an open subset of X such that $W_\alpha \cap S = H_\alpha$. Now let $\beta \in I$ be arbitrary and define $\mathscr{G} = (G_\alpha)_{\in \alpha I}$ as follows: set

$$G_\beta = (V_\beta \cap W_\beta) \cup X - S;$$

and if $\alpha \neq \beta$, let $G_\alpha = V_\alpha \cap W_\alpha$. Then one easily verifies that $\mathscr{G}$ is a uniformly locally finite open cover of X such that $G_\alpha \cap S = H_\alpha$ for each $\alpha \in I$.

(4) *implies* (5). Let S be a closed subset of X and suppose that $\mathscr{F} = (F_\alpha)_{\alpha \in I}$ is a uniformly locally finite family of subsets of S. If $I = \varnothing$, then clearly (5) holds. Therefore assume that $I \neq \varnothing$. Since S is normal and $I \neq \varnothing$, there exists a uniformly locally finite open cover $\mathscr{H} = (H_\alpha)_{\alpha \in I}$ of S such that $\mathscr{F}$ strongly screens $\mathscr{H}$. By (4) there exists a uniformly locally finite open cover $\mathscr{G} = (G_\alpha)_{\alpha \in I}$ of X such that $G_\alpha \cap S = H_\alpha$ for each $\alpha \in I$. Since S is closed in X and $F_\alpha \subset S$, it follows that $\mathrm{cl}_S F_\alpha = \mathrm{cl}_X F_\alpha$ and hence $\mathrm{cl}_X F_\alpha \subset G_\alpha$. Therefore, by 12.2, $\mathscr{F}$ is uniformly locally finite in X.

(5) *implies* (1). Suppose that $\mathscr{F} = (F_\alpha)_{\alpha \in I}$ is a discrete family of closed subsets of X and let $S = \bigcup_{\alpha \in I} F_\alpha$. Observe that $\mathscr{F}$ is a locally finite family of open subsets of S such that $\mathrm{cl}_S F_\alpha \subset F_\alpha$ for each $\alpha \in I$. By 12.2, $\mathscr{F}$ is uniformly locally finite in S, so, by (5), $\mathscr{F}$ is uniformly locally finite in X. Therefore, by 12.2, there exists a locally finite family $\mathscr{G}$ of open subsets of X such that $\mathscr{G}$ expands $\mathscr{F}$. Moreover, by 10.8, there exists a family $\mathscr{H}$ of open subsets of X such that $\mathscr{F}$ screens $\mathscr{H}$ and $\mathscr{H}$ strongly screens $\mathscr{G}$ and such that $\mathscr{F}$ is similar to $(\mathrm{cl}\ H)_{H \in \mathscr{H}}$. It follows that $\mathscr{H}$ is a family of mutually disjoint open subsets of X expanding $\mathscr{F}$, whence X is collectionwise normal. The proof is now complete. □

12.5 Remark Theorem 12.4 can be modified to give equivalent conditions for γ-collectionwise normality. Because of the complexity of the proof we choose not to include this additional modification. However we will make use of this observation in later sections and the reader is urged to check that the cardinality adjustments may readily be made.

13 Uniform spaces, normal covers, and pseudometrics

In Section 3, our approach for introducing uniformities on a non-empty set X was to utilize a particular filter on the power set of $X \times X$ or equivalently, particular neighborhoods of the diagonal. Each member or entourage U of the uniformity $\mathscr{U}$ determines a pseudometric, in fact a uniformly continuous pseudometric as will be shown in this section. In addition each entourage U also determines a cover of X, namely $\{U(x): x \in X\}$. In fact this cover enjoys a 'Lebesgue property' and will be called a 'uniform cover'. Moreover for any uniform space $(X, \mathscr{U})$, the collection of all uniformly continuous pseudometrics or the collection of all 'uniform covers' generates a unique uniformity $\mathscr{V}$. The uniformity $\mathscr{V}$ is precisely the uniformity $\mathscr{U}$. Consequently, depending upon the problem at hand and the appropriateness of one of the uniformity constructions, free use may and will be made of any of the approaches defining a uniformity.

The non-topological notion of completeness which is appropriate in the metric space structure is also appropriate in the generalization to the uniform space structure. Every Tychonoff space has a compactification and a realcompletion. Analogously every such space has a Hausdorff completion. Not so analogous though, is the uniqueness of this completion. On the other hand it resembles the uniqueness of the Stone–Čech compactification and the Hewitt–Nachbin realcompletion.

Properties possessed by a compactification or realcompletion of a space X yield information about X itself. In a similar fashion the completion is discussed. It is possible to state when the completion is compact, when it is paracompact, when it is Lindelöf or when it is realcomplete.

To begin let us consider the equivalent formulations mentioned. Let $\mathscr{P}$ be a collection of pseudometrics on a non-empty set X. For $\epsilon > 0$ and d in $\mathscr{P}$, let

$$U(d, \epsilon) = \{(x, y) \in X \times X: d(x, y) < \epsilon\}.$$

The collection $U(\mathscr{P}) = \{U(d, \epsilon): d \in \mathscr{P}, \epsilon > 0\}$ is a subbase for a unique uniformity on X.

13.1 DEFINITION The unique uniformity of the preceding paragraph is called the *uniformity on X generated by $\mathscr{P}$* and is denoted by $\mathscr{U}(\mathscr{P})$. If confusion is not likely, this uniformity $\mathscr{U}(\mathscr{P})$ will be designated by simply $\mathscr{U}$. If (X, d) is a pseudometric space then the uniformity generated by $\mathscr{P} = \{d\}$ is called the *pseudometric uniformity* and is denoted by $\mathscr{U}_d$. The set $U(d, \epsilon)$ defined above is called a *uniform d-sphere* of *radius* ϵ.

If X is a topological space, the functions in $C(X)$ can be used to define various uniformities on X. For each $f \in C(X)$, let ψ_f be the pseudometric on X defined as

$$\psi_f(x, y) = |f(x) - f(y)| \quad (x, y \in X).$$

Note that $\psi_f = d \circ (f \times f)$ where d is the usual metric on $\mathbf{R}$. Thus ψ_f is a continuous pseudometric on X by 2.14.

13.2 DEFINITION The peudometric ψ_f defined above is called the *pseudometric associated with f*. If $\mathscr{A}$ is a subset of $C(X)$, then the *uniformity on X generated by $\mathscr{A}$* is the uniformity on X generated by $\{\psi_f : f \in \mathscr{A}\}$.

Thus if a topological space has non-constant real-valued continuous functions as Tychonoff or completely regular spaces do, then the space also has numerous non-trivial continuous pseudometrics. Moreover the space has many uniform structures defined on it. In fact for any collection $\mathscr{P}$ of continuous pseudometrics on X, the uniformity $\mathscr{U}(\mathscr{P})$ is a continuous uniformity on X (that is, $\mathscr{T}(\mathscr{U}(\mathscr{P})) \subset \mathscr{T}$).

13.3 PROPOSITION *Suppose that $(X, \mathscr{T})$ is a topological space and that $\mathscr{U}$ is the uniformity generated by a collection $\mathscr{P}$ of continuous pseudometrics on X (in particular by a subset $\mathscr{A}$ of $C(X)$). Then $\mathscr{U}$ is a continuous uniformity on X.*

Proof. By 3.12, a basic set of $\mathscr{T}(\mathscr{U})$ is of the form $U(x_0)$ where $U = U(d, \epsilon)$, x_0 is a fixed point of X, and $\epsilon > 0$. But

$$U(x_0) = S(x_0; d, \epsilon)$$

and $S(x_0; d, \epsilon)$ is in $\mathscr{T}_d$. Since d is continuous, $\mathscr{T}_d \subset \mathscr{T}$ (2.12) and therefore $\mathscr{T}(\mathscr{U}) \subset \mathscr{T}$. □

Let us now consider a characterization for a uniformly continuous pseudometric. Since a pseudometric is a function on the product set, its uniform continuity or its continuity is in reference to the corresponding structure for the Cartesian product. But more internal is the following statement which relates it to the uniform structure. The proof is left as an exercise.

13.4 PROPOSITION *Suppose that* $(X, \mathscr{U})$ *is a uniform space and that* d *is a pseudometric on* X. *Then* d *is uniformly continuous on* $X \times X$ *(with respect to the product uniformity on* $X \times X$*) if and only if* $U(d, \epsilon) \in \mathscr{U}$ *for all* $\epsilon > 0$.

13.5 COROLLARY *If* $\mathscr{P}$ *is a collection of pseudometrics on a uniform space* $(X, \mathscr{U})$ *and if* $\mathscr{U} = \mathscr{U}(\mathscr{P})$, *then* $\mathscr{P}$ *is a collection of uniformly continuous pseudometrics on* $X \times X$.

Now for the sake of brevity it will be said that d is *uniformly continuous on* X (instead of 'on $X \times X$'). For this the following definition is given.

DEFINITION Suppose that d is a pseudometric on a uniform space $(X, \mathscr{U})$. The pseudometric d is a *uniformly continuous pseudometric on* $(X, \mathscr{U})$ if d is uniformly continuous on $X \times X$ where $X \times X$ is given the product uniformity (see 3.16).

From 13.5 the following question naturally arises. If the uniformity $\mathscr{U}$ on X is equivalent to a uniformity $\mathscr{U}(\mathscr{P})$, then must $\mathscr{P}$ be the collection of all uniformly continuous pseudometrics on X? We can now determine the collection of all uniformly continuous pseudometrics on X. It is the intention to show that this collection generates the original uniformity on X. The structural details are begun in the next proposition which we leave as an exercise. What is pertinent to its solution is Section 8 and in particular 8.2. It says that every entourage contains a 'small' uniform d-sphere for some uniformly continuous pseudometric d.

13.6 PROPOSITION *Suppose that* $(X, \mathscr{U})$ *is a uniform space and that* $U \in \mathscr{U}$. *Then there exists a uniformly continuous pseudometric* d *on* X *such that* $U(d, \frac{1}{4}) \subset U$.

This proposition is originally due to Weyl (see [XXIX]). It suggests that the collection $U(\mathscr{P})$ is a base for $\mathscr{U}$ where $\mathscr{P}$ is the collection of all pseudometrics uniformly continuous with respect to $\mathscr{U}$. Indeed it is. Even more so $\mathscr{P}$ exhibits two structural properties which when taken as axioms for any collection of pseudometrics they immediately yield a uniform structure.

13.7 PROPOSITION *Let $\mathscr{P}$ be the collection of all uniformly continuous pseudometrics defined on the non-empty uniform space $(X, \mathscr{U})$. Then $\mathscr{P}$ has the following properties.*

(P. 1) *If $d_1 \in \mathscr{P}$ and $d_2 \in \mathscr{P}$ then $d_1 \vee d_2 \in \mathscr{P}$.*

(P. 2) *If e is any pseudometric on X such that for all $\epsilon > 0$ there is a $d \in \mathscr{P}$ and a $\delta > 0$ with $d(x,y) < \delta$ implying $e(x,y) < \epsilon$, then also $e \in \mathscr{P}$.*

Conversely if $\mathscr{P}$ is any collection of pseudometrics on a non-empty set X satisfying (P. 1) *then $U(\mathscr{P})$ is a base for a unique uniformity $\mathscr{U}$ on X. If $\mathscr{P}$ satisfies also* (P. 2) *then $\mathscr{P}$ is precisely the collection of all pseudometrics uniformly continuous with respect to $\mathscr{U}$.*

Proof. First assume $\mathscr{P}$ to be the collection of all uniformly continuous pseudometrics on the space $(X, \mathscr{U})$. Since

$$(1) \qquad U(d_1 \vee d_2, \epsilon) = U(d_1, \epsilon) \cap U(d_2, \epsilon)$$

for all d_1 and d_2 in $\mathscr{P}$ and $\epsilon > 0$, it follows from (U. 4) and 13.4 that (P. 1) is satisfied. If e is any pseudometric satisfying the condition of (P. 2), then for a given $\epsilon > 0$, there exist $\delta > 0$ and $d \in \mathscr{P}$ such that

$$(2) \qquad U(d, \delta) \subset U(e, \epsilon).$$

Consequently it follows from (U. 5) and 13.4, that the pseudometric e is in $\mathscr{P}$.

On the other hand if $\mathscr{P}$ is any collection of pseudometrics on the topological space $(X, \mathscr{T})$ satisfying (P. 1), one looks at the sets $U(d, \epsilon)$ for $d \in \mathscr{P}$ and $\epsilon > 0$. The axioms of 3.4 for a base may now be verified. The axioms (B. 1) and (B. 2) are trivially satisfied. Axiom (B. 3) follows from the triangle equality which gives us

$$U(d, \tfrac{1}{2}\epsilon) \circ U(d, \tfrac{1}{2}\epsilon) \subset U(d, \epsilon)$$

for any $\epsilon > 0$ and $d \in \mathscr{P}$. Axiom (B. 4) follows directly from (1) and (P. 1). Thus by 3.4, $U(\mathscr{P})$ is a base for a unique uniformity $\mathscr{U} = \mathscr{U}(\mathscr{P})$. Let $\mathscr{P}^*$ be all pseudometrics on X that are uniformly continuous with respect to $\mathscr{U}(\mathscr{P})$. Since $U(\mathscr{P}) \subset \mathscr{U}(\mathscr{P}^*)$, it is clear by 13.4 that $\mathscr{P} \subset \mathscr{P}^*$. If $e \in \mathscr{P}^*$ and if $\epsilon > 0$ then $U(e, \epsilon) \in \mathscr{U}$ by 13.4. Since $U(\mathscr{P})$ is a base for $\mathscr{U}(\mathscr{P})$, there is a $d \in \mathscr{P}$ and $\delta > 0$ such that

$$U(d, \delta) \subset U(e, \epsilon).$$

Consequently if $\mathscr{P}$ also satisfies (P. 2) then it follows that $e \in \mathscr{P}$ and therefore $\mathscr{P} = \mathscr{P}^*$. Our proof is now complete. □

13.8 COROLLARY *If $(X, \mathscr{U})$ is a uniform space and if $\mathscr{P}$ is the collection of all pseudometrics uniformly continuous with respect to $\mathscr{U}$, then $\mathscr{U} = \mathscr{U}(\mathscr{P})$.*

Proof. By 13.6 for each $U \in \mathscr{U}$ there is a $d \in \mathscr{P}$ such that $U(d, \frac{1}{4}) \subset U$. Consequently $U(\mathscr{P})$ is a base for $\mathscr{U}$. □

We have thus indicated that there is a one-to-one correspondence between collections of pseudometrics $\mathscr{P}$ satisfying (P. 1) and (P. 2) and uniform structures satisfying the conditions of 3.1.

As in the case of functions, the countable sum of uniformly continuous pseudometrics is uniformly continuous. From this it follows that a uniformity with a countable base yields a pseudometrizable uniform topology.

13.9 PROPOSITION *Suppose that $(X, \mathscr{U})$ is a uniform space and that $(d_n)_{n \in \mathbf{N}}$ is a sequence of uniformly continuous pseudometrics on X. Then the pseudometric $d = \sum_{n=1}^{\infty} (d_n \wedge 1)/2^n$ is a uniformly continuous pseudometric on $(X, \mathscr{U})$.*

Proof. It is easy to show that d is indeed a pseudometric on X. To show uniform continuity let $\epsilon > 0$ and choose $k \in \mathbf{N}$ such that $\sum_{m=k}^{\infty} 1/2^m < \frac{1}{2}\epsilon$. For each $m = 1, \ldots, k-1$, by 13.4, $U(d_m, \epsilon/2k) \in \mathscr{U}$. Hence

$$U = U(d_1, \epsilon/2k) \cap \ldots \cap U(d_{k-1}, \epsilon/2k)$$

is in $\mathscr{U}$. Moreover, if $(x, y) \in U$, then

$$d(x, y) = \sum_{m=1}^{k-1} \frac{1}{2^m} d_m(x, y) + \sum_{m=k}^{\infty} \frac{1}{2^m} d_m(x, y) < \frac{(k-1)\epsilon}{2k} + \frac{\epsilon}{2} < \epsilon,$$

whence $U \subset U(d, \epsilon)$. Consequently $U(d, \epsilon) \in \mathscr{U}$. Therefore d is uniformly continuous by 13.4. □

For any collection $\mathscr{P}$ of pseudometrics, the uniformity $\mathscr{U}(\mathscr{P})$ generated by $\mathscr{P}$ has been discussed. Certain specific uniformities are now relabeled for the sake of brevity. The collections $C(X)$ and $C^*(X)$ of functions on X determine pseudometrics and therefore uniformities. For each $f \in C^*(X)$ the pseudometric ψ_f is totally bounded (see Exercise 5).

In the following definition and subsequent results, the work of Aquaro is pertinent and should be referred to by the reader (see [23], [24] and [398] to [403]).

DEFINITION Suppose that X is a completely regular topological space. Let $\mathscr{U}_0(X)$ be the uniformity generated by the collection of all continuous pseudometrics on X. The uniformity $\mathscr{U}_0(X)$ is called the *universal uniformity on* X. The uniformity generated by the collection $C(X)$ will be denoted by $\mathscr{C}(X)$ and the uniformity generated by the collection $C^*(X)$ by $\mathscr{C}^*(X)$. Let $\mathscr{U}_T(X)$ be the uniformity on X generated by the collection of all totally bounded continuous pseudometrics on X. For each infinite cardinal number γ, let $\mathscr{U}_\gamma(X)$ denote the uniformity on X generated by the collection of all continuous γ-separable pseudometrics on X. The uniformity $\mathscr{U}_\gamma(X)$ is called the *γ-uniformity on* X whereas $\mathscr{U}_T(X)$ is called the *T-uniformity on* X.

For completely regular spaces, all of the uniformities just defined yield completely regular topologies.

13.10 PROPOSITION *If X is a completely regular space then $\mathscr{U}_0(X)$, $\mathscr{C}(X)$, $\mathscr{C}^*(X)$, $\mathscr{U}_T(X)$ and $\mathscr{U}_\gamma(X)$ for all infinite cardinal numbers γ are admissible uniformities on X. Moreover, $\mathscr{U}_0(X)$ is the largest (finest) admissible uniformity on X.*

Proof. It is shown that $\mathscr{C}^*(X)$ is admissible, leaving the proof for $\mathscr{U}_0(X)$ and $\mathscr{C}(X)$ as an exercise.

For every f is $C^*(X)$, ψ_f is a continuous pseudometric (see Exercise 6) and thus $\mathscr{T}_{\psi_f} \subset \mathscr{T}$. Consequently, the uniform topology $\mathscr{T}(\mathscr{C}^*(X))$ equals $\cup \{\mathscr{T}_{\psi_f}: f \in C^*(X)\}$ and is contained in $\mathscr{T}$. If G is any $\mathscr{T}$-open set and if x is in G, by hypothesis there is an f in $C^*(X)$ such that $f(y) = 1$ if $y \in X - G$ and $f(x) = 0$. Then $x \in S(x; \psi_f, 1) \subset G$. It follows that $\mathscr{T} = \mathscr{T}(\mathscr{C}^*(X))$.

Clearly $\mathscr{U}_0(X) \supset \mathscr{U}_\gamma(X)$ for any infinite cardinal number γ. Moreover, for each $f \in C(X)$, the pseudometric ψ_f on X associated with f is $\aleph_0$-separable (see Exercise 6) and hence γ-separable. It follows that $\mathscr{U}_\gamma(X) \supset \mathscr{C}(X)$. Now by Exercise 7 both $\mathscr{U}_0(X)$ and $\mathscr{C}(X)$ are admissible uniformities and one can easily verify that $\mathscr{T}(\mathscr{U}_0(X)) \supset \mathscr{T}(\mathscr{U}_\gamma(X)) \supset \mathscr{T}(\mathscr{C}(X))$. It follows that $\mathscr{U}_\gamma(X)$ is admissible. Moreover note that $\mathscr{U}_{\aleph_0}(X) \supset \mathscr{U}_T(X) \supset \mathscr{C}^*(X)$ and $\mathscr{T}(\mathscr{U}_{\aleph_0}(X)) \supset \mathscr{T}(\mathscr{U}_T(X)) \supset \mathscr{T}(\mathscr{C}^*(X))$. Consequently it follows also that $\mathscr{U}_T(X)$ is admissible. By 13.8 and Exercise 7 it follows that $\mathscr{U}_0(X)$ is the largest admissible uniformity on X. This completes the proof. □

Thus completely regular spaces have admissible uniformities. In fact as the next theorem shows it is only the completely regular spaces which have even one such uniformity.

13.11 Theorem *A topological space $(X, \mathscr{T})$ is completely regular if and only if it is uniformizable. It is Tychonoff if and only if it admits a Hausdorff uniformity.*

Proof. If X is completely regular then by 13.10, X is uniformizable. Conversely, suppose that $\mathscr{U}$ is a uniformity on X such that $\mathscr{T}(\mathscr{U}) = \mathscr{T}$. If F is a closed subset of X and if x_0 is a point not in F then by 3.12, there exists $U \in \mathscr{U}$ such that $U(x_0) \cap F = \varnothing$. But then 13.6 implies there is a uniformly continuous pseudometric d on X such that $U(d, \frac{1}{4}) \subset U$. It now follows that d^F is a continuous real-valued function (2.13). Moreover, for any point x not in the closed set F we have $d^F(x) \neq 0$. Let $f = \dfrac{1}{d^F(x_0)} \cdot d^F$ and observe that f is a real-valued continuous function that assumes the value of 1 at x_0 and the value 0 on F. Hence the uniform topology is completely regular. The second statement is immediate from the above and 3.13. □

The function d^F constructed in the previous theorem can actually be shown to be uniformly continuous. Consequently the following results.

13.12 COROLLARY *Suppose that X is a completely regular topological space and that $\mathscr{U}$ is an admissible uniformity on X. If x is a point not in a closed set F then there exists a uniformly continuous real-valued function $f \in C(X, E)$, $f(x) = 0$, and $f(y) = 1$ if $y \in F$.*

Thus the interrelationship between pseudometrics and uniformities has been established. Since a uniformity may be considered as a collection of pseudometrics satisfying (P. 1) and (P. 2) it is natural to consider the metric space concepts of completeness and Cauchy sequences. To generalize these concepts to uniform spaces, the notion of filters is useful. Frequent use will be made of results and concepts from the theory of metric spaces.

DEFINITION Let $(X, \mathscr{U})$ be a uniform space and let $\mathscr{F}$ be a collection of subsets of X. If $U \in \mathscr{U}$ and if $A \subset X$ then the subset A is *small of order* U if $A \times A \subset U$. The collection $\mathscr{F}$ contains *arbitrarily small sets* if for every $U \in \mathscr{U}$ there exists $F \in \mathscr{F}$ such that F is small of order U (that is, $F \times F \subset U$). If $\mathscr{F}$ is a $\mathscr{Z}$-filter, then $\mathscr{F}$ is a *Cauchy $\mathscr{Z}$-filter* if $\mathscr{F}$ contains arbitrarily small sets. The uniform space $(X, \mathscr{U})$ is said to be *complete* if every family of closed sets with the finite intersection property that also contains arbitrarily small sets, has a non-empty intersection. A *completion* of the uniform space $(X, \mathscr{U})$ is a complete uniform space that contains X homeomorphically as a dense subspace.

It is immediate from the definition that every compact space is complete with respect to any admissible uniformity on the space. For completely regular spaces as 13.10 has shown, the universal uniformity $\mathscr{U}_0(X)$, is the largest admissible uniformity on X. In fact it is the only admissible one if X happens to be also compact.

13.13 PROPOSITION *The only admissible uniformity on a compact completely regular space X is the universal uniformity $\mathscr{U}_0(X)$.*

Proof. Let $\mathscr{U}$ be any admissible uniformity on X generated by the collection $\mathscr{P}$ of all uniformly continuous (with respect to $\mathscr{U}$) pseudometrics on X. We wish to show that any continuous pseudometric d on X (and therefore a pseudometric that belongs to the generators of $\mathscr{U}_0(X)$) is uniformly continuous with respect to $\mathscr{U}$. Let $\epsilon > 0$. Since d is continuous on X, for each $x \in X$ there is a $d_x \in \mathscr{P}$ and $\delta_x > 0$ such that $S(x; d_x, \delta_x) \subset S(x; d, \epsilon)$. Since X is compact there is a finite set $F \subset X$ such that X is the union of the spheres $S(x; d_x, \delta_x)$ for $x \in F$. However the pseudometric e that is the supremum of the pseudometrics d_x, for $x \in F$, is an element in $\mathscr{P}$ since $\mathscr{P}$ satisfies (P. 1) of 13.7. If δ is the infimum of the δ_x, for $x \in F$, then $e(x, y) < \delta$ implies that $d(x, y) < \epsilon$. Consequently by (P. 2) of 13.7, $d \in \mathscr{P}$. Hence $\mathscr{U}_0(X) \subset \mathscr{U}$. □

13.14 COROLLARY† *Every continuous map from a compact uniform space into a uniform space is uniformly continuous.*

13.15 PROPOSITION *Suppose that $\mathscr{F}$ is a collection of subsets of the uniform space $(X, \mathscr{U})$ and that $\mathscr{P}$ is the collection of all uniformly continuous pseudometrics on X. Then $\mathscr{F}$ contains arbitrarily small sets if and only if for every $d \in \mathscr{P}$ and for all $\epsilon > 0$, $\mathscr{F}$ contains a set of d-diameter less than ϵ.*

The easy proof is left as an exercise.

It is necessary to check now that the notions introduced here are compatible with the usual pseudometric space notions of Cauchy sequences and completeness. Firstly every pseudometric space (X, d) is a uniform space where $\mathscr{U}_d$ is the admissible uniformity generated by d. Consequently the 'uniform concepts' must be analyzed to see if they yield the proper topological concepts for $\mathscr{T}(\mathscr{U}_d) = \mathscr{T}_d$. In particular for pseudometric spaces the statement would be: every Cauchy sequence converges if and only if every Cauchy $\mathscr{Z}$-filter converges if and only if every family of closed sets which has the finite intersection property and which contains arbitrarily small sets is fixed. This is left to the exercises with sufficient hints to assist in the development. Once this is done it is then possible to use results from the theory of pseudometric spaces for our considerations here.

For pseudometric spaces the following are known and are assumed.

(1) A closed subspace of a complete pseudometric space is complete.

† Can one first prove this corollary and then obtain the proposition, itself, as a corollary?

(2) Every pseudometric space has a completion (which is again a pseudometric space).

The Tychonoff product of an uncountable collection of non-trivial pseudometric spaces is in general not a pseudometric space. Each pseudometric space is a uniform space and as such may be complete. The concept of completeness is naturally couched in uniform spaces. The Tychonoff product of pseudometric spaces, however, has the admissible product uniformity on it. Consequently to check the completeness of the Tychonoff product of pseudometric spaces one must consider the product as a uniform structure. Then it is possible to use 13.15 or an equivalent formulation. This will be done in the exercises. For now it is recorded here.

(3) A product of complete pseudometric spaces is complete.

The statements (1) and (3) may also be shown for uniform spaces in general as is done in 13.19. Statement (2) may also be generalized to uniform spaces once it is shown that every uniform space can be homeomorphically embedded as a subset of a product of pseudometric spaces. It should be noted that the existence of a completion for a uniform space may be proved by a direct construction. However, here the intention is to utilize as much as possible the results of pseudometric spaces.

13.16 THEOREM *Every uniform space $(X, \mathscr{U})$ is uniformly isomorphic to a subspace of a product of pseudometric spaces.*

Proof. Let $\mathscr{P}$ be the collection of all uniformly continuous pseudometrics on X. Let Y be the Cartesian product of the pseudometric spaces (X_d, d) for each d in $\mathscr{P}$ where $X_d = X$ for all d. The set Y is a uniform space when it is given the product uniformity (see 3.16), that is the smallest uniformity such that the projections are uniformly continuous. Let f be the one-to-one map of X into Y defined at each of its d-coordinates by $(f(x))_d = x$. Let $\bar{x} = f(x)$ for x in X. Now the mapping $\mathrm{pr}_d \circ f$ for each d in $\mathscr{P}$ is just the identity map of X onto the dth coordinate space X_d and is uniformly continuous. Consequently by 3.17 the function f is uniformly continuous *onto* a subspace Y^* of Y. Now the mapping f^{-1} of Y^* into X agrees with each projection pr_d for d in $\mathscr{P}$, that is $f^{-1}(\bar{x}) = \mathrm{pr}_d(\bar{x}) = x$ for each d in $\mathscr{P}$. Thus for any d

in $\mathscr{P}$, $d\circ(f^{-1}\times f^{-1}) = d\circ \mathrm{pr}_d \times \mathrm{pr}_d$ on Y^*. But pr_d is uniformly continuous with respect to $\mathscr{U}$ for each d in $\mathscr{P}$. Consequently by Exercise 6, f^{-1} is uniformly continuous. Therefore f is a uniform isomorphism of X into Y^* a subspace of Y. □

13.17 COROLLARY *Every uniform space $(X, \mathscr{U})$ is uniformly isomorphic to a subspace of a product of complete pseudometric spaces.*

Proof. Every pseudometric space is isometric (and therefore uniformly isomorphic!) to a dense subspace of a complete pseudometric space. The conclusion follows from the theorem. □

13.18 COROLLARY *Every Hausdorff uniform space is uniformly isomorphic to a subspace of a product of complete metric spaces.*

Proof. Exercise. □

13.19 THEOREM *The following statements are true.*

(*a*) *A closed subspace of a complete uniform space $(X, \mathscr{U})$ is complete.*

(*b*) *The uniform topological product of uniform spaces*

$$((X_\alpha, \mathscr{U}_\alpha))_{\alpha\in I}$$

is complete if and only if $(X_\alpha, \mathscr{U}_\alpha)$ is complete for all $\alpha \in I$.

(*c*). *A complete subspace of a Hausdorff uniform space is closed.*

Proof. Suppose that $(X, \mathscr{U})$ is a complete uniform space and that F is a closed subspace of X. If $\mathscr{F}$ is a collection of closed sets in F with the finite intersection property and which also contains arbitrarily small sets in F, then $\mathscr{F}$ has these same properties in the complete space X and the conclusion follows. The slightly (and only slightly) more difficult (*b*) is left as an exercise, as well as (*c*). □

The point has arrived where one can now show quickly that every uniform space has a completion. Then, what is more surprising, it can be shown that the completion is unique up to a uniform isomorphism.

13.20 THEOREM *Every uniform space $(X, \mathscr{U})$ has a completion.*

Proof. Applying 13.17 to $(X, \mathscr{U})$ and then taking the closure of the image space, the conclusion follows by 13.19(*a*). □

As a notational convention the completion of the uniform space $(X, \mathscr{U})$ will be represented as $(\gamma X, \gamma\mathscr{U})$ or when confusion is clearly not attainable simply by the symbol γX.

To show the 'uniqueness' of the completion, the following result is needed.

13.21 THEOREM *If X is a dense uniform subspace of the uniform space $(Y, \mathscr{U})$ then every uniformly continuous mapping f of X into a complete Hausdorff uniform space $(Z, \mathscr{V})$ has an extension to a uniformly continuous mapping f^* of Y into Z.*

Proof. Let $\mathscr{P}_Y$ and $\mathscr{P}_Z$ be the collection of all uniformly continuous pseudometrics on Y and Z respectively. By 13.8 these generate the uniformities $\mathscr{U}$ and $\mathscr{V}$ respectively. Let y be any point in Y and let $\mathscr{F}_y$ be the $Z(Y)$-filter generated by the collection of closed sets

$$F_{d,\delta} = \{x \in X : d(x, y) \leqslant \delta\}$$

for all $d \in \mathscr{P}_y$ and $\delta > 0$. By construction and by 13.15 $\mathscr{F}_y$ contains arbitrarily small sets. The filter $\mathscr{F}_y$ determines a $Z(Z)$-filter on Z, namely

$$Z(\mathscr{F}_y) = \{B \in Z(Z) : f^{-1}(B) \in \mathscr{F}_y\}.$$

Since f is uniformly continuous, for each $d \in \mathscr{P}_Z$, one has $d \circ f \times f \in \mathscr{P}_y | X \times X$ (Exercise 6). Consequently by 13.15 $Z(\mathscr{F}_y\}$ is a Cauchy filter on the complete space Z. Since Z is Hausdorff $Z(\mathscr{F}_y)$ converges to a unique point z_y of Z (Exercise 8). Define a function f^* from Y into Z by $f^*(y) = z_y$ for each $y \in Y$. By 13.12 and Exercise 3, if y is in X then $\mathscr{F}_y$ converges to y and $Z(\mathscr{F}_y)$ converges to $f(y)$. Consequently $f^*(y) = f(y)$ (Exercise 8). It remains to show that f^* is a uniformly continuous function on Y. For this the reader may make use of Exercise 6. □

Let us now observe that the completion of a uniform space is unique.

13.22 COROLLARY *If Y_1 and Y_2 are two completions of a uniform space X, then there is a uniform isomorphism of Y_1 onto Y_2 that leaves X pointwise fixed.*

A Tychonoff space X may have many compactifications as well as many realcompactifications. Such is clearly not the case with the completion of a completely regular space. On the other hand the Stone–Čech compactification of a Tychonoff space X is unique in the sense of 13.22 coupled with the fact that X is C^*-embedded in βX. This is the proper analogue for the completion in the hierarchy of compactifications of X. This will be made clearer below.

In [XI], Gilman and Jerison have constructed the completion γX directly from βX. The uniform structure on X was considered as a collection of pseudometrics $\mathscr{P}$ satisfying (P. 1) and (P. 2). Since X is C^*-embedded in βX but $X \times X$ is not in general C^*-embedded in $\beta X \times \beta X$, each $d \in \mathscr{P}$ was extended in two steps, one variable at a time, to a suitable subspace cX of βX. The completion γX is a quotient space of the subspace cX. This approach has the technical advantage of establishing immediately the relationship between βX and γX. The intention here has been to take a more direct and more simplified approach. However the relationship among γX, υX and βX can also be formalized here. Namely this requires a closer look at the uniformities $\mathscr{C}(X)$ and $\mathscr{C}^*(X)$ on a completely regular space X.

The following proposition will give us the insight we will need to demonstrate this relationship.

13.23 PROPOSITION *If a Tychonoff space X is realcompact then it is complete in the uniform structure $\mathscr{C}(X)$.*

Proof. Let $\mathscr{F}$ be a $Z(X)$-ultrafilter on X that is not real. By 5.21(14), there is a function $f \in C(X)$ such that f is unbounded on each $Z \in \mathscr{F}$. But then $\mathscr{F}$ contains no set of finite ψ_f-diameter. Hence by 13.15, $\mathscr{F}$ is not Cauchy. Thus if X is realcomplete then every Cauchy (with respect to $\mathscr{C}(X)$) $Z(X)$-ultrafilter is fixed. Thus $(X, \mathscr{C}(X))$ is complete. □

13.24 THEOREM *Let X be a Tychonoff space.*

(a) The completion of X in the uniform structure $\mathscr{C}(X)$ is $(\upsilon X, \mathscr{C}(\upsilon X))$.

(b) The completion of X in the uniform structure $\mathscr{C}^*(X)$ is $(\beta X, \mathscr{C}^*(\beta X))$.

Proof. By 13.23 the space $(\upsilon X, \mathscr{C}(\upsilon X))$ is complete. Moreover $\mathscr{C}(\upsilon X)|X \times X = \mathscr{C}(X)$ and X is dense in υX. Thus by 13.22 the completion of X in $\mathscr{C}(X)$ is $(\upsilon X, \mathscr{C}(\upsilon X))$.

Part (*b*) follows readily with a similar argument since every compact space is complete. □

This last theorem is interesting in its own right. Nachbin ([295]) called a uniform space X *saturated* if it is complete under the weakest uniform structure such that for all $f \in C(X)$, f is uniformly continuous. We have just seen that these are precisely the realcomplete spaces. Consequently the above gives another way of introducing the Hewitt–Nachbin realcompletion (also making more appropriate realcomplete over realcompact (see [XI])).

Having the relationship between pseudometrics and entourages, let us now consider the relationships that exist between these and covers of the space. Let us begin with a uniform space $(X, \mathscr{U})$. The covers of X on which we will focus our attention are the 'uniform covers' of X as defined in the next definition.

To motivate the definition let us recall Lebesgue's lemma (see 2.19(11)). If $\mathscr{A}$ is an open cover of a compact subset K of a pseudometric space (X, d), then there is an $\epsilon > 0$ such that the family $(S(x, \epsilon))_{x \in K}$ refines $\mathscr{A}$.

Let ϵ be the positive real number obtained from the open cover $\mathscr{A}$ of the compact set K. Then the subset $U(d, \epsilon)$ of $X \times X$ is an open subset containing the diagonal of $X \times X$. Consequently the preceding implies the following purely topological result. If $\mathscr{A}$ is an open cover of a compact pseudometric space X then there is a neighborhood U of the diagonal of $X \times X$ such that the family $(U(x))_{x \in X}$ refines $\mathscr{A}$. In [XVIII], Kelley has pointed out that this result is correct for compact regular spaces in general.

Since a pseudometric space is a concrete realization of a uniform space the concept may be 'lifted' to the abstraction.

DEFINITION Let $\mathscr{V}$ be a collection of subsets of X. By $\mathscr{V} \times \mathscr{V}$ is meant the family $\{V \times U \colon V, U \in \mathscr{V}\}$. By $\mathscr{V} \Delta \mathscr{V}$ is meant that subcollection of $\mathscr{V} \times \mathscr{V}$ consisting of all $V \times V$ for $V \in \mathscr{V}$.

13.25 Definition A cover $\mathscr{V}$ of a uniform space $(X, \mathscr{U})$ is said to be a *uniform cover* of X if there is a $U \in \mathscr{U}$ such that the family $(U(x))_{x \in X}$ refines $\mathscr{V}$. In particular, if $U \in \mathscr{U}$, the family $(U(x))_{x \in X}$ is a uniform cover of X and is called the *uniform cover of X associated with U*. Now if μ is a collection of covers of the non-empty uniform space $(X, \mathscr{U})$, then the covers in μ generate the uniformity $\mathscr{U}$ if the collection

$$\{\cup(\mathscr{V} \Delta \mathscr{V}): \mathscr{V} \in \mu\}$$

is a base for $\mathscr{U}$.

The collection of uniform covers of a uniform space $(X, \mathscr{U})$ (like the collection of uniformly continuous pseudometrics) enjoy some interesting properties. It will be shown that uniform covers are always normal covers of the space. In showing this the proof utilizes the fact that every uniform cover of $(X, \mathscr{U})$ has an *open uniform* refinement. This is reminiscent of the fact that every entourage contains an open entourage and thus every uniformity defined in terms of entourages has a base consisting of open neighborhoods of the diagonal.

Moreover the first three conclusions of the next theorem may be taken as axioms for a collection of covers of a space $(X, \mathscr{T})$ to yield an entourage uniformity.

13.26 Theorem *Suppose that μ is the collection of all uniform covers of the uniform space $(X, \mathscr{U})$. Then μ has the following properties.*

(1) *If $\mathscr{W} \in \mu$ then there exists $\mathscr{V} \in \mu$ such that $\mathscr{V} <^* \mathscr{W}$.*

(2) *If $\mathscr{V}$ and $\mathscr{W}$ are in μ then $\mathscr{V} \wedge \mathscr{W}$ is in μ (see 1.23).*

(3) *If $\mathscr{V} \in \mu$ and if $\mathscr{W}$ is a cover of X such that $\mathscr{V} < \mathscr{W}$ then $\mathscr{W} \in \mu$.*

(4) *Every element of μ is a normal cover.*

Proof. The proof of (2) and (3) are clear. To prove (1) let $\mathscr{W} \in \mu$ and let $U \in \mathscr{U}$ such that $(U(x))_{x \in X} < \mathscr{W}$. Let $V \in \mathscr{U}$ such that V is symmetric and $V^3 \subset U$. Note that $\mathscr{V} = (V(x))_{x \in X}$ is in μ. To show that $\mathscr{V} <^* \mathscr{W}$ suppose that $z \in \operatorname{st}(V(x), \mathscr{V})$. Then there exists $y \in X$ such that $z \in V(y)$ and $V(y) \cap V(x) \neq \varnothing$. Let $w \in V(y) \cap V(x)$. Thus (y, z), (y, w) and (x, w) are elements of V and therefore $(x, z) \in V^3 \subset U$. Hence $z \in U(x)$.

Finally for (4) it is necessary first to show that every element of μ has an open uniform refinement. If $\mathscr{V} \in \mu$ then there exists $U \in \mathscr{U}$ such that $(U(x))_{x \in X}$ refines $\mathscr{V}$. By 3.10, there exists an open subset V of $X \times X$ such that $V \subset U$. Now $V(x)$ is open in X. It follows then that $(V(x))_{x \in X}$ is an open uniform cover of X that refines $\mathscr{V}$.

Now suppose that $\mathscr{V}$ is in μ. By the above there exists an open uniform cover $\mathscr{V}_1$ of X such that $\mathscr{V}_1 < \mathscr{V}$. By (1) there exists $\mathscr{W} \in \mu$ such that $\mathscr{W} <^* \mathscr{V}_1$ and by the above there exists an open uniform cover $\mathscr{V}_2$ such that $\mathscr{V}_2 < \mathscr{W}$. Continuing inductively a sequence $(\mathscr{V}_n)_{n \in \mathbf{N}}$ of open covers of X is obtained such that $\mathscr{V}_{n+1} <^* \mathscr{V}_n$ for each $n \in \mathrm{N}$ and $\mathscr{V}_1 < \mathscr{V}$. It follows that $\mathscr{V}$ is normal. □

13.27 THEOREM *Let μ be a collection of covers on a non-empty set X and consider the following conditions.*

(μ. 1) *If $\mathscr{W}$ is in μ then there is a $\mathscr{V} \in \mu$ such that $\mathscr{V} <^* \mathscr{W}$.*

(μ. 2) *If $\mathscr{V}$ and $\mathscr{W}$ are in μ then $\mathscr{V} \wedge \mathscr{W}$ is in μ.*

(μ. 3) *If $\mathscr{V}$ is in μ and if $\mathscr{W}$ is a cover of X such that $\mathscr{V} < \mathscr{W}$ then $\mathscr{W}$ is in μ.*

If μ satisfies (μ. 1) and (μ. 2) then μ generates a unique (entourage) uniformity $\mathscr{U}(\mu)$ on X. If μ also satisfies (μ. 3) then μ is precisely the collection of all covers of X that are uniform with respect to $\mathscr{U}(\mu)$.

Proof. Theorem 3.4 is utilized to show that

$$\mathscr{B} = \{\cup(\mathscr{V} \Delta \mathscr{V}) : \mathscr{V} \in \mu\}$$

is a base for a unique uniformity on X. Conditions (B. 1) and (B. 2) clearly hold. The axiom (B. 3) follows from condition (μ. 2). Thus $\mathscr{B}$ is a base for a unique uniformity $\mathscr{U}(\mu)$ on X.

Let μ^* be the collection of all covers of X uniform with respect to $\mathscr{U}(\mu)$. Now if $\mathscr{W}$ is in μ then by (μ. 1) there is a $\mathscr{V}$ in μ such that $\mathscr{V} <^* \mathscr{W}$. Now $B = \cup(\mathscr{V} \Delta \mathscr{V})$ is in $\mathscr{U}(\mu)$, and the family $(B(x))_{x \in X}$ refines $\mathscr{W}$. Consequently $\mu \subset \mu^*$. On the other hand if $\mathscr{W}$ is in μ^* then since $\mathscr{W}$ is uniform with respect to $\mathscr{U}(\mu)$ there is a B in $\mathscr{B}$ such that the family $(B(x))_{x \in X}$ refines $\mathscr{W}$. By definition of $\mathscr{B}$ there is a $\mathscr{V}$ in μ such that $B = \cup(\mathscr{V} \Delta \mathscr{V})$. To show that $\mathscr{V}$ refines $\mathscr{W}$ let $V \in \mathscr{V}$ and choose $x \in V$. Then there exists an element W_x in $\mathscr{W}$ such that $B(x) \subset W_x$ and one easily shows that $V \subset W_x$. If condition (μ. 3) holds, we then have $\mu^* \subset \mu$ and therefore they are equal. □

13.28 COROLLARY *If $(X, \mathscr{U})$ is a uniform space (in the sense of entourages) and if μ is the collection of all covers of X uniform with respect to μ then $\mathscr{U} = \mathscr{U}(\mu)$.*

Proof. For each U in $\mathscr{U}$ there is a symmetric V in $\mathscr{U}$ such that $V \circ V \subset U$. Moreover the family $\mathscr{V} = (V(x))_{x \in X}$ is a uniform cover of X, that is $\mathscr{V}$ is in μ. If $B = \cup(\mathscr{V} \Delta \mathscr{V})$, then $B \subset U$. Consequently $\mathscr{U}(\mu)$ is a base for $\mathscr{U}$. By 13.26, μ satisfies all the conditions of 13.27, whence the result follows. □

Thus there is a one-to-one correspondence between collections of covers on a non-empty set X satisfying conditions (μ. 1), (μ. 2), and (μ. 3) and entourage uniformities just as the latter was put in one-to-one correspondence with collections of pseudometrics satisfying (P. 1) and (P. 2).

Let us repeat that a uniform cover $\mathscr{A}$ of a uniform space $(X, \mathscr{U})$ is just a normal cover of X, that is, there exists an appropriate sequence $(\mathscr{A}_n)_{n \in \mathbf{N}}$ of *open* covers of X. Every pseudometric d on X, also defines a 'normal' (not necessarily open) cover $\mathscr{A}$ on X. Notice that a normal cover is a purely topological notion whereas uniform covers and pseudometrics are not. It is interesting then to note that in view of 13.27 and its corollary, the uniform structure may be determined entirely by the topological structure emanating from collections of normal covers satisfying (μ. 1), (μ. 2), and (μ. 3).

The question as to when a normal cover of a uniform space is a uniform cover is an interesting one. In fact, a normal cover of a compact subset of a uniform space $(X, \mathscr{U})$ is a uniform cover. This follows from the more general result that any open cover of a compact subset of $(X, \mathscr{U})$ is a uniform cover. Recall our motivation for this concept when the discussion centered on the closed unit interval and compact pseudometric spaces. Thus the result should not at all be surprising. Is it possible to do better for normal covers? We answer this question in part by characterizing the γ-uniformities in terms of a class of normal covers. To this end let μ_γ be the collection of all normal *open* covers of power at most γ. It is easily seen that μ_γ satisfies (μ. 1) and (μ. 2) and thus μ_γ generates a unique uniformity $\mathscr{U}(\mu_\gamma)$ on X. Our aim is to now prove that this uniformity is precisely the γ-uniformity $\mathscr{U}_\gamma$. But first a lemma.

13.29 LEMMA *Let $(X,\mathscr{T})$ be a topological space and let $\mathscr{V}$ be a normal open cover of X of power at most γ. Then there exists a γ-separable continuous pseudometric d on X such that*

$$U(d,\tfrac{1}{4}) \subset \cup(\mathscr{V}\Delta\mathscr{V}).$$

Proof. If $\mathscr{V}$ is a normal open cover of power at most γ then by 10.6, there exists a normal sequence $(\mathscr{V}_n)_{n\in\mathbf{N}}$ of open covers, each of power at most γ. Hence by 8.9, there exists a γ-separable continuous pseudometric d on X that is associated with $(\mathscr{V}_n)_{n\in\mathbf{N}}$. Thus if $d(x,y) < \frac{1}{4}$ then $x\in\operatorname{st}(y,\mathscr{V}_1)$ and so there is an element of $\mathscr{V}_1$ that contains both x and y. But $\mathscr{V}_1$ refines $\mathscr{V}$ and therefore there exists $V\in\mathscr{V}$ such that $(x,y)\in V\times V$. □

13.30 THEOREM *Suppose that X is a topological space and that γ is an infinite cardinal number. Then the γ-uniformity $\mathscr{U}_\gamma(X)$ is equal to the uniformity generated by μ_γ.*

Proof. Let $\mathscr{U}(\mu_\gamma)$ be the uniformity generated by μ_γ and let $\mathscr{P}_\gamma$ be the collection of all continuous γ-separable pseudometrics on X. To show that $\mathscr{U}_\gamma \subset \mathscr{U}(\mu_\gamma)$ let $d\in\mathscr{P}_\gamma$ and let $\epsilon > 0$. We prove that $U(d,\epsilon)\in\mathscr{U}(\mu_\gamma)$. Now the family $(S(x; d, \frac{1}{2}\epsilon))_{x\in X}$ is an open cover of the paracompact space $(X,\mathscr{T}_d)$ and hence by 8.8 has a locally finite open refinement $\mathscr{V}$ of power at most γ. Since $(X,\mathscr{T}_d)$ is normal, $\mathscr{V}$ is normal (10.5). But $\mathscr{T}_d\subset\mathscr{T}$ implies that $\mathscr{V}$ is a normal open cover of $(X,\mathscr{T})$ of power at most γ. If

$$(x,y)\in\cup(\mathscr{V}\Delta\mathscr{V})$$

then x and y are in some V in $\mathscr{V}$ and $V\subset S(z; d, \frac{1}{2}\epsilon)$ for some $z\in X$. It follows that $d(x,y) < \epsilon$ and therefore

$$\cup(\mathscr{V}\Delta\mathscr{V})\subset U(d,\epsilon),$$

whence $U(d,\epsilon)\in\mathscr{U}(\mu_\gamma)$.

Conversely suppose that $\mathscr{V}\in\mu_\gamma$ so that $\mathscr{V}$ is a normal open cover of power at most γ. By 13.29, there exists a γ-separable continuous pseudometric d on X such that $U(d, \frac{1}{4}) \subset \cup(\mathscr{V} \Delta \mathscr{V})$ and therefore $\mathscr{U}(\mu_\gamma) \subset \mathscr{U}_\gamma$. □

13.31 Exercises

(1) Prove 13.4, 13.6, 13.15, 13.18, and 13.19(*b*), (*c*).

(2) If $\mathscr{P}$ is a collection of pseudometrics on a non-empty set X, prove that $U(\mathscr{P})$ is a subbase for a unique uniformity on X.

(3) Let X be a uniform subspace of Y and let $\mathscr{P}_X$ and $\mathscr{P}_Y$ be the collections of all bounded uniformly continuous pseudometrics on X and Y respectively. Then $\mathscr{P}_Y | X \times X = \mathscr{P}_X$.

(4) The sum of two uniformly continuous pseudometrics is uniformly continuous. A uniformity with a countable base is a pseudometrizable uniform topology.

(5) If $f \in C^*(X)$ then ψ_f is totally bounded. In fact a continuous pseudometric d on X is totally bounded if and only if for each $\epsilon > 0$, X is a finite union of zero-sets of diameter at most ϵ. (For $\epsilon > 0$, choose $k \in \mathrm{N}$ such that $f(x) \leqslant (k+1) \cdot \epsilon$ for all $x \in S$. Using this, define zero sets for $n = 1, \ldots, k$.)

(6) If $f \in C(X)$ then ψ_f is a separable continuous pseudometric. If f is a uniformly continuous real valued function on the uniform space $(X, \mathscr{U})$ then ψ_f is also uniformly continuous. More generally if f is a continuous map from the uniform space $(X, \mathscr{U})$ into the uniform space $(Y, \mathscr{V})$ and if $\mathscr{P}_X$ and $\mathscr{P}_Y$ are the collections of all uniformly continuous functions on X and Y, then f is uniformly continuous if and only if for $d \in \mathscr{P}_Y$, $d \circ (f \times f) \in \mathscr{P}_X$.

(7) If X is completely regular then $\mathscr{U}_0(X)$ and $\mathscr{C}(X)$ are admissible uniformities on X. If $\mathscr{P}$ is any collection of pseudometrics uniformly continuous with respect to the (admissible) uniformity $\mathscr{U}$ then $\mathscr{U}(\mathscr{P}) \subset \mathscr{U}$.

(8) A filter $\mathscr{F}$ converges to a point x of a space $(X, \mathscr{T})$ if and only if for every open set G containing x there is an $F \in \mathscr{F}$ such that $F \subset G$. If the uniform space is Hausdorff then a filter converging to a point converges to a unique point. A uniform space is complete if and only if every Cauchy $\mathscr{Z}$-filter converges to a point x of X. For metric spaces every Cauchy sequence converges if and only if every Cauchy $\mathscr{Z}$-filter converges if and only if every family of closed sets with the finite intersection property and containing arbitrary closed sets is fixed.

(9) Prove that a product of complete pseudometric spaces is complete.

(10) Prove that an open cover of a compact subset of a uniform space is a uniform cover.

(11) Let S be a subset of a completely regular space X and let $\mathscr{U}(S)$ and $\mathscr{V}(S)$ be uniformities on S such that $\mathscr{U}(S) \subset \mathscr{V}(S)$. If $\mathscr{V}(S)$ has a continuous extension to $\mathscr{V}^*(X)$ then $\mathscr{U}(S)$ has a continuous extension $\mathscr{U}^*(X)$ and $\mathscr{U}^*(X) \subset \mathscr{V}^*(X)$.

(12) A linear topological space that is normable with norm p is called a *Banach space* if it is complete in its metric topology $\mathscr{T}_{d_p}$.

(*a*) For any space X, the set $C^*(X)$ is a Banach space if we define the norm p as follows

$$p(g) = \sup |g(x)| \quad g \in C^*(X),$$

where the supremum is taken over all the real numbers $|g(x)|$ for $x \in X$. On the other hand, if X is taken to be the closed real interval $[a, b]$, then $\mathscr{C} = C([a, b])$ is not a Banach space if the metric d is defined as

$$d(f, g) = \int_a^b |f(t) - g(t)| dt \quad (f, g \in \mathscr{C}).$$

(*b*) A complete metrizable locally convex space is called a *Fréchet space*. Every complete normable space and hence every Banach space is a Frechet space.

(13) Let $(X, \mathscr{U})$ be a uniform space and let $\mathscr{P}$ be the collection of all uniformly continuous pseudometrics on X. For $d \in \mathscr{P}$ if every d-discrete subspace is realcompact, then every real $Z(X)$-ultrafilter $\mathscr{F}$ on X is a Cauchy $Z(X)$-filter. (For $d \in \mathscr{P}$, construct a d-discrete subset A of X by selecting one point z_α from each Z_α satisfying (*d*) through (*g*) of 4.27(6). A real $Z(A)$-ultrafilter $\mathscr{F}_A$ on A is then defined as follows: for B a subset of A, $B \in \mathscr{F}_A$ if and only if $\cup \{Z_\alpha : z_\alpha \in A\} \in \mathscr{F}$. Thus for any $d \in \mathscr{P}$ and $\epsilon > 0$, there is an α for which $\{z_\alpha\} \in \mathscr{F}_A$ or $Z_\alpha \in \mathscr{F}$. See 13.15.)

The question as to the existence of measurable cardinals has been discussed previously. Thus to be able to say that every real $Z(X)$-ultrafilter is a Cauchy $Z(X)$-filter one must exclude the existence of d-discrete subspaces of measurable cardinal.

(14) (Shirota, [344]) Let X be a Tychonoff space in which every closed discrete subspace has non-measurable cardinal. A necessary and sufficient condition that X admits a complete uniform structure is that X be realcompact. (Using the hypothesis, necessity follows immediately from Exercise 13. Sufficiency follows from 13.24.) Thus the hypothesis is only needed for necessity. See comment at end of Exercise 13 above.

(15) If $(X, \mathscr{U})$ is dense in a Hausdorff uniform space $(Y, \mathscr{V})$ then every uniformly continuous pseudometric on X can be extended to a uniformly continuous pseudometric on Y. Compare with 13.21.

IV. *Normality and pseudometrics*

Characterizations of normal spaces have been given in terms of various types of covers. Many of the original covers or the refinements were normal covers. In particular recall that every countable cozero-set cover is a normal cover. Also recall that to each normal open cover there is associated a continuous pseudometric (see 8.4).

In Section 6 one of the most important characterizations of a normal space was given. A space is normal if and only if every closed set is C-embedded (or equivalently C^*-embedded) in X. It is natural to ask then how do our characterizations of normal spaces using covers apply to the extension to the whole space of various covers from closed subspaces? Continuing, it would be natural to ask how can normal spaces be characterized in terms of the extension of appropriate continuous pseudometrics on the closed subspaces?

Of course questions of this type motivate a notion which is next to be presented. That is, to consider subspaces of a space for which every continuous pseudometric extends or for which a particular class of pseudometrics extend. From consideration of these notions will come new characterizations of normal spaces (Sections 16 and 17) and new characterizations of collectionwise normal spaces (Section 15). Also from this arises an interesting equivalence for Ulam's problem concerning the existence of measurable cardinals (Section 15).

14 P^{γ}-embedding and equivalent formulations

Our development is begun with the consideration of the extension of certain classes of continuous pseudometrics.

Arens (see 14.4) first showed that for a closed subset S of a metric space X, every continuous pseudometric on S can be extended to a continuous pseudometric on X. This is the foundation of our work. Before progressing further, it is necessary to make the following definitions.

14.1 DEFINITION Suppose that S is a subset of a topological space X, and that γ is an infinite cardinal number. The subset S is *P^γ-embedded in* X if every γ-separable continuous pseudometric on S can be extended to a γ-separable continuous pseudometric on X. The subset S is *P-embedded in* X if every continuous pseudometric on S can be extended to a continuous pseudometric on X.

In this section characterizations for extending continuous pseudometrics are given in terms of locally finite cozero-set covers, normal cozero-set covers, normal open covers, and locally finite normal open covers. The main theorem of this section is stated as 14.5 and this theorem characterizes P^γ-embedding, while 14.6 gives characterizations of P^γ-embedding, that involve additional hypothesis on the given topological space X. In 14.7 and 14.8 P-embedding is characterized.

First let us see how the extension of a single pseudometric is characterized in terms of covers.

14.2 THEOREM *Suppose that S is a subset of a topological space X, that γ is an infinite cardinal number and that d is a continuous γ-separable pseudometric on S. Then the following statements are equivalent.*

(1) *The pseudometric d can be extended to a γ-separable continuous pseudometric on X.*

(2) *Every uniform cover of $(S, \mathscr{U}_d)$ has a refinement that can be extended to a normal open cover of X.*

(3) *For each $\epsilon > 0$, $\mathscr{G}_\epsilon = (S(x; d, \epsilon))_{x\in S}$ has a refinement that can be extended to a normal open cover of X of power at most γ.*

For the time being our immediate goal will be to prove 14.2 which concerns itself with the characterization, in terms of covers, for the extension of a single pseudometric. The foundation of this work was laid in Hausdorff's theorem (14.3). We will give a proof of this theorem due to Arens (see [27]) that is much shorter and shows also how a metric space may be isometrically embedded in its own space of bounded continuous functions. To prepare the way for this, the following 'somewhat geometrical' ideas are needed.

LEMMA 1 *If A is a closed subset of a metric space (X, d), then there is a locally finite family $\mathscr{F} = (f_\alpha)_{\alpha \in I}$ of non-negative real-valued functions on X and a family $F = (a_\alpha)_{\alpha \in I}$ of points in A such that*

(*a*) *for each $\alpha \in I$, $f_\alpha(x) = 0$ for all $x \in A$;*

(*b*) $\sum_{\alpha \in I} f_\alpha(x) = 1$ *for all $x \in X - A$;*

(*c*) *for each a in A and for each $\alpha \in I$, if $f_\alpha(x) > 0$ then*

$$d(a, a_\alpha) < 3d(a, x) \quad \textit{and} \quad d(a, x) < d(a, a_\alpha) + 2d(x, A).$$

Proof. Let $\mathscr{V} = (V_\alpha)_{\alpha \in I}$ be a locally finite open refinement of the cover $\{S(x; d, \epsilon_x) : x \in X - A\}$ of the paracompact space $X - A$ where $\epsilon_x = \frac{1}{4}d(x, A)$. For each $V_\alpha \in \mathscr{V}$ there is a point $x \in X - A$ such that $V_\alpha \subset S(x; d, \epsilon_x)$. Call this point x_α and let ϵ_x be written as ϵ_α. Now select a point $a_\alpha \in A$ such that $d(a_\alpha, x_\alpha) < \frac{5}{4}d(x_\alpha, A)$. For each $\alpha \in I$ let $g_\alpha = d^{X - V_\alpha}$. Since $\mathscr{V}$ is locally finite, each x in X has a neighborhood on which all but finitely many of the g_α vanish. Thus

$$s(x) = \sum_{\alpha \in I} g_\alpha(x)$$

is both finite and continuous. Since $s(x)$ is never zero, the functions $f_\alpha = g_\alpha / s$ $(\alpha \in I)$ are continuous non-negative real-valued functions satisfying (*a*) and (*b*) above.

For part (*c*) let us assume that $f_\alpha(x) > 0$, that is $x \in V_\alpha$. Since $V_\alpha \subset S(x_\alpha; d, \epsilon_\alpha)$,

$$d(a_\alpha, x) \leqslant d(a_\alpha, x_\alpha) + d(x_\alpha, x) < \tfrac{5}{4}d(x_\alpha, A) + \tfrac{1}{4}d(x_\alpha, A).$$

Moreover for $a \in A$,

$$d(x_\alpha, A) \leqslant d(x_\alpha, a) \leqslant d(x_\alpha, x) + d(x, a) < \tfrac{1}{4}d(x_\alpha, A) + d(x, a).$$

So $\frac{3}{4}d(x_\alpha, A) < d(x, a)$, or $d(a_\alpha, x) < 2d(x, a)$ and thus

$$d(a, a_\alpha) < d(a, x) + d(x, a_\alpha) < d(a, x) + 2d(x, a) = 3d(a, x).$$

Now note that $d(x, A) > \frac{3}{4}d(x_\alpha, A)$. On the other hand, since $d(a_\alpha, x) < \frac{6}{4}d(x_\alpha, A)$ then $d(a_\alpha, x) < 2d(x, A)$. Finally

$$d(a, x) < d(a, a_\alpha) + d(a_\alpha, x) < d(a, a_\alpha) + 2d(x, A). \quad \square$$

Using this result we may consider a Tietze–Urysohn type extension theorem for continuous functions on a metric space whose range of values lie in a convex (locally convex) topological linear space L. This is done without enlarging the convex hull of the image. This result was first shown by J. Dugundji in [104], however the proof is that of [27].

LEMMA 2 *Let K be a convex subset of a locally convex linear topological space L and let A be a closed subset of a pseudometric space (X,d). If f is a continuous function on A with values in K, then f may be extended continuously to a function f^* defined on X with values in K.*

Proof. To prove this result we make use of Lemma 1 and its notation. Thus we define the extension f^* of the function f by

$$f^*(x) = \sum_{\alpha\in I} f_\alpha(x) f(a_\alpha) \quad \text{for} \quad x \in X - A$$

and

$$f^*(x) = f(x) \quad \text{for} \quad x \in A.$$

It is now necessary to show that f^* is a continuous function on X.

The topology $\mathcal{T}$ of L is determined by the family $\mathcal{P}$ of all continuous pseudonorms on L (see 2.19(16)), that is by the family of all neighborhoods

$$S(0; d_p, r) = \{x : p(x) < r\}$$

of the zero vector where d_p is the pseudometric defined by the pseudonorm $p \in \mathcal{P}$ and for $r > 0$.

Let p be a fixed pseudonorm in $\mathcal{P}$ and let $\epsilon > 0$. Since f is continuous on A for each $a \in A$ there is a positive number r such that $f(S(a; d, r)) \subset S(f(a); d_p, \epsilon)$. If $x \in S(a; d, \tfrac{1}{3}r)$, then for those *finitely* many f_α such that $f_\alpha(x)$ is not zero we have $d(a, a_\alpha) < r$ (Lemma 1(c)). Thus (the only point x of interest here is $x \in X - A$) we have

$$\begin{aligned} d_p(f^*(x), f(a)) &= p(f^*(x) - f(a)) = p\Big(\sum_{\alpha\in I} (f_\alpha(x) f(a_\alpha) - f(a))\Big) \\ &= p\Big(\sum_{\alpha\in I} f_\alpha(x) f(a_\alpha) - \sum_{\alpha\in I} f_\alpha(x) f(a)\Big) \\ &\leqslant \sum_{\alpha\in I} f_\alpha(x)\, p(f(a_\alpha) - f(a)) < \epsilon. \end{aligned}$$

We emphasize again the 'finiteness' of the family f_α which here permits the penultimate inequality as well as the fact that $\sum_{\alpha \in I} f_\alpha(x) = 1$ for $x \in X - A$ and for those f_α. More succinctly, for $a \in A$, we have $f^*(S(a; d, \frac{1}{3}r)) \subset S(f^*(a); d_p, \epsilon)$ which is the continuity of f^* for $a \in A$. If $x \in X - A$ then there is a neighborhood N_x of x and a finite subset F of the indexing set I such that $N_x \cap (X - Z(f_\alpha)) \neq \varnothing$ only if $\alpha \in F$. Thus for $y \in N_x$

$$f^*(y) = \Sigma\{f_\alpha(y) f(a_\alpha) \colon \alpha \in I\} = \Sigma\{f_\alpha(y) f(a_\alpha) \colon \alpha \in F\}$$

or f^* is continuous on N_x. It follows then that f^* is continuous on all of X and that f^* is a continuous extension of f as required. □

We are now ready to give Hausdorff's theorem (see [171]) and the shorter proof alluded to previously. This proof is that found in [27].

14.3 Theorem (*Hausdorff* [171]) *Let f be a continuous map from a closed subset A of a metric space (X, d) into another metric space (B, r). Then the metric space B can be isometrically embedded in a metric space Y such that there is a continuous function F from X into Y with $F|A = f$, B is a closed subset in Y, and f is a homeomorphism of $X - A$ with $Y - B$.*

Proof. Without loss of generality we may assume that d is a bounded metric on X. We will consider $C^*(X)$ as a Banach space with norm ρ defined by $\rho(g) = \sup\{|g(x)| \colon x \in X\}$ for $g \in C^*(X)$ and for any space X.

For each b in B, define the function r_b from X into $\mathbf{R}$ by

$$r_b(c) = r(b, c) \quad (c \in B). \tag{1}$$

In general r_b is not bounded, however the difference function $r_b - r_c$ is always in $C^*(B)$ for b and c in B. In fact,

$$\rho(r_b - r_c) = r(b, c) \tag{2}$$

for

$$r_b(c) - r_c(c) = r_b(c) = r(b, c)$$

and for each x in B

$$\rho(r_b - r_c) \geqslant r(b, x) - r(c, x) \geqslant r(b, c).$$

Let k be a fixed point in A. We now map A into $C^*(B)$ by the function ϕ defined for a in A by

$$\phi_a = r_{f(a)} - r_{f(k)}. \tag{3}$$

The map ϕ is continuous for

$$(4) \qquad \rho(\phi_a - \phi_c) = \rho(r_{f(a)} - r_{f(c)}) = r(f(a), f(c)).$$

Note that $\phi_a \in C^*(B)$ for each $a \in A$. By Lemma 2, ϕ may be extended in a continuous fashion to all of X. Let ϕ represent also this extension, that is, ϕ maps X into $C^*(B)$.

Let L be the product space $C^*(B) \times \mathbf{R} \times C^*(X)$ with the norm ρ_L defined by

$$(5) \qquad \rho_L((h, j, k)) = \max(\rho(h), |j|, \rho(k))$$

for $h \in C^*(B)$, $j \in \mathbf{R}$ and $k \in C^*(X)$. Note that the functions d^x for $x \in X$ are in the Banach space $C^*(X)$. Let us now define the function F mapping X into L by

$$(6) \qquad F(x) = (\phi_x, d^A(x), d^A(x)\, d^x) \quad (x \in X).$$

Since each of the elements in the triple is continuous, the function F is continuous. Let $Y = F(X)$ and let $B_1 = F(A)$. Of course B_1 is a closed subset of Y. In fact, if $a \in A$, then $F(a) = (\phi_a, 0, 0)$ for $d^A(a) = d(a, A) = 0$. Thus if $y \in Y$ is the limit of any sequence $(y_n)_{n \in \mathbf{N}}$ in $F(A)$ (where $(a_n)_{n \in \mathbf{N}}$ is a sequence in A such that $y_n = F(a_n) = (\phi_{a_n}, 0, 0)$ for each $n \in \mathbf{N}$) then the projections $\mathrm{pr}_2(y)$ and $\mathrm{pr}_3(y)$ must be zero also. Thus there is an $x \in X$ such that $y = F(x) = (\phi_x, 0, 0)$. Since A is closed, x must be in A, for otherwise $d(x)$ must be positive.

The continuous function F actually defines an isometry between B and B_1 for if a and c are in A then as in (4)

$$(7) \quad \rho_L(F(a) - F(c)) = \rho(\phi_a - \phi_c) = \rho(r_{f(a)} - r_{f(c)}) = r(f(a), f(c)).$$

Thus identifying B with B', F becomes an extension of f, continuous on all of X.

If $y \in X - A$ and if $F(x) = F(y)$ then $d^A(x) = d^A(y) > 0$ and d^x must equal d^y (cancelling!). Consequently $x = y$. Thus on $Y - B_1$, F has an inverse. It remains to show that here we have a homeomorphism. If $y \in X - A$ and if the sequence $(F(x_n))_{n \in \mathbf{N}}$ converges to $F(y)$ then the sequence $(d^A(x_n))_{n \in \mathbf{N}}$ converges to the positive d^y. In turn the sequence $(d^A(x_n)\, d^{x_n})_{n \in \mathbf{N}}$ converges to $d^A(y)\, d^y$. Thus the sequence $(d^{x_n})_{n \in \mathbf{N}}$ of bounded functions in $C^*(X)$ converges to d^y. But this just says that (as in (4)) the sequence

$d(x_n, y) = \rho(d^{x_n}, d^y)$ converges to zero. Thus on $Y - B_1$, the inverse of F is also continuous. □

Let us mention again, here, how the proof of Hausdorff's result made use of the embedding of a metric space into a space of bounded continuous functions. Let A be the metric space of 14.3. Then A was embedded, via ϕ, into $C^*(B)$ where (4) in the proof 14.3 showed that the image of A under ϕ is isometric to the image of A under f in B. On the other hand, $B_1 = F(A)$ was just the set of triples $(\phi_a, 0, 0)$ for $a \in A$ and (7) in 14.3 showed that B_1 is isometric to the image of B under f. Thus the image of A under ϕ is isometric to $B_1 = F(A)$. If f is a homeomorphism of A onto B then F may be taken to be a homeomorphism also (see Exercise 3). Thus the above shows how the metric space A may be isometrically embedded in its own space of bounded continuous functions.

14.4 THEOREM (*Arens* [27]) *Suppose that S is a closed subset of a metric space X. If d is a continuous pseudometric on S, then d can be extended to a continuous pseudometric on X.*

Proof. Define a relation R on S as follows:

$$x\,R\,y \text{ in case } d(x, y) = 0 \quad (x, y \in S).$$

Observe that R is an equivalence relation on S. Let $S^* = S/R$ be the quotient space of S modulo R and let $f: S \to S^*$ be the canonical map. Then the formula $d^*(f(x), f(y)) = d(x, y)$ $(x, y \in S)$ determines a well defined map d^* on $S^* \times S^*$.

One easily verifies that (S^*, d^*) is a metric space, that $\mathscr{T}_{d^*}$ is the quotient topology on S^*, and that the canonical map f from S into S^* is an isometry. Thus, by 14.3 there exist a metric space (Y, m) and a continuous map $\bar{f}$ from X into Y such that S^* is isometrically embedded as a closed subset of Y and such that $\bar{f}|S = f$. Let $\bar{d} = m \circ (\bar{f} \times \bar{f})$. Then $\bar{d}$ is a continuous pseudometric (2.14) and $\bar{d}|S \times S = d$. The proof is now complete. □

With these results all is ready to prove 14.2. As one might expect, the difficult part of the proof will be in constructing a continuous pseudometric on X from condition (3) of 14.2.

Proof of 14.2. (1) *implies* (2). Suppose that $\mathscr{V}$ is a uniform cover of $(S, \mathscr{U}_d)$. Then by definition there exists $W \in \mathscr{U}_d$ such that $(W(x))_{x \in S}$ refines $\mathscr{V}$. Then there exists $\epsilon > 0$ such that

$$U(d, \epsilon) \subset W.$$

Note that $U(d,\epsilon)(x) = \{y \in X: d(x,y) < \epsilon\} = S(x; d, \epsilon)$. Since d can be extended to a γ-separable continuous pseudometric d^* on X, $\mathscr{W}^* = (S(x; d^*, \epsilon))_{x \in X}$ is an open cover of X such that $\mathscr{W}^*|S < \mathscr{V}$. Since (X, d^*) is paracompact, $\mathscr{W}^*$ has a normal open refinement $\mathscr{V}^*$ (of power at most γ). Clearly $\mathscr{V}^*|S$ refines $\mathscr{V}$, and therefore (2) holds.

(2) *implies* (3). If $\epsilon > 0$, if A is γ-separable dense subset of S, and if $\mathscr{G}_{\epsilon/2} = (S(a; d, \epsilon/2))_{a \in A}$, then $U(d,\epsilon)(x) = S(x; d, \epsilon)$ implies therefore $\mathscr{G}_{\epsilon/2}$ is a uniform cover of S. The result now follows from (2) and 10.13.

(3) *implies* (1). For each $m \in \mathbf{N}$, let

$$\mathscr{G}_m = (S(x; d, 2^{-(m+3)}))_{x \in S}.$$

By (3), there exists a refinement of $\mathscr{G}_m$ that extends to a normal open cover $\mathscr{V}^m$ of X of power at most γ. Then, by 10.12 there exists a normal sequence $(\mathscr{V}_i^m)_{i \in \mathbf{N}}$ of open covers of X such that $\mathscr{V}_1^m$ refines $\mathscr{V}^m$ and such that, for each $i \in \mathbf{N}$, the cardinality of $\mathscr{V}_i^m$ is at most γ. Now for all $i, m \in \mathbf{N}$, let

$$\mathscr{U}^m = \bigwedge_{j=1}^{m} \mathscr{V}^j$$

and

$$\mathscr{U}_i^m = \bigwedge_{j=1}^{m} \mathscr{V}_i^j.$$

The assertion is that, for all $i, m \in \mathbf{N}$,

(i) $\mathscr{U}^m$ and $\mathscr{U}_i^m$ are open covers of X of power at most γ.
(ii) $\mathscr{U}_{i+1}^m <^* \mathscr{U}_i^m$ and $\mathscr{U}_1^m < \mathscr{U}^m$,
(iii) $\mathscr{U}_i^{m+1} < \mathscr{U}_i^m$ and $\mathscr{U}^{m+1} < \mathscr{U}^m$, and
(iv) $\mathscr{U}^m|S < \mathscr{G}_m$.

To see that (i) holds, note that $\mathscr{V}^j$ and $\mathscr{V}_i^j$ are open covers of X for all $i \in \mathbf{N}$, and that both $\mathscr{U}^m$ and $\mathscr{U}_i^m$ have power at most $\gamma^m = \gamma$ (see 1.23). Moreover, (ii) is true by 1.24 and (iii) holds by 1.25. Finally $\mathscr{U}^m < \mathscr{V}^m$ (see 1.23) and $\mathscr{V}^m|S < \mathscr{W}^m < \mathscr{G}_m$, so (iv) is valid.

Now again consider any $m \in \mathbf{N}$. It follows from (i) and (ii) that $(\mathscr{U}_i^m)_{i\in\mathbf{N}}$ is a normal sequence of open covers of X. Then, by 8.4, and 8.9, there exists a γ-separable continuous pseudometric r_m on X that is associated with $(\mathscr{U}_i^m)_{i\in\mathbf{N}}$. Thus for each $x, y \in X$,

$$r_m(x,y) < 2^{-(i+1)} \quad \text{implies} \quad x \in \operatorname{st}(y, \mathscr{U}_i^m)$$

and $$x \in \operatorname{st}(y, \mathscr{U}_i^m) \quad \text{implies} \quad r_m(x,y) < 2^{-(i-3)}.$$

Also note that

(*) $$\textit{if} \quad x, y \in S \quad \textit{and if} \quad r_m(x,y) < 2^{-3}$$
$$\textit{then} \quad d(x,y) < 2^{-(m+2)}.$$

For suppose that x and y are in S and that $r_m(x,y) < 2^{-3}$. Then $x \in \operatorname{st}(y, \mathscr{U}_2^m)$. Moreover, by (ii) and (iv), one has $\mathscr{U}_2^m <^* \mathscr{U}_1^m < \mathscr{U}^m$ and $\mathscr{U}^m|S < \mathscr{G}_m$, so there exists $z \in S$ such that

$$\operatorname{st}(y, \mathscr{U}_2^m) \cap S \subset S(z; d, 2^{-(m+3)}).$$

Thus x and y are in $S_d(z, 2^{-(m+3)})$ and therefore $d(x,y) < 2^{-(m+2)}$.

Define a mapping r from $X \times X$ into $\mathbf{R}^+$ by

$$r(x,y) = \sum_{m\in\mathbf{N}} 2^{-m} r_m(x,y) \quad (x, y \in X).$$

One easily verifies that r is a continuous pseudometric on X. In fact, r is γ-separable in X. For each $m \in \mathbf{N}$, r_m is γ-separable in X, so there exists $A_m \subset X$ such that $|A_m| \leqslant \gamma$ and A_m is r_m-dense in X. Let $A = \bigcup_{m\in\mathbf{N}} A_m$ and note that $|A| \leqslant \aleph_0 \cdot \gamma = \gamma$. Let $x \in X$, let $\epsilon > 0$, and choose $k \in \mathbf{N}$ such that $2^{-k} < \frac{1}{2}\epsilon$. Then there exists $y \in S(x; r_k, 2^{-(k+4)}) \cap A_k$. Since $r_k(x,y) < 2^{-(k+4)}$, it follows that $x \in \operatorname{st}(y, \mathscr{U}_{k+3}^k)$. Moreover, if $1 \leqslant m \leqslant k$, then by (iii) we have $\mathscr{U}_{k+3}^k < \mathscr{U}_{k+3}^m$ and therefore $x \in \operatorname{st}(y, \mathscr{U}_{k+3}^m)$. Hence

$$r_m(x,y) < 2^{-k} < \tfrac{1}{2}\epsilon$$

whenever $1 \leqslant m \leqslant k$. Then

$$r(x,y) = \sum_{m=1}^{k} 2^{-m} r_m(x,y) + \sum_{m=k+1}^{\infty} 2^{-m} r_m(x,y)$$
$$< \sum_{m=1}^{k} 2^{-m}(\tfrac{1}{2}\epsilon) + \sum_{m=k+1}^{\infty} 2^{-m} < \epsilon,$$

so $y \in S(x; r, \epsilon) \cap A$. Thus A is r-dense in X and therefore r is γ-separable in X.

The next result needed is

$$(**) \quad \textit{if} \quad x, y \in S, \quad \textit{if} \quad i > 3, \quad \textit{and if} \quad r(x,y) < 2^{-i}$$
$$\textit{then} \quad d(x,y) < 2^{-(i-1)}.$$

If $r(x,y) = 0$, then $r_m(x,y) < 2^{-3}$ for all $m \in \mathbf{N}$, so, by (*), $d(x,y) < 2^{-(m+2)}$ for all $m \in \mathbf{N}$, whence $d(x,y) = 0$. Suppose therefore that $r(x,y) > 0$ and let k be the smallest positive integer such that $2^{-(k+1)} \leqslant r(x,y)$. Then $k > i-1 \geqslant 3$ and we have $2^{-(k-3)} \cdot r_{k-3}(x,y) \leqslant r(x,y) < 2^{-k}$. Thus $r_{k-3}(x,y) < 2^{-3}$ and therefore, by (*), $d(x,y) < 2^{-(k-1)} \leqslant 2^{-(i-1)}$.

Define a relation R on X as follows:

$$x \, R \, y \text{ in case } r(x,y) = 0 \quad (x, y \in X).$$

Observe that R is an equivalence relation on X. Let $X^* = X/R$ be the quotient space of X modulo R and let $\sigma: X \to X^*$ be the canonical map. Then the formula

$$r^*(\sigma(x), \sigma(y)) = r(x,y) \; (x, y \in X)$$

determines a well defined map r^* on $X^* \times X^*$. One easily verifies that (X^*, r^*) is a metric space, that $\mathscr{T}_{r^*}$ is the quotient topology on X^* and that the canonical map $\sigma: X \to X^*$ is an isometry. Since X^* is a continuous image of X (that is, with respect to $\mathscr{T}_r$ and $\mathscr{T}_{r^*}$) and since X has a dense subset A with $|A| \leqslant \gamma$, then X^* also has a dense subset (namely $\sigma(A)$) with $|\sigma(A)| \leqslant \gamma$.

Let $S^* = \sigma(S)$ and observe that if $a, a', b, b' \in S$ with

$$\sigma(a) = \sigma(a') \quad \text{and} \quad \sigma(b) = \sigma(b'),$$

it follows that $r(a,a') = 0$ and $r(b,b') = 0$. Thus $r(a,a') < 2^{-i}$ for all $i \in \mathbf{N}$, so, by (**), $d(a,a') < 2^{-(i-1)}$ for all $i \in \mathbf{N}$, and hence $d(a,a') = 0$. Similarly $d(b,b') = 0$. Moreover,

$$d(a,b) \leqslant d(a,a') + d(a',b') + d(b',b) = d(a',b').$$

Therefore $d(a,b) \leqslant d(a',b')$. Similarly $d(a',b') \leqslant d(a,b)$ and therefore $d(a,b) = d(a',b')$. From this it follows that one can define a map d^* from $S^* \times S^*$ into $\mathbf{R}^+$ as follows:

$$d^*(\sigma(a), \sigma(b)) = d(a,b) \quad (a, b \in S).$$

Then one easily verifies that d^* is a pseudometric on S^*.

Let $\mathscr{D}^*$ be the metric uniformity on $\operatorname{cl} S^*$ generated by the single metric $r^*|\operatorname{cl} S^* \times \operatorname{cl} S^*$ (13.1). Let us show that d^* is uniformly continuous on $S^* \times S^*$ relative to the product subspace uniformity. If $\epsilon > 0$, choose $i \in \mathbf{N}$, $i > 2$, such that $2^{-i} < \epsilon$. If $\sigma(x), \sigma(y) \in S^*$ such that $r^*(\sigma(x), \sigma(y)) \leqslant 2^{-(i+1)}$, then

$$r(x, y) < 2^{-(i+1)},$$

so, by (**), $d^*(\sigma(x), \sigma(y)) = d(x, y) < 2^{-i} \leqslant \epsilon$. Thus

$$U(r^*|S \times S, 2^{-(i+1)}) \subset U(d^*, \epsilon)$$

and therefore $U(d^*, \epsilon) \in \mathscr{D}^*|S^* \times S^*$. By 13.4, d^* is uniformly continuous on $S^* \times S^*$ and hence by 13.21 d^* has an extension to a uniformly continuous function from $\operatorname{cl} S^* \times \operatorname{cl} S^*$ into $\mathbf{R}$. One easily verifies that this extension (which shall again be denoted by d^*) is a pseudometric on $\operatorname{cl} S^*$.

It now follows by 14.4 that d^* can be extended to a continuous pseudometric $\overline{d^*}$ on X^*. Since (X^*, r^*) is a pseudometric space with r^* γ-separable, $\overline{d^*}$ is γ-separable in X^*. Define $\bar{d}$ on $X \times X$ by $\bar{d} = \overline{d^*} \circ (\sigma \times \sigma)$. Since σ is continuous relative to the given topology on X, $\bar{d}$ is a continuous pseudometric on X (2.14). Moreover, if x, $y \in S$, then $\bar{d}(x, y) = \overline{d^*}(\sigma(x), \sigma(y)) = d^*(\sigma(x), \sigma(y)) = d(x, y)$. Therefore $\bar{d}|S \times S = d$. Since $\bar{d}$ is continuous relative to $\mathscr{T}_r$ and since r is γ-separable, it follows that $\bar{d}$ is γ-separable. Therefore, (1) holds. □

Now for subsets which are P^{γ}-embedded, the main results evolve. Notice that the pseudometric being bounded does not influence the result.

14.5 THEOREM *Suppose that S is a subset of a topological space X, and that γ is an infinite cardinal number. Then the following statements are equivalent.*

(1) *The subset S is P^{γ}-embedded in X.*

(2) *Every γ-separable bounded continuous pseudometric on S can be extended to a γ-separable bounded continuous pseudometric on X.*

(3) *Every γ-separable bounded continuous pseudometric on S can be extended to a continuous pseudometric on X.*

(4) *Every locally finite cozero-set cover of S of power at most γ has a refinement that can be extended to a locally finite cozero-set cover of X.*

(5) *Every normal cozero-set cover of S of power at most γ has a refinement that can be extended to a normal cozero-set cover of X.*

(6) *Every normal open cover of S of power at most γ has a refinement that can be extended to a normal open cover of X.*

(7) *Every locally finite normal open cover of S of power at most γ has a refinement that can be extended to a locally finite normal open cover of X.*

Furthermore, the above conditions are also equivalent to the conditions obtained from (4) *through* (7) *by requiring that the extended cover be of power at most γ.*

Before proving this theorem some parenthetical remarks are appropriate. It is clear that condition (4) of 14.5 is equivalent to the following.

(4′) *If $\mathscr{U}$ is a locally finite cozero-set cover of S of power at most γ, then there exists a locally finite cozero-set cover of X of power at most γ such that $\mathscr{V}|S < \mathscr{U}$.*

Using 10.13 one easily verifies that (4) is also equivalent to the following.

(4″) *If $\mathscr{U}$ is a locally finite cozero-set cover of S of power at most γ, then there exists a locally finite cozero-set cover $\mathscr{V}$ of X such that $\mathscr{V}|S$ screens $\mathscr{U}$.*

Similar equivalences hold for conditions (5), (6), and (7), as well as for similar conditions in future theorems that characterize P^γ-embedding. Whenever it is convenient, free use of these alternative formulations will be made.

If S is a subset of a topological space X and if γ is an infinite cardinal number, then one can define S to be *P^*-embedded* (respectively, *$P^{\gamma*}$-embedded*) in X in case every bounded (respectively, γ-separable bounded) continuous pseudometric on S can be extended to a bounded (respectively, γ-separable bounded) continuous pseudometric on X. However, in 14.5 it is shown that S is P^γ-embedded in X if and only if S is $P^{\gamma*}$-embedded in X. Moreover, in 14.7 it is shown that S is P-embedded in X if and only if S is P^*-embedded in X. Thus it is unnecessary to define the concepts of P^*-embedding and $P^{\gamma *}$-embedding since they do not really differ from P-embedding and P^γ-embedding respectively.

Finally, the first discussion of P^γ-embedding appears to be in Arens [27, 28]. Arens called the concept 'γ-normally embedding' and was able to prove the equivalence of (1) and (4) of 14.5 for the case in which X is a normal topological space and S is a closed subset of X [28, Theorem 2.4].

Proof of 14.5. (1) *implies* (2). Let d be a γ-separable bounded continuous pseudometric on S. By (1), there exists a γ-separable continuous pseudometric e on X such that $e|S \times S = d$. Now $d \leqslant \boldsymbol{\alpha}$ for some $\alpha \in \mathbf{R}^+$. Then one easily verifies that $\bar{d} = e \wedge \boldsymbol{\alpha}$ is a γ-separable bounded continuous pseudometric on X such that $\bar{d}|S \times S = d$. Thus (2) holds.

(2) *implies* (3). This implication is immediate.

(3) *implies* (4). If X is empty, then clearly (4) holds. We may therefore assume that S is non-empty. Let $\mathscr{T}$ be the given topology on X and suppose that $\mathscr{U} = (U_\alpha)_{\alpha \in I}$ is a locally finite cozero-set cover of S of power at most γ. By 10.10, $\mathscr{U}$ is normal, so there exists a normal sequence $(\mathscr{V}_i)_{i \in \mathbf{N}}$ of open covers of S such that $\mathscr{V}_1 < \mathscr{U}$ and such that, for each $i \in \mathbf{N}$, $\mathscr{V}_i$ has power at most γ (10.12). Then, by 8.4 and 8.9 there exists a γ-separable continuous pseudometric d on S that is associated with $(\mathscr{V}_i)_{i \in \mathbf{N}}$. Therefore, by (3), there is a continuous pseudometric $\bar{d}$ on X such that $\bar{d}|S \times S = d$. Let $\mathscr{W}' = (S(x; \bar{d}, 2^{-4}))_{x \in X}$. Since $(X, \bar{d})$ is a pseudometric space, it is paracompact, so there is a locally finite open cover $\mathscr{W}$ of X such that $\mathscr{W} < \mathscr{W}'$. By 2.16, it follows that $\mathscr{W}$ is a locally finite cozero-set cover of X relative to $\mathscr{T}$. It remains to show that $\mathscr{W}|S < \mathscr{U}$. Let W be a member of the family $\mathscr{W}$ and suppose that $x \in X$ such that $W \subset S(x; \bar{d}, 2^{-4})$. If

$$S(x; \bar{d}, 2^{-4}) \cap S = \varnothing,$$

then $W \cap S \subset U_\alpha$ for all $\alpha \in I$. So suppose that there exists $u \in S(x; \bar{d}, 2^{-4}) \cap S$. Since $\mathscr{V}_2 <^* \mathscr{V}_1 < \mathscr{U}$, there exists $\alpha \in I$ such that $\operatorname{st}(u, \mathscr{V}_2) \subset U_\alpha$. If $y \in W \cap S$, then

$$\bar{d}(y, u) \leqslant \bar{d}(y, x) + \bar{d}(x, u) < 2^{-4} + 2^{-4} = 2^{-3}.$$

Thus $d(y, u) = \bar{d}(y, u) < 2^{-3}$, so $y \in \operatorname{st}(u, \mathscr{V}_2) \subset U_\alpha$. Therefore $W \cap S \subset U_\alpha$, whence $\mathscr{W}|S < \mathscr{U}$, so (4) holds.

(4) *implies* (5). Suppose that $\mathscr{U}$ is a normal cozero-set cover of S of power at most γ. By 10.10 and 10.13, $\mathscr{U}$ has a locally finite cozero-set refinement of power at most γ. Then, by (4), there exists a locally finite, cozero-set cover $\mathscr{W}$ of X such that $\mathscr{W}|S < \mathscr{V}$. But then $\mathscr{W}$ is normal (10.10), so (5) holds.

(5) *implies* (6). Let $\mathscr{U}$ be a normal open cover of S. By 10.10, $\mathscr{U}$ has a locally finite cozero-set refinement $\mathscr{V}$ of power at most γ. Since $\mathscr{V}$ is normal (10.10), it follows, by (5), that $\mathscr{V}$ has a refinement that can be extended to a normal cozero-set cover of X. Therefore (6) holds.

(6) *implies* (7). Suppose that $\mathscr{U}$ is a locally finite normal open cover of S of power at most γ. By (6), $\mathscr{U}$ has a refinement that can be extended to a normal open cover $\mathscr{V}$ of X. Then, by 10.10, there exists a locally finite normal open refinement of $\mathscr{V}$. Therefore (7) holds.

(7) *implies* (1). Suppose that d is a γ-separable continuous pseudometric on S, let $\epsilon > 0$ and let $\mathscr{G}_\epsilon = (S(x; d, \epsilon))_{x \in S}$. Then $\mathscr{G}_\epsilon$ is an open cover of $(S, \mathscr{T}_d)$ and hence has a locally finite normal open refinement $\mathscr{W}$ of power at most γ. Condition (7) now implies 14.2(3) and it follows that d can be extended to a continuous γ-separable pseudometric on X. Therefore S is P^γ-embedded in X.

The final assertion of the theorem follows from 14.2. The proof is now complete. □

If X is a normal space and if S is a closed subset of X then we can characterize P^γ-embedding in additional ways as follows.

14.6 THEOREM *Suppose that S is a closed subset of a normal topological space X, and that γ is an infinite cardinal number. Then the following statements are equivalent.*

(1) *The subset S is P^γ-embedded in X.*

(2) *Every locally finite open cover of S of power at most γ has a refinement that can be extended to a locally finite open cover of X of power at most γ.*

(3) *If $\mathscr{F}$ is a γ-uniformly locally finite collection of subsets of S then $\mathscr{F}$ is γ-uniformly locally finite in X.*

(4) *If $\mathscr{F}$ (respectively, $\mathscr{H}$) is a locally finite family of closed*

(respectively, open) subsets of S *of power at most* γ *such that* $\mathscr{F}$ *screens* $\mathscr{H}$*, then there exists a locally finite family* $\mathscr{G}$ *of open subsets of* X *such that* $\mathscr{G}|S$ *expands* $\mathscr{F}$ *and screens* $\mathscr{H}$*.*

(5) *If* $(H_\alpha)_{\alpha \in I}$ *is a uniformly locally finite open cover of* S *of power at most* γ*, then there exists a uniformly locally finite open cover* $(G_\alpha)_{\alpha \in I}$ *of* X *such that* $G_\alpha \cap S = H_\alpha$ *for each* $\alpha \in I$*.*

Proof. (1) *implies* (2). This is an immediate consequence of 14.5, 10.5, and 10.10.

(2) *implies* (3). By hypothesis $\mathscr{F} = (F_\alpha)_{\alpha \in I}$ is a γ-uniformly locally finite family of subsets of S, so there exists a locally finite open cover $\mathscr{U} = (U_\beta)_{\beta \in J}$ of S of power at most γ such that $\mathscr{F}$ is finite with respect to $\mathscr{U}$. By (2) and 1.8, there exists a locally finite open cover $\mathscr{V} = (V_\beta)_{\beta \in J}$ of X such that $V_\beta \cap S \subset U_\beta$ for each $\beta \in J$. Clearly $\mathscr{V}$ has power at most γ, so it will suffice to show that $\mathscr{F}$ is finite with respect to $\mathscr{V}$. Let $\beta \in J$. Then there exists a finite subset K of I such that $U_\beta \cap F_\alpha = \varnothing$ if $\alpha \notin K$. But

$$V_\beta \cap F_\alpha \subset U_\beta \cap F_\alpha,$$

so $V_\beta \cap F_\alpha = \varnothing$ if $\alpha \notin K$. Thus $\mathscr{F}$ is finite with respect to $\mathscr{V}$.

(3) *implies* (4). Suppose that $(F_\alpha)_{\alpha \in I}$ (respectively, $(H_\alpha)_{\alpha \in I}$) is a locally finite family of closed (respectively, open) subsets of S of power at most γ such that $F_\alpha \subset H_\alpha$ for each $\alpha \in I$. Then, by 12.2, $(F_\alpha)_{\alpha \in I}$ is uniformly locally finite in S. Therefore, by (3), $(F_\alpha)_{\alpha \in I}$ is uniformly locally finite in X. Hence, by 12.2, there exists a locally finite family $(G'_\alpha)_{\alpha \in I}$ of open subsets of X such that $\mathrm{cl}_X F_\alpha \subset G'_\alpha$ for each $\alpha \in I$. Let $(H'_\alpha)_{\alpha \in I}$ be a family of open subsets of X such that $H'_\alpha \cap S = H_\alpha$ for each $\alpha \in I$. Set $G_\alpha = G'_\alpha \cap H'_\alpha$ $(\alpha \in I)$. Then one easily verifies that $(G_\alpha)_{\alpha \in I}$ is a locally finite family of open subsets of X such that $F_\alpha \subset G_\alpha \cap S \subset H_\alpha$ for each $\alpha \in I$.

(4) *implies* (5). This implication follows from 12.4 and 12.5.

(5) *implies* (1). Suppose that $\mathscr{U} = (U_\alpha)_{\alpha \in I}$ is a locally finite normal open cover of S of power at most γ. Since S is normal, there exists an open cover $\mathscr{H} = (H_\alpha)_{\alpha \in I}$ of S such that

$$\mathrm{cl}_S H_\alpha \subset U_\alpha$$

for each $\alpha \in I$ (10.1). Moreover, $\mathscr{H}$ is uniformly locally finite in S (12.2), so, by (5), there exists a uniformly locally finite open cover $\mathscr{G} = (G_\alpha)_{\alpha \in I}$ of X such that $G_\alpha \cap S = H_\alpha$ for each $\alpha \in I$. Thus $\mathscr{G}$ is a locally finite open cover of X of power at most γ such that $\mathscr{G}|S < \mathscr{U}$. Since $\mathscr{G}$ is normal (10.5), it follows that 14.5(7) holds, whence S is P^γ-embedded in X.

The proof is now complete. □

The equivalence of (1) and (2) in 14.6 is a generalization of Theorem 2.4 of Arens [28]. There the equivalence of (1) and (2) was proved for the case of X a normal space and S a closed subset.

The concept of P-embedding will now drop the cardinality conditions attached to the covers of 14.5. The next two theorems will be concerned with the P-embedding of a subset.

14.7 Theorem *If X is a topological space and if $S \subset X$, then the following statements are equivalent.*

(1) *The subset S is P-embedded in X.*

(2) *The subset S is P^γ-embedded in X for all infinite cardinal numbers γ.*

(3) *Every bounded continuous pseudometric on S can be extended to a bounded continuous pseudometric on X.*

(4) *Every locally finite cozero-set cover of S has a refinement that can be extended to a locally finite cozero-set cover of X.*

(5) *Every normal cozero-set cover of S has a refinement that can be extended to a normal cozero-set cover of X.*

(6) *Every normal open cover of S has a refinement that can be extended to a normal open cover of X.*

(7) *Every locally finite normal open cover of S has a refinement that can be extended to a locally finite normal open cover of X.*

Proof. (1) *implies* (3). Suppose that d is a bounded continuous pseudometric on S. By (1), there exists a continuous pseudometric e on X such that $e|S \times S = d$. Now $d \leqslant \boldsymbol{\alpha}$ for some $\alpha \in \mathbf{R}^+$. Then one easily verifies that $\bar{d} = e \wedge \boldsymbol{\alpha}$ is a bounded continuous pseudometric on X such that $\bar{d}|S \times S = d$.

(3) *implies* (2). If (3) holds, then 14.5(3) is valid and it follows that S is P^γ-embedded in X for all cardinal numbers γ.

(2) *implies* (1). This implication is immediate.

The equivalence of (2), (4), (5), (6), and (7) follows from 14.5. The proof is now complete. □

14.8 THEOREM *Suppose that S is a closed subset of a normal topological space X. Then the following statements are equivalent.*

(1) *The subset S is P-embedded in X.*

(2) *If $(F_\alpha)_{\alpha\in I}$ is a uniformly locally finite family of subsets of S, then $(F_\alpha)_{\alpha\in I}$ is uniformly locally finite in X.*

(3) *If $\mathscr{F}$ (respectively $\mathscr{H}$) is a locally finite collection of closed (respectively, open) subsets of S such that $\mathscr{F}$ screens $\mathscr{H}$, then there exists a locally finite collection $\mathscr{G}$ of open subsets of X such that $\mathscr{G}|S$ expands $\mathscr{F}$ and screens $\mathscr{H}$.*

(4) *If $(H_\alpha)_{\alpha\in I}$ is a uniformly locally finite open cover of S, then there exists a uniformly locally finite open cover $(G_\alpha)_{\alpha\in I}$ of X such that $G_\alpha \cap S = H_\alpha$ for each $\alpha\in I$.*

This result is an immediate consequence of 14.6 and the fact that S is P-embedded in X if and only if S is P^γ-embedded in X for all infinite cardinal numbers γ. □

Let us now turn our attention to characterizing the extending of γ-separable continuous pseudometrics in terms of uniformities.

14.9 THEOREM *Suppose that S is a subset of a completely regular topological space $(X, \mathscr{T})$, that γ is an infinite cardinal number and that d is a γ-separable continuous pseudometric on S. Then the following statements are equivalent.*

(1) *The pseudometric d can be extended to a (γ-separable) continuous pseudometric on X.*

(2) *There exists a continuous uniformity $\mathscr{V}$ on X such that $\mathscr{V}|S\times S = \mathscr{U}_d$.*

Proof. Assuming (1), there exists a continuous pseudometric d^* on X such that $d^*|S\times S = d$. Let $\mathscr{V} = \mathscr{U}_d$. Since $d^*|S\times S = d$, $\mathscr{V}|S\times S = \mathscr{U}_d$. Moreover, since d^* is continuous, $\mathscr{T}(\mathscr{V}) \subset \mathscr{T}$.

On the other hand, if (2) is assumed, let $d\in\mathscr{U}_d$. By (2), there exists $V\in\mathscr{V}$ such that $V\cap(S\times S) = U$. Let $\mathscr{A} = (V(x))_{x\in X}$ and note that $\mathscr{A}$ is a uniform open cover of $(X,\mathscr{V})$. Hence $\mathscr{A}$ is a normal open cover relative to $\mathscr{T}(\mathscr{V})$. Since $\mathscr{T}(\mathscr{V}) \subset \mathscr{T}$, 14.2(2) holds and therefore d can be extended to a continuous (γ-separable) pseudometric on X. □

14.10 PROPOSITION (*Gantner* [143]) *Suppose that S is a subset of a completely regular topological space X, and that γ is an* *infinite cardinal number. The following statements are then equivalent.*

(1) *The subset S is P^{γ}-embedded in X.*

(2) $\mathscr{U}_{\gamma}(X)|S \times S = \mathscr{U}_{\gamma}(S)$.

(3) $\mathscr{U}_{\gamma}(S)$ *has an admissible extension to* X.

(4) $\mathscr{U}_{\gamma}(S)$ *has a continuous extension to* X.

Proof. (1) *implies* (2). Note that $\mathscr{U}_{\gamma}(X)|S \times S$ is always contained in $\mathscr{U}_{\gamma}(S)$. To prove the converse, let $U(d, \epsilon)$ be a subbasic element of $\mathscr{U}_{\gamma}(S)$, where d is a γ-separable continuous pseudometric on S. By (1), there exists a continuous γ-separable pseudometric e on X such that $e|S \times S = d$. But then

$$U(e, \epsilon) \in \mathscr{U}_{\gamma}(X)$$

and therefore $U(d, \epsilon) = U(e, \epsilon) \cap (S \times S) \in \mathscr{U}_{\gamma}(X)|S \times S$. It follows that (2) holds.

(2) *implies* (3). This implication is clear.

That (3) *implies* (4) is obviously true, so let us show that (4) *implies* (1). Let d be a γ-separable continuous pseudometric on the subset S. Now the uniformity $\mathscr{U}_d(S)$, generated by d, is a subcollection of the uniformity $\mathscr{U}_{\gamma}(S)$. By (4), $\mathscr{U}_{\gamma}(S)$ has a continuous extension $\mathscr{U}^*_{\gamma}(X)$. Consequently by 13.31(11) there is a continuous extension $\mathscr{U}^*_d(X)$ of $\mathscr{U}_d(S)$ to X and $\mathscr{U}^*_d(X) \subset \mathscr{U}^*_{\gamma}(X)$. Statement (1) then follows from 14.9. □

14.11 Proposition *If S is a subset of a completely regular space X, then the following statements are equivalent.*

(1) *The subset S is P-embedded in X.*

(2) $\mathscr{U}_0(S) = \mathscr{U}_0(X)|S \times S$.

(3) $\mathscr{U}_0(S)$ *has an admissible extension to* X.

(4) $\mathscr{U}_0(S)$ *has a continuous extension to* X.

Proof. First, (1) *implies* (2) is immediate by 14.9 and (2) *implies* (3) follows from the observation that $\mathscr{U}_0(S)$ is an admissible uniformity on S.

That (3) *implies* (4) is trivial.

The proof that (4) *implies* (1) is similar to the proof that (4) *implies* (1) in 14.10 and is left as an exercise. □

Finally, let us show that extending pseudometrics can be

characterized in terms of σ-locally finite and σ-discrete covers. First the following preliminary results are needed.

14.12 THEOREM *Suppose that X is a topological space and that γ is an infinite cardinal number. If $\mathscr{U}$ is a σ-locally finite cozero-set cover of cardinality at most γ, then $\mathscr{U}$ admits a locally finite cozero-set refinement of cardinality at most γ.*

Proof. Suppose that $\mathscr{U}$ is a σ-locally finite cozero-set cover of X. Thus $\mathscr{U} = \bigcup_{n\in \mathbf{N}} \mathscr{U}_n$ where each $\mathscr{U}_n = (U^n_\alpha)_{\alpha\in I_n}$ is a family of locally finite cozero-sets of X. For each $n\in \mathbf{N}$, let

$$U_n = \bigcup_{\alpha\in I_n} U^n_\alpha$$

and note that by 2.8, each U_n is a cozero-set in X. Therefore $(U_n)_{n\in\mathbf{N}}$ is a countable cozero-set cover of X and hence, by 11.3, there exists a locally finite cozero-set cover $(V_n)_{n\in\mathbf{N}}$ of X such that $V_n \subset U_n$ for each $n\in\mathbf{N}$. Now let $M = \{(n,\alpha)\colon n\in\mathbf{N} \text{ and } \alpha\in I_n\}$ and for each $(n,\alpha)\in M$, let

$$W_{n\alpha} = V_n \cap U^n_\alpha.$$

It is asserted that $\mathscr{W} = (W_{n\alpha})_{(n,\alpha)\in M}$ is a locally finite cozero-set cover of X that refines $\mathscr{U}$. To see that $\mathscr{W}$ is locally finite, let $x\in X$. Since $(V_n)_{n\in\mathbf{N}}$ is locally finite there exist a neighborhood G_0 of x and a finite subset F of $\mathbf{N}$ such that $G_0 \cap V_n = \varnothing$ if $n\notin F$. Moreover, since $\mathscr{U}_n$ is locally finite, for each $n\in F$, there exist a neighborhood G_n of x and a finite subset K_n of I_n such that $G_n \cap U^n_\alpha = \varnothing$ if $\alpha\notin K_n$. Let $G = G_0 \cap (\bigcap_{n\in F} G_n)$ and let

$$P = \{(n,\alpha)\in M\colon n\in F \quad\text{and}\quad \alpha\in K_n\}.$$

Clearly G is a neighborhood of x and P is a finite subset of M and one easily verifies that $G\cap W_{n\alpha} = \varnothing$ if $(n,\alpha)\notin P$. Note that the cover $\mathscr{W}$ has cardinality at most $\aleph_0\cdot\gamma = \gamma$. Since the other assertions are obvious, the proof is now complete. □

14.13 LEMMA *Suppose that f is a continuous function from a topological space X into a topological space Y. If $\mathscr{U}$ is a discrete collection of cozero-subsets of Y, then $f^{-1}(\mathscr{U})$ is a discrete collection of cozero-subsets of X.*

The straightforward proof is omitted. □

14.14 Theorem *Suppose that X is a topological space and that γ is an infinite cardinal number. If $\mathscr{U}$ is a locally finite cozero-set cover of cardinality at most γ, then $\mathscr{U}$ admits a σ-discrete cozero-set refinement of cardinality at most γ.*

Proof. Suppose that $\mathscr{U}$ is a locally finite cozero-set cover of X of cardinality at most γ. By 10.10 and 1.8 there exist a metric space (Y,d), a continuous function f from X into (Y,d), and an open cover $\mathscr{V}$ of cardinality at most γ of (Y,d) such that $f^{-1}(\mathscr{V})$ refines $\mathscr{U}$. Now an open cover in a metric space has a σ-discrete cozero-set refinement and this refinement has cardinality at most γ (2.17). Therefore, there exists a cover $\mathscr{N}$ of Y such that $\mathscr{N} = \bigcup_{n\in\mathbf{N}} \mathscr{N}_n$, each $\mathscr{N}_n$ is a collection of discrete cozero-sets in Y, and $\mathscr{N}$ refines $\mathscr{V}$. Note that $f^{-1}(\mathscr{N}) = \bigcup_{n\in\mathbf{N}} f^{-1}(\mathscr{N}_n)$, and by 14.13, each $f^{-1}(\mathscr{N}_n)$ is a discrete collection of cozero-sets of X. Since $f^{-1}(\mathscr{N})$ clearly refines $\mathscr{U}$ the proof is now complete. □

The main result may now be made evident.

14.15 Theorem *Suppose that S is a subset of a topological space X and that γ is an infinite cardinal number. Then the following statements are equivalent.*

(1) *The subset S is P^γ-embedded in X.*

(2) *Every normal locally finite cozero-set cover of S of cardinality at most γ has a refinement that can be extended to a σ-discrete cozero-set cover of X.*

(3) *Every σ-locally finite cozero-set cover of S of cardinality at most γ has a refinement that can be extended to a locally finite cozero-set cover of X of cardinality at most γ.*

(4) *Every σ-locally finite cozero-set cover of S of cardinality at most γ has a refinement that can be extended to a σ-discrete cozero-set cover of X.*

(5) *Every σ-discrete cozero-set cover of S of cardinality at most γ has a refinement that can be extended to a σ-locally finite cozero-set cover of X of cardinality at most γ.*

Proof. (1) *implies* (2). If $\mathscr{U}$ is a normal locally finite cozero-set cover of S of cardinality at most γ, then, by 14.5, $\mathscr{U}$ has a refinement that extends to a normal open cover $\mathscr{V}$. By 10.10 every normal open cover has a locally finite cozero-set refinement $\mathscr{N}$ and by 14.14, $\mathscr{N}$ has a σ-discrete refinement $\mathscr{A}$. Clearly $\mathscr{A}|S$ refines $\mathscr{U}$, hence (2) holds.

(2) *implies* (4). This implication is clear.

(4) *implies* (1). If $\mathscr{U}$ is a normal locally finite cozero-set cover of S of cardinality at most γ, then clearly $\mathscr{U}$ is σ-locally finite and therefore by (4) there exists a σ-discrete cozero-set cover $\mathscr{V}$ of X such that $\mathscr{V}|S$ refines $\mathscr{U}$. By 14.12 $\mathscr{V}$ has a locally finite cozero-set refinement $\mathscr{N}$ and clearly $\mathscr{N}|S$ refines $\mathscr{U}$. Therefore by 14.5 S is P^γ-embedded in X. The proof that (1), (3) and (5) are equivalent proceeds in a similar manner and we omit it. □

Since a subspace S of a topological space X is P-embedded in X if and only if it is P^γ-embedded in X for all infinite cardinal numbers γ, it is clear that 14.15 could be stated in terms of P-embedding. We leave this formulation as an exercise.

14.16 Exercises

(1) If S is a subset of a topological space X, then the following statements are equivalent.

(*a*) The subset S is paracompact normal and P-embedded in X.

(*b*) Every open cover of S has a refinement that can be extended to a locally finite cozero-set cover of X.

(*c*) Every open cover of S has a refinement that can be extended to a normal cozero-set cover of X.

(*d*) Every open cover of S has a refinement that can be extended to a normal open cover of X.

(*e*) Every open cover of S has a refinement that can be extended to a locally finite normal open cover of X.

(2) Suppose that X is a completely regular topological space. The space X is called *almost compact* in case $|X-\beta X| \leqslant 1$. We say that X is *absolutely P-embedded* in case every embedding of X in a completely regular space, is a P-embedding.

If X is a completely regular topological space then the following statements are equivalent.

(*a*) X is almost compact.

(*b*) If $X \subset Y$ and if Y is completely regular then $\beta X \subset \beta Y$.

(*c*) Every embedding of X in a completely regular space is a C-embedding.
(*d*) The space X admits a unique uniform structure.
(*e*) The space X is absolutely P-embedded.

(3) In 14.3, refine Hausdorff's theorem to show that if f (as in 14.3) is a homeomorphism of A with B, then it can be arranged that F also is a homeomorphism. (Keep F as constructed in the proof. Check that if the sequence $(F(x_n))_{n\in\mathbf{N}}$ converges to $F(a)$ where $x_n \in X - A$ for all n and where $a \in A$, then $(x_n)_{n\in\mathbf{N}}$ converges to a. Consider the function $h \in C^*(B)$ defined by

$$h = \phi_{x_n} - \phi_a = \sum_{\alpha\in I} f_\alpha(x_n)\,(r_{f(a_\alpha)} - r_{f(a)}),$$

where the f_α and a_α are as in Lemma 1. Then $\rho(h)$ converges to 0. Consider $h(f(a))$ and make use of part (*c*) of Lemma 1 and the fact that f is a homeomorphism on A.)

(4) If r is a pseudometric defined on a closed subset of a metric space (X, d), then r may be extended to a pseudometric r^* on all of X in such a way that
(*a*) in $X - A$, r^* is equivalent to the metric d of X;
(*b*) if $x \in X - A$ then for some positive $\epsilon > 0$, $r(x, y) < \epsilon$ implies $y \in X - A$;
(*c*) if r is a metric equivalent to d on A, then the extension r^* is equivalent to d on all of X.

(We use Exercise 3 and Lemma 2. In A form the equivalence classes for the relation $r(x, y) = 0$ and obtain the obvious metric space B. The natural mapping of A into B satisfies 14.3. If m is the metric on Y, then $m(F(x), F(y))$ extends $r(x, y)$.) Notice how this is related to 14.4!

(5) Let d be a pseudometric defined on a closed subset A of a paracompact Hausdorff space. Then there is a pseudometric d^* defined on all of X such that for x and y in A and for $k = 4, 5, \ldots$, if $d^*(x, y) < 2^{-k}$ then $d(x, y) < 2^{-k}$.

(Note that this result can be used to show that a pseudometric d defined on a closed subset of a paracompact Hausdorff space may be extended to a pseudometric on X (see [27]). Later we will see this result when we consider the more general class of collectionwise normal spaces.)

(Partition X, using the pseudometric d^*, into a set X^* of equivalence classes x^* given by $d^*(x,y) = 0$. Then define $\bar{d}(x^*, y^*) = d^*(x,y)$. The natural map is a continuous function of X into the metric space X^*. If A^* is the closure in X^* of the image of A, then d may be carried over (in a unique way) to A^*.)

(6) (*Exercise on simultaneous extension.*) Let K be a linear (or possibly just convex) subset of a locally convex linear space and let A be a closed subset of the metric space X. If F is the collection of all continuous functions on A with values in K, then each f in F may be extended by an f^* (as in Lemma 2) in such a way that, for $f_1, \ldots, f_n$ in F and $c_1, \ldots, c_n$ real numbers (non-negative with sum 1 when K is merely convex) we have

$$(c_1 f_1 + \ldots + c_n f_n)^* = c_1 f_1^* + \ldots + c_n f_n^*.$$

(This is a generalization of a result in Kakutani [215] and is due to Arens in [27].)

15 Extending continuous pseudometrics and other topological concepts

In this section the concept of P^γ-embedding is discussed. It is related to the existence of a measurable cardinal (Ulam's problem), to C-, C^*-, and z-embedding, to pseudocompact spaces and to collectionwise normal spaces.

In 15.2 it is shown that a P-embedded subset of a space X is necessarily C-embedded in X. In 15.3 it is shown that the converse holds if the cardinality of the *set* X is non-measurable and if S is (topologically) dense in the *space* X. From this, a purely topological formulation of Ulam's set theoretical problem is given in 15.5.

Urysohn's lemma for normal spaces shows that the separation conditions imposed on the open sets are sufficient to permit the existence of non-trivial continuous pseudometrics. The Tietze–Urysohn extension theorem then characterized normal spaces as those in which all closed subsets are C-embedded. In 15.7, collectionwise normal spaces are shown to be those spaces in which all closed subsets are P-embedded. It would be interesting if an analogous characterization could be given for completely regular spaces.

Every Tychonoff space is C^*-embedded in its Stone–Čech compactification. If the underlying set has non-measurable cardinal, 15.12 shows that pseudocompact Tychonoff spaces are precisely those which satisfy the stronger condition of being P-embedded in βX. Moreover any pseudocompact Tychonoff space which satisfies the stronger separation condition of normality enjoys the even stronger separation condition of collectionwise normality as is shown in 15.15. In consideration of another 'compactness' or 'countability' axiom, it will evolve (15.18) that Lindelöf, C-embedded subsets of Tychonoff spaces are P-embedded.

Finally z-embedded subsets are considered. They are characterized (15.20) in terms of normal covers. Some results concerning extensions of pseudometrics are related in 15.21.

With the fact that ψ_f is an $\aleph_0$-separable pseudometric for $f \in C(X)$ and with the help of the following lemma it is possible to show that a subset being P-embedded is definitely stronger than it being C-embedded in the space.

15.1 Lemma *Suppose that X is a topological space and that S is a $P^{\aleph_0}$-embedded subset of X. If $f \in C(S)$, if $Z_S(f) \neq \varnothing$, and if $f \geqslant \mathbf{0}$, then there exists $g \in C(X)$ such that $g|S = f$.*

Proof. Let $f \in C(S)$ and suppose that $f \geqslant \mathbf{0}$ and that $Z = Z_S(f) \neq \varnothing$. Then by 13.31(6), ψ_f is a continuous $\aleph_0$-separable pseudometric on X. Since S is $P^{\aleph_0}$-embedded in X, there exists a continuous pseudometric d on X such that $d|S \times S = \psi_f$. Then $d^Z \in C(X)$ (2.13). Moreover, if $x \in S$, then

$$d^Z(x) = \psi_f(x, Z) = \inf_{z \in Z} |f(x) - f(z)| = f(x)$$

and it follows that $d^Z|S = f$. The proof is now complete. □

15.2 Theorem *Suppose that X is a topological space. If S is $P^{\aleph_0}$-embedding in X, then S is C-embedded in X.*

Proof. If S is empty, then S is clearly C-embedded in X. Now assume that $S \neq \varnothing$ and let $f \in C(S)$. By 2.2, it may be assumed that $f \geqslant \mathbf{0}$. Let $a \in S$ be arbitrary and let $f(a) = \alpha$. Then $f \vee \boldsymbol{\alpha}$ and

$f \wedge \alpha$ are continuous functions on S. Let $g = (f \vee \alpha) - \alpha$ and $h = -((f \wedge \alpha) - \alpha)$. Since $g \geqslant 0, h \geqslant 0, Z_S(g) \neq \varnothing$, and $Z_S(h) \neq \varnothing$, it follows by 15.1 that there exist $\bar{g}, \bar{h} \in C(X)$ such that $\bar{g}|S = g$ and $\bar{h}|S = h$. Let $k = (\bar{g} - \bar{h}) + \alpha$. Then one easily verifies that $k \in C(X)$ and that $k|S = f$. The proof is now complete. □

Shortly it will be shown that, in general, the converse does not hold (see 15.8). Thus, having seen that if S is P-embedded in X then S is C-embedded in X and having remarked that, in general, the converse does not hold, it is natural to ask under what conditions the converse does hold. It is now possible to give some sufficient conditions for the converse to hold.

15.3 Theorem *Suppose that X is a completely regular space, that the cardinality of X is non-measurable, and that S is a dense subspace of X. Then the subspace S is C-embedded in X if and only if it is P-embedded in X.*

Proof. For the necessity let $\mathscr{P}(S)$ denote the universal uniformity on S (see 13.1) and let $\gamma(S, \mathscr{P}(S))$ denote the completion of $(S, \mathscr{P}(S))$. Since X is non-measurable, it follows by 13.31(14) that $\gamma(S, \mathscr{P}(S))$ is realcomplete, and hence $\gamma(S, \mathscr{P}(S)) = \upsilon S$. Since S is C-embedded in X, $\mathrm{cl}_{\upsilon X} S = \upsilon S$ (5.21(13c)). Moreover, $X = \mathrm{cl}_X S \subset \mathrm{cl}_{\upsilon X} S = \upsilon S$ and it follows that $\upsilon S = \upsilon X$ (6.9(2)). Therefore, $\gamma(S, \mathscr{P}(S)) = \upsilon S = \upsilon X$.

Suppose that d is a continuous pseudometric on S. By 13.5, d is uniformly continuous on $(S, \mathscr{P}(S))$ and hence by 13.31(15), d can be extended to a continuous pseudometric $\bar{d}$ on $\gamma(S, \mathscr{P}(S)) = X$. Let $e = \bar{d}|X \times X$. Then e is a continuous pseudometric on X and $e|S \times S = d$. Therefore S is P-embedded in X.

Conversely since a P-embedded set is $P^{\aleph_0}$-embedded, the result follows from 15.2. □

15.4 Theorem *For discrete topological spaces X, the cardinality of the set X is non-measurable if and only if the subspace X is P-embedded in υX.*

Proof. The necessity is immediate by 15.3. On the other hand suppose that $|X|$ is measurable. Since X is a discrete topological space, $X \neq \upsilon X$ (5.21(15c)). Let $p \in \upsilon X - X$ and let $\mathscr{U} = (\{x\})_{x \in X}$. Then $\mathscr{U}$ is a locally finite cozero-set cover of X and therefore by

14.7, $\mathscr{U}$ has a refinement, which must be $\mathscr{U}$ itself, that can be extended to a locally finite cozero-set cover of υX. Therefore $(\{x\})_{x\in X}$ is locally finite in υX, so there exists a neighborhood U of p in υX such that $U \cap X$ is finite. Thus $U \cap X$ is closed in υX and it follows that $(\upsilon X - (U \cap X)) \cap U$ is a neighborhood of p in υX. Since X is dense in υX, there exists an $x \in X$ such that

$$x \in (\upsilon X - (U \cap X)) \cap U.$$

Therefore $x \in X \cap U$. But $x \in \upsilon X - (U \cap X)$ implies that $x \notin U \cap X$, a contradiction. The proof is now complete. □

Now an interesting fact is related. The question in set theory of the existence of a measurable cardinal is related to the existence of a C-embedded dense subset S of a completely regular topological space and a continuous pseudometric d on S which does not extend continuously to X. On the other hand if one is to accept a negative answer to Ulam's question, then for dense subsets of completely regular spaces, C- and P-embedding are the same.

15.5 THEOREM *The following statements are equivalent.*

(1) *Every cardinal number is non-measurable.*

(2) *If X is completely regular and if S is a dense C-embedded subset of X, then S is P-embedded in X.*

Proof. (1) *implies* (2). This is immediate by 15.3.

(2) *implies* (1). Suppose that there exists a set X such that $|X|$ is measurable. Let X have the discrete topology and note that X is dense and C-embedded in X. But, by 15.4, X is not P-embedded in υX. The proof is now complete. □

At first glance it would seem that one should be able to prove 15.4 more directly. In particular, if S is a dense C-embedded subset of X and if $\mathscr{U}$ is a locally finite cozero-set cover of S, since S is C-embedded, each element of $\mathscr{U}$ may be extended to a cozero-set and since S is dense, this extension is a locally finite cover of X. Theorem 15.5 shows us that this is not the case. The situation is further confused because this technique does work for other special cases (see 16.2).

(2) *implies* (3). This result follows from 15.2 and the fact that if f is a uniformly continuous real-valued function on $(S, \mathscr{C}(S))$ then ψ_f is a uniformly continuous pseudometric on $(S, \mathscr{C}(S))$.

(3) *implies* (1). This result follows from 17.3. □

As an immediate result of 17.6 and known characterizations of normal spaces, the following theorem shows the germane character of normal spaces in a manner similar to 17.6.

17.7 THEOREM *If X is a completely regular space, then the following statements are equivalent.*

(1) *The space X is normal.*

(2) *For every closed subset S of X, $\mathscr{C}(S) = \mathscr{C}(X)|S \times S$.*

(3) *The space $(S, \mathscr{C}(S))$ is almost $P^{\aleph_0}$-embedded in X for every closed subset S of X.*

(4) *The space $(S, \mathscr{C}(S))$ is almost C-embedded in X for every closed subset S of X.*

We have seen in the consideration of continuous real-valued functions that $\aleph_0$-separable continuous pseudometrics were of significant assistance. We will see in the consideration of bounded continuous real-valued functions that totally bounded continuous pseudometrics are of significant assistance.

17.8 DEFINITION A subset S of a topological space X is *T-embedded in X* in case every totally bounded continuous pseudometric on S has a totally bounded continuous pseudometric extension to X.

Our first order of business will be to characterize T-embedding in terms of finite cozero-set covers, finite normal cozero-set covers, and finite normal open covers. To do this we must go through a proof similar to 14.2 and make the appropriate modifications. In 17.9 not all of the possible equivalences to T-embedding are stated. It is a worthwhile exercise to establish them however.

17.9 THEOREM *If S is a subspace of a topological space X, then the following statements are equivalent.*

(1) *The subset S is T-embedded in X.*

(2) *Every totally bounded continuous pseudometric on S can be extended to a continuous pseudometric on X.*

(3) *Every finite normal cozero-set cover of S has a refinement that can be extended to a normal open cover of X.*

(4) *Every finite normal open cover of S has a refinement that can be extended to a finite normal cozero-set cover of X.*

Proof. The implication (1) *implies* (2) is immediate so let us consider (2) *implies* (3). Assume (2) and suppose that $\mathscr{U}$ is a finite normal cozero-set cover of S. By 10.12 there exists a normal sequence $(\mathscr{V}_i)_{i\in\mathbf{N}}$ of open covers of S such that $\mathscr{V}_1$ refines $\mathscr{U}$ and such that, for each $i\in\mathbf{N}$, $\mathscr{V}_i$ is finite. Then, by 8.4, there exists a continuous pseudometric d on S that is associated with $(\mathscr{V}_i)_{i\in\mathbf{N}}$ and, by 8.9, d is totally bounded. Therefore, by (2), there is a continuous pseudometric $\bar{d}$ on X such that $\bar{d}|S\times S = d$. Let $\mathscr{W}' = (S_{\bar{d}}(x, 2^{-4}))_{x\in X}$. Since $(X, \bar{d})$ is a pseudometric space, it is paracompact, so there is a locally finite open cover $\mathscr{W}$ of X such that $\mathscr{W}$ refines $\mathscr{W}'$. By 2.16 and the fact that a locally finite cozero-set cover is normal (10.10), it follows that $\mathscr{W}$ is a normal open cover of X relative to the given topology on X and one easily verifies that $\mathscr{W}|S$ refines $\mathscr{U}$.

The implication that (3) *implies* (4) follows from 10.10. Finally assuming (4), let $\mathscr{T}$ be the given topology on X and suppose that d is a totally bounded continuous pseudometric on S. We wish to show (1).

For each $m\in\mathbf{N}$ there exists a finite subset F_m of S such that

$$\mathscr{S}_m = (S_d(x, 2^{-(m+3)}))_{x\in F_m}$$

covers S.

Now consider any $m\in\mathbf{N}$ and note that $\mathscr{S}_m$ is a normal cover relative to $\mathscr{T}_d$. Therefore, by (4), there exists a refinement of $\mathscr{S}_m$ that extends to a finite normal cozero-set cover $\mathscr{V}^m$ of X. By 10.12, there exists a normal sequence $(\mathscr{V}_i^m)_{i\in\mathbf{N}}$ of open covers of X such that $\mathscr{V}_1^m$ refines $\mathscr{V}^m$ and such that $\mathscr{V}_i^m$ is finite for each $i\in\mathbf{N}$. The proof now follows the proof (3) *implies* (1) in 14.2. We will make comments, however, where there is some modification. Then as in that proof, one easily verifies that (i) $\mathscr{U}^m$ and $\mathscr{U}_i^m$ (as defined there) are finite open covers of X, and that (ii), (iii), and (iv) also hold as given there.

Now again consider any $m\in\mathbf{N}$. It follows from (i) and (ii) that $(\mathscr{U}_i^m)_{i\in\mathbf{N}}$ is a normal sequence of finite open covers of X. Then, by 8.4, there exists a continuous pseudometric r_m on X that is associated with $(\mathscr{U}_i^m)_{i\in\mathbf{N}}$ and by 8.9, r_m is totally bounded. By (ii) and (iv), it is also possible to have statement (*) on p. 171. Then defining the continuous pseudometric r as in 14.2, it is seen that r is totally bounded. In fact if $\epsilon > 0$, let $k \in \mathbf{N}$ such that $2^{-(k+1)} < \epsilon$. Since r_k is totally bounded, there exists a finite subset F of X such that $\bigcup_{x\in F} S(x; r_k, 2^{-(k+4)}) = X$. If $z \in X$, then $z \in S(x; r_k\, 2^{-(k+4)})$ for some $x \in F$ and it follows that $z \in \mathrm{st}\,(x, \mathscr{U}_{k+3}^k)$. Furthermore, if $1 \leqslant m \leqslant k$, then by (iii) we have $\mathscr{U}_{k+3}^k$ refines $\mathscr{U}_{k+3}^m$ and hence $z \in \mathrm{st}\,(x, \mathscr{U}_{k+3}^m)$. Hence $r_m(x, y) < 2^{-k} < \frac{1}{2}\epsilon$ whenever $1 \leqslant m \leqslant k$. Then

$$r(x,z) = \sum_{m=1}^{k} 2^{-m} r_m(x,z) + \sum_{m=k+1}^{\infty} 2^{-m} r_m(x,z)$$
$$< \sum_{m=1}^{k} 2^{-m}(\tfrac{1}{2}\epsilon) + \sum_{m=k+1}^{\infty} 2^{-m} < \epsilon.$$

The statement (**) of p. 172 holds and it is easy to show that the resulting pseudometrics r^* and d^* as defined there are totally bounded.

Let $\mathscr{U}^*$ be the uniformity on X^* generated by the metric r^*. Let $\mathscr{V}^* = \mathscr{U}^*|S^* \times S^*$ and note that $(S^*, \mathscr{V}^*)$ is a uniform subspace of $(X^*, \mathscr{U}^*)$. Using (**) and the fact that $\sigma: X \to X^*$ is an isometry one easily shows that d^* is a uniformly continuous pseudometric on S^*. Therefore, by 17.12(2), there exists a uniformly continuous bounded pseudometric e on X^* such that $e|S^* \times S^* = d^*$. To show that e is totally bounded, let $\epsilon > 0$ and let

$$B^* = \{(x^*, y^*) \in X^* \times X^* \colon e(x^*, y^*) < \epsilon\}.$$

Since e is uniformly continuous, $B^* \in \mathscr{U}^*$ and therefore, there exists $\delta > 0$ such that

$$\{(x^*, y^*) \in X^* \times X^* \colon r^*(x^*, y^*) < \delta\} \subset B^*.$$

Now r^* is totally bounded, so there is a finite subset F^* of X^* such that $\bigcup_{x^*\in F^*} S(x^*; r^*, \delta) = X$. One easily shows that

$$\bigcup_{x^*\in F^*} S(x^*; e, \epsilon) = X^*$$

and it follows that e is totally bounded.

Define $\bar{d}$ on $X \times X$ by $\bar{d} = e \circ (\sigma \times \sigma)$. Since σ is continuous relative to $\mathscr{T}$, $\bar{d}$ is a continuous pseudometric on X (2.14). Moreover, if $x, y \in S$, then

$$\bar{d}(x,y) = e(\sigma(x),\sigma(y)) = d^*(\sigma(x),\sigma(y)) = d(x,y).$$

Therefore, $\bar{d}|S \times S = d$. Since σ is an isometry and since e is totally bounded, it follows that $\bar{d}$ is totally bounded. Therefore (1) holds.

The proof is now complete. □

The appropriateness of T-embedding lies in the fact that it is equivalent to C^*-embedding. Here we see how this is possible.

17.10 **Theorem** *If S is a subset of a topological space X, then the following statements are equivalent.*

(1) *The space S is C^*-embedded in X.*

(2) *The space S is T-embedded in X.*

Proof. (1) *implies* (2). Suppose that $(U_1, \ldots, U_n)$ is a finite cozero-set cover of S. By 10.11, there exists a zero-set cover $(V_1, \ldots, V_n)$ of S such that $V_i \subset U_i$ for $i = 1, \ldots, n$. Moreover, for each $i = 1, \ldots, n$, V_i and $S - U_i$ are completely separated in S (6.2) and therefore by 6.6 they are completely separated in X. Thus there exists $f_i \in C(X,E)$ such that $f_i(x) = 0$ if $x \in V_i$ and $f_i(x) = 1$ if $x \in S - U_i$. Let $A_i = \{x \in X : f_i(x) \leqslant \frac{1}{2}\}$ and let

$$B_i = \{x \in X : f_i(x) < 1\}.$$

Note that $G = X - (A_1 \cup \ldots \cup A_n)$ is a cozero-set in X. Let $\mathscr{W} = (W_1, \ldots, W_n)$ be defined as follows: set $W_1 = G \cup B_1$ and $W_i = B_i$ for $i = 2, \ldots, n$. Clearly $\mathscr{W} = (W_1, \ldots, W_n)$ is a finite cozero-set cover of X such that $\mathscr{W}|S$ refines $\mathscr{U}$. Therefore 17.9 holds and S is T-embedded in X.

(2) *implies* (1). The proof is exactly that of 15.2 once it is observed that the pseudometric ψ_f, for $f \in C^*(S)$, is totally bounded (see 13.31(5)). □

Reformulating the characterizations for normal spaces in terms of C^*-embedding, we add (as a consequence of 17.9, 17.10, and 11.7) to the normal space characterizations the following.

17.11 THEOREM *If X is a topological space then the following statements are equivalent.*

(1) *The space X is normal.*

(2) *Every closed subset of X is T-embedded in X.*

(3) *For every closed subset S of X, every totally bounded continuous pseudometric on S can be extended to a continuous pseudometric on X.*

(4) *For every closed subset S of X, every finite normal cozero-set cover of S has a refinement that can be extended to a normal open cover of X.*

(5) *For every closed subset S of X, every finite normal open cover of S has a refinement that can be extended to a finite normal cozero-set cover of X.*

17.12 EXERCISES

(1) If X and Y are topological spaces, if m is a totally bounded continuous pseudometric on Y, and if $f\colon X \to Y$ is a continuous function, then $m \circ (f \times f)$ is a totally bounded continuous pseudometric on X.

(2) Suppose that $(X, \mathscr{U})$ is a uniform space and that S is a uniform subspace of X. Then every bounded uniformly continuous pseudometric on S can be extended to a bounded uniformly continuous pseudometric on X. (If d is a bounded uniformly continuous pseudometric on S, let $U_n = \{(x,y) \in S \times S : d(x,y) < 2^n\}$ for all $n \in \mathbf{N}$. Then there exists $V_n \in \mathscr{U}$ such that V_n is symmetric and $V_n \cap (S \times S) = U_n$. Use 13.6 to show for all n in $\mathbf{N}$ there exists a uniformly continuous pseudometric r_n on X such that $r_n \leqslant 1$ and $\{(x, y) \in X \times X : r_n(x, y) < 1/4\} \subset V_n$. For all n in $\mathbf{N}$ let $e_n = 2^{n+2} r_n$ and set $e = \sum_{i=0}^{k} e_i$. Then e is a bounded uniformly continuous pseudometric (13.9) such that $e|S \times S \geqslant d$. For $(x, y) \in X \times X$, let $f(x, y) = \inf\{e(x, a) + d(a, b) + d(b, y) : a, b \in S\}$ and $r(x, y) = \min\,(d(x, y), e(x, y))$. Then r is a bounded uniformly continuous pseudometric on X such that $r|S \times S = d$.)

V. *Normality and uniformities*

Extensions of collections of continuous real-valued functions and the extensions of collections of continuous pseudometrics have had appropriate applications in normal and collectionwise normal spaces. In Chapter III, it was shown that the concept of a uniform space could be studied in terms of collections of pseudometrics, or collections of covers, or collections of entourages. With this in mind it seems appropriate, to consider extensions of uniformities. Here this will be done.

Uniformities generated by various collections of continuous pseudometrics will be considered. The uniformities suitable to normal spaces and the ones suitable to collectionwise normal spaces will be exhibited in a fashion analogous to our previous work. For example, a space will be normal if and only if for every closed subset F, every precompact (prerealcomplete) admissible uniformity on F may be extended to a precompact (prerealcomplete) continuous uniformity on X. A space X is collectionwise normal if and only if for every closed subset F, every admissible uniformity may be extended to a continuous uniformity on X.

18 Uniformities generated by γ-separable continuous pseudometrics

Particular attention has been placed on pseudometrics whose associated topologies are γ-separable where γ is an infinite cardinal number. Since any collection of pseudometrics generates a uniformity (as in Section 13) let us first consider those uniformities which are generated by all γ-separable continuous pseudometrics. We will then drop the requirement that the uniformities be generated by γ-separable pseudometrics to investigate the case for any admissible uniformity. In a manner analogous to our approach with P-embedding, concepts may now be defined for extending uniformities from a subspace to the whole space.

18.1 DEFINITION Let S be a subspace of a completely regular topological space X and let γ be an infinite cardinal number. The subset S is *γ-uniformly embedded* in X if every admissible uniformity on S generated by a collection of continuous γ-separable pseudometrics on S has a continuous extension to X. The subset S is *u_γ-embedded in X* if it is γ-uniformly embedded in X where the extension uniformity on X is, in addition, admissible. The subset S is *uniformly embedded in X* if every admissible uniformity on S can be extended to a continuous uniformity on X. The subset S is said to be *u-embedded in X* if it is uniformly embedded in X and if the extension uniformity is also admissible.

It is clear that every u_γ-embedded subset is also γ-uniformly embedded. By 14.10 *the subset S is P^γ-embedded in X if and only if $\mathscr{U}_\gamma(S) = \mathscr{U}_\gamma(X)|S \times S$.* Now $\mathscr{U}_\gamma(S)$ is an admissible uniformity on S generated by a collection of γ-separable continuous pseudometrics (see 13.10). From this will follow the last result in the next proposition.

Also it is clear that every u-embedded subset is uniformly embedded. Since the universal uniformity $\mathscr{U}_0(S)$ is an admissible uniformity on S, 14.11 shows that every uniformly embedded subset is also P-embedded. In fact the last two concepts are equivalent (see 18.9(3)).

18.2 PROPOSITION *For a completely regular space X and for γ an infinite cardinal number, every u_γ-embedded subset is γ-uniformly embedded and every γ-uniformly embedded subset is P^γ-embedded. Moreover every u-embedded subset is uniformly embedded and every uniformly embedded subset is P-embedded.*

Actually a subset being P^γ-embedded in a space is equivalent to the subspace being γ-uniformly embedded in X. However these are not equivalent to the subset being u_γ-embedded in X.

To see this let X be any Tychonoff pseudocompact space of

non-measurable cardinality such that X is not almost compact. An example of such a space is the space $\Lambda = \beta\mathbf{R} - (\beta\mathbf{N} - \mathbf{N})$ which was constructed in Section 9 (see 9.5). Another example of such a space X may be obtained by taking X to be a Σ-product of the product space $A^{\mathbf{R}}$, where A is a two-point discrete space (see 9.7). Now if X is pseudocompact then X is P-embedded in βX by 15.12. On the other hand, since X is not almost compact there are at least two distinct admissible uniformities on X (see 14.16(2)). But βX is compact and so there is a unique admissible uniformity on βX that cannot extend the two uniformities on X. Therefore X is not $u_{\aleph_0}$-embedded in βX.

By the way, the above example also shows that not every P-embedded subset is u-embedded. However it will be possible to show that the two concepts are equivalent for closed subsets.

Having demonstrated this counter-example let us now show that in terms of characterizing P^{γ}-embedding, the word 'continuous' plays the key role as opposed to 'admissible' when one considers the extension uniformity. That is, every P^{γ}-embedded subset is also γ-uniformly embedded. Using this result it is then possible to show how every closed P^{γ}-embedded subset is also a u_{γ}-embedded subset.

18.3 THEOREM *Suppose that S is a subset of a completely regular topological space $(X, \mathcal{T})$, and that γ is an infinite cardinal number. Then the following statements are equivalent.*

(1) *The subset S is P^{γ}-embedded in X.*

(2) *The subset S is γ-uniformly embedded in X.*

Proof. To show that (1) *implies* (2) let us suppose that $\mathcal{U}$ is an admissible uniformity on S generated by a collection $\mathcal{P}$ of γ-separable continuous pseudometrics on S. By 13.5 without loss of generality it is possible to assume that $\mathcal{P}$ is the collection of all uniformly continuous γ-separable pseudometrics on $(S, \mathcal{U})$. Let $\mathcal{P}^*$ be the collection of all continuous γ-separable pseudometrics d on X such that $d|S \times S \in \mathcal{P}$ and let $\mathcal{U}(\mathcal{P}^*)$ be the uniformity generated by $\mathcal{P}^*$. Clearly $\mathcal{P}^*|S \times S \subset \mathcal{P}$. On the other hand, if $d \in \mathcal{P}$ then since S is P^{γ}-embedded in X there exists a continuous γ-separable pseudometric e on X such that $e|S \times S = d$ and therefore $d \in \mathscr{P}^*|S \times S$. We thus have $\mathscr{P}^*|S \times S = \mathscr{P}$ and hence $\mathscr{U}(\mathscr{P}^*)|S \times S = \mathscr{U}$. To complete the proof observe that

$$\mathscr{U}(\mathscr{P}^*) \subset \mathscr{U}_\gamma(X)$$

and therefore $\mathscr{T}(\mathscr{U}(\mathscr{P}^*)) \subset \mathscr{T}(\mathscr{U}_\gamma(X)) = \mathscr{T}$. □

In [143], Gantner showed that for closed subsets of completely regular spaces, P^γ-embedding and u_γ-embedding are equivalent.

18.4 THEOREM (*Gantner*) *Suppose that S is a closed subset of a completely regular topological space $(X, \mathscr{T})$ and that γ is an infinite cardinal number. If S is P^γ-embedded in X then S is u_γ-embedded in X.*

Proof. Using the notation in 18.3 it is necessary to prove that $\mathscr{T} \subset \mathscr{T}(\mathscr{U}(\mathscr{P}^*))$. Suppose that $G \in \mathscr{T}$ and that $x \in G$. Since X is completely regular let us assume that G is a cozero-set in X. There are two cases to be considered.

If $x \notin S$ then $G \cap (X - S)$ is a neighborhood of x in X so there exists $f \in C(X, E)$ such that $f(x) = 1$ and $f(y) = 0$ for

$$y \in S \cup (X - G).$$

Then the pseudometric ψ_f on X associated with f is a continuous $\aleph_0$-separable pseudometric on X. If $(x, y) \in S \times S$ then

$$\psi_f(x, y) = |f(x) - f(y)| = 0.$$

Consequently, ψ_f is a continuous extension to X of the pseudometric $\mathbf{0}$ on S. Since $\mathbf{0}$ is obviously a uniformly continuous γ-separable pseudometric on $(S, \mathscr{U})$ one has $\mathbf{0} \in \mathscr{P}$. Therefore by definition $\psi_f \in \mathscr{P}^*$. Now the set $W = \{y \in X: \psi_f(y, x) < 1\}$ is an element of $\mathscr{T}(\mathscr{U}(\mathscr{P}^*))$ and $x \in W \subset G$.

Now suppose $x \in S$. Since G is a cozero-set neighborhood of x in X there exists a zero-set neighborhood Z of x in X such that $Z \subset G$ (see 4.3). Then Z and $X - G$ are disjoint zero-sets in X and so there exists a $g \in C(X, E)$ such that $g(x) = 0$ for x in Z and $g(y) = 1$ for $y \in X - G$ (see 6.2). Since Z is a neighborhood of x in X, $Z \cap S$ is a neighborhood of x in S. Since $\mathscr{U}$ is an admissible uniformity on S and since $S - \operatorname{int}(Z \cap S)$ is a closed set in S disjoint from $\{x\}$, there exists a $\mathscr{U}$-uniformly continuous real-valued function

f in $C(S, E)$ such that $f(x) = 0$, $f(y) = 1$ for $y \in S - \operatorname{int}(S \cap Z)$ (13.12). But then the pseudometric ψ_f on S associated with f is a uniformly continuous γ-separable pseudometric on $(S, \mathscr{U})$. It follows that $\psi_f \in \mathscr{P}$. Thus since S is P^γ-embedded in X, there exists $d \in \mathscr{P}^*$ such that $d|S \times S = \psi_f$. Then the function $d^{\{x\}}$ is continuous (2.13) and $d^{\{x\}}|S = f$. Let $h = d^{\{x\}} \vee g$ and observe that h is a real-valued continuous function on X such that $h(y) = f(y)$ for all $y \in S$. In fact if $y \in S \cap Z$ then $g(y) = 0$. Since $h \geqslant \mathbf{0}$, one has $h(y) = f(y)$. If $y \in S - (S \cap Z)$ then $f(y) = 1$ and since $g \leqslant \mathbf{1}$, $h(y) = f(y)$. Thus the pseudometric ψ_h on X associated with h is a continuous $\aleph_0$-separable pseudometric on X and $\psi_h|S \times S = \psi_f$. Thus $\psi_h \in \mathscr{P}^*$. Finally let us set $W = \{y \in X: \psi_h(y, x) < 1\}$. Then $x \in W$ and $W \in \mathscr{T}(\mathscr{U}(\mathscr{P}^*))$. Moreover if $y \in W$ then $h(x) = f(x) = 0$ implies that $g(y) \leqslant h(y) = \psi_h(y, x) < 1$, whence $y \in G$.

Therefore in either case, $W \in \mathscr{T}(\mathscr{U}(\mathscr{P}^*))$ and $x \in W \subset G$. It follows that $G \in \mathscr{T}(\mathscr{U}(\mathscr{P}^*))$ and so $\mathscr{T} \subset \mathscr{T}(\mathscr{U}(\mathscr{P}^*))$. □

COROLLARY *Suppose that S is a closed subset of a completely regular topological space X. If S is P-embedded in X then S is u-embedded in X.*

Proof. Let $\gamma = |\operatorname{cl} S| + \aleph_0$ and suppose that $\mathscr{U}$ is an admissible uniformity on S. Then by 13.8, $\mathscr{U}$ is generated by a collection $\mathscr{P}$ of continuous pseudometrics on S. Clearly every element in $\mathscr{P}$ is γ-separable. Since S is $\mathscr{P}^\gamma$-embedded in X, we may apply 18.4 to conclude that $\mathscr{U}$ can be extended to an admissible uniformity on X. Therefore S is u-embedded in X. □

The problem of extending a γ-separable continuous pseudometric from the closure of a subset S to the whole space X is related to the extension of such pseudometrics from the subset. The reader is asked to give the easy proof of the following two results.

18.5 PROPOSITION *Let X be a Tychonoff space. If S is P^γ-embedded in X then $\operatorname{cl} S$ is P^γ-embedded in X.*

Using 18.4 now and this proposition, it is easy to show that $\operatorname{cl} S$ is always u_γ-embedded in X whenever S is P^γ-embedded in X. Consequently, the real problem in showing that S is u_γ-embedded in X whenever S is P^γ-embedded in X comes between S and its closure.

18.6 THEOREM *Suppose that X is a Tychonoff space and that γ is an infinite cardinal number. If S is a P^γ-embedded subset of X then* $\operatorname{cl} S$ *is u_γ-embedded in X.*

It is now possible to show that if an admissible uniformity generated by a collection of γ-separable continuous pseudometrics (let us call such a uniformity a γ-separable uniformity) may be extended to an admissible (continuous) uniformity then it may be extended to an admissible (continuous) uniformity generated by a collection of γ-separable continuous pseudometrics.

18.7 THEOREM *Suppose that S is a subset of a completely regular topological space $(X, \mathscr{T})$ and that γ is an infinite cardinal number. If $\mathscr{U}$ is an admissible uniformity on S generated by a collection of continuous γ-separable pseudometrics on S and if $\mathscr{U}$ can be extended to an admissible (respectively, continuous) uniformity on X then there exists an admissible (respectively, continuous) uniformity $\mathscr{U}^*$ on X generated by a collection of continuous γ-separable pseudometrics on X such that $\mathscr{U}^* | S \times S = \mathscr{U}$.*

Proof. Suppose that $\mathscr{U}$ is an admissible uniformity on S generated by a collection $\mathscr{P}$ of γ-separable continuous pseudometrics on S. By 13.5 one may assume without loss of generality, that $\mathscr{P}$ is the collection of all uniformly continuous γ-separable pseudometrics on $(S, \mathscr{U})$. By hypothesis there exists an admissible (respectively, continuous) uniformity $\mathscr{V}$ on X such that $\mathscr{V} | S \times S = \mathscr{U}$. Let $\mathscr{P}^*$ be the collection of all continuous γ-separable pseudometrics d on X such that $d | S \times S \in \mathscr{P}$ and let $\mathscr{U}(\mathscr{P}^*)$ be the uniformity generated by $\mathscr{P}^*$. Clearly

$$\mathscr{P}^* | S \times S \subset \mathscr{P}.$$

On the other hand if $d \in \mathscr{P}$, then d is a γ-separable uniformly continuous pseudometric on $(S, \mathscr{U})$, $\mathscr{U}_d \subset \mathscr{U}$, and $(S, \mathscr{U})$ is a uniform subspace of $(X, \mathscr{V})$. Therefore by 17.4, there exists a continuous γ-separable pseudometric e on X such that $e | S \times S = d$. It follows that $d \in \mathscr{P}^* | S \times S$. Thus one has $\mathscr{P}^* | S \times S = \mathscr{P}$ and hence $\mathscr{U}(\mathscr{P}^*) | S \times S = \mathscr{U}$.

To complete the proof one must show that $\mathscr{T}(\mathscr{U}(\mathscr{P}^*)) = \mathscr{T}$. Observe that $\mathscr{U}(\mathscr{P}^*) \subset \mathscr{U}_\gamma(X)$ and therefore

$$\mathscr{T}(\mathscr{U}(\mathscr{P}^*)) \subset \mathscr{T}(\mathscr{U}_\gamma(X)) = \mathscr{T}.$$

Conversely suppose that $G \in \mathscr{T}$ and that $x \in G$. By 13.12 there exists an $f \in C(X, E)$ which is uniformly continuous on $(X, \mathscr{U})$ such that $f(x) = 0$ and $f(y) = 1$ for y in $X - G$. But then ψ_f, the pseudometric associated with f, is $\aleph_0$-separable and uniformly continuous on $(X, \mathscr{V})$. Hence $\psi_f | S \times S$ is a uniformly continuous γ-separable pseudometric on $(S, \mathscr{U})$. Therefore by definition $\psi_f \in \mathscr{P}^*$. Let $W = \{y \in X: \psi_f(x, y) < 1\}$. Observe that W is an element of $\mathscr{T}(\mathscr{U}(\mathscr{P}^*))$ and that $x \in W \subset G$ whence

$$\mathscr{T} \subset \mathscr{T}(\mathscr{U}(P^*)). \square$$

Using our characterizations of γ-collectionwise normality in terms of P^γ-embedding, some characterizations of the former concept in terms of uniformity extensions may now be given.

18.8 THEOREM *Suppose that X is a completely regular topological space and that γ is an infinite cardinal number. Then the following statements are equivalent.*

(1) *The space X is γ-collectionwise normal.*

(2) *Every closed subset of X is γ-uniformly embedded in X.*

(3) *Every closed subset of X is u_γ-embedded in X.*

(4) *For every closed subset F of X, $\mathscr{U}_\gamma(F)$ has an admissible extension to X.*

(5) *For every closed subset F of X, $\mathscr{U}_\gamma(F) = \mathscr{U}_\gamma(X) | F \times F$.*

(6) *Every discrete family of closed subsets of X of cardinality* at *most γ is uniformly discrete in $(X, \mathscr{U}_\gamma(X))$.*

Proof. In showing (1) *implies* (2) let X be a γ-collectionwise normal space and let S be a closed subset of X. By 15.6, S is P^γ-embedded in X. Therefore by 18.3, S is γ-uniformly embedded in X.

By 18.2, 18.3, and 18.4, it follows that (2) and (3) are equivalent.

Since $\mathscr{U}_\gamma(F)$ is an admissible uniformity on F, (3) *implies* (4) is trivial, and (4) *implies* (5) follows from 14.10.

To show that (5) *implies* (6) suppose that $(F_\alpha)_{\alpha \in I}$ is a discrete family of closed subsets of X. Let $F = \bigcup_{\alpha \in I} F_\alpha$ and let d be a pseudometric on F defined by

$$\begin{aligned} d(x,y) &= 1 \quad \text{if } x \in F_\alpha,\ y \in F_\beta \text{ and } \alpha \neq \beta, \\ &= 0 \quad \text{otherwise.} \end{aligned}$$

Then d is a continuous γ-separable pseudometric on F and therefore by definition of $\mathscr{U}_\gamma(F)$, $U(d,1) \in \mathscr{U}_\gamma(F)$. Then by (5), there exists $W \in \mathscr{U}_\gamma(X)$ such that $W \cap (F \times F) = U(d,1)$. Let $V \in \mathscr{U}_\gamma(X)$ be a symmetric entourage such that $V \circ V \subset W$. Now the family $(V(F_\alpha))_{\alpha \in I}$ is pairwise disjoint. For suppose that $\alpha, \beta \in I$ and that there exists $z \in V(F_\alpha) \cap V(F_\beta)$. Then there are points $x \in F_\alpha$ and $y \in F_\beta$ such that $(x,z) \in V$ and $(y,z) \in V$ and hence $(x,y) \in W$. Since $(x,y) \in F_\alpha \times F_\beta \subset F \times F$, it follows that $(x,y) \in U(d,1)$. But then $d(x,y) < 1$ and therefore $\alpha = \beta$.

Finally for (6) *implies* (1) suppose $(F_\alpha)_{\alpha \in I}$ is a discrete family of closed subsets of X of power at most γ. By (6) there exists $U \in \mathscr{U}_\gamma(X)$ such that the family $(U(F_\alpha))_{\alpha \in I}$ is pairwise disjoint. By 3.10, $V = \operatorname{int} U$ is in $\mathscr{U}_\gamma(X)$. Moreover $F_\alpha \subset V(F_\alpha) \subset U(F_\alpha)$. Also since $V(F_\alpha) = \bigcup_{x \in F_\alpha} V(x)$ and since $V(x)$ is open it follows that $V(F_\alpha)$ is open. Thus $(V(F_\alpha))_{\alpha \in I}$ is a family of pairwise disjoint open sets. Therefore X is γ-collectionwise normal. □

Utilizing 18.8 one can now give characterizations of collectionwise normal spaces. The proofs are obtained by some modifications of those in 18.8 in conjunction with the technique developed in the proof of the corollary to 18.4, that is, let $\gamma = |X| + \aleph_0$ wherever necessary. Consequently these proofs are left as exercises.

Corollary *If X is a completely regular topological space, then the following statements are equivalent.*

(1) *The space X is collectionwise normal.*

(2) *Every closed subset of X is uniformly embedded in X.*

(3) *Every closed subset of X is u-embedded in X.*

(4) *For every closed subset F of X, $\mathscr{U}_0(F)$ can be extended to an admissible uniformity on X.*

(5) *For every closed subset F of X,*

$$\mathscr{U}_0(F) = \mathscr{U}_0(X) | F \times F.$$

(6) *Every discrete family of closed subsets of X is uniformly discrete in $(X, \mathscr{U}_0(X))$.*

18.9 Exercises

(1) Suppose that X is a completely regular topological space, that $A \subset B \subset X$, and that γ is an infinite cardinal number. Then the following statements hold.
(*a*) If A is u_γ-embedded in X then A is u_γ-embedded in B.
(*b*) If A is u_γ-embedded in B and if B is u_γ-embedded in X, then A is u_γ-embedded in X.

(2) Let S be a realcomplete subset of a Tychonoff space X and let γ be an infinite cardinal number. Then S is u_γ-embedded in X, if and only if S is closed and P^γ-embedded in X. (Use the fact that $(S, \mathscr{C}(S))$ is a complete uniform subspace of $(X, \mathscr{U})$ whenever S is realcomplete and $\mathscr{U}$ is an extension of $\mathscr{C}(S)$. Use also 18.2 and 18.6.)

(3) For a completely regular space X, a subset is P-embedded in X if and only if it is uniformly embedded in X (modify 18.3).

(4) Suppose that X is a completely regular space and that $A \subset B \subset X$. Then the following statements are true.
(*a*) If A is u-embedded in X, then A is u-embedded in B.
(*b*) If A is u-embedded in B and if B is u-embedded in X, then A is u-embedded in X.

(5) If S is a P-embedded subset of a completely regular space X, then $\operatorname{cl} S$ is u-embedded in X.

(6) Let S be a topologically complete subspace of a Tychonoff space X. Then S is u-embedded in X if and only if S is closed and P-embedded in X (a complete uniform subspace of a Hausdorff uniform space is closed).

19 Extending uniformities generated by continuous functions

Normal spaces have been characterized in terms of C- and C^*-embedding. Every normal space is completely regular and completely regular spaces are the appropriate abstraction of metric spaces in which one may consider uniform concepts. Consequently the question arises as to how one may characterize normal spaces in terms of uniform concepts. The similar question for collectionwise normal spaces has already been answered (see corollary to 18.8).

19.1 Definition Let X be a completely regular space. A uniformity $\mathscr{U}$ on X is said to be *precompact* if it is generated by a collection of bounded continuous real-valued functions. The uniformity is said to be *prerealcomplete* if it is generated by a collection of continuous real-valued functions.

Let S be a subset of X. The subset S is *precompact uniformly embedded* in X if every precompact admissible uniformity on S has a continuous extension to X. It is *prerealcomplete uniformly embedded* in X if every prerealcomplete admissible uniformity on S has a continuous extension to X. It is *c-embedded* (respectively, *c*-embedded*) in X if it is prerealcomplete (respectively, precompact) uniformly embedded in X and if the extension uniformity is admissible.

The obvious relationships among these concepts are given in the following diagram.

19.2 Proposition *Let S be a subspace of a completely regular space. The following diagrammatic implications hold for S.*

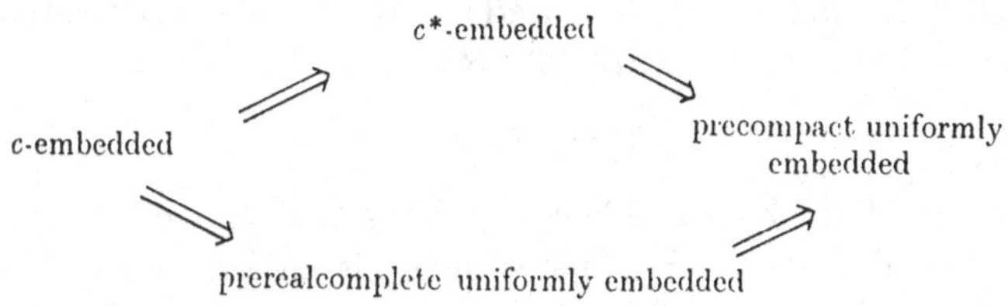

For pseudocompact spaces, c- and c^*-embedding are the same.

19.3 Proposition *If S is a pseudocompact subspace of a completely regular space X then S is c-embedded in X if and only if S is c*-embedded in X.*

Proof. If $\mathscr{U}$ is an admissible prerealcomplete uniformity then S pseudocompact implies that $C(S) = C^*(S)$ and therefore $\mathscr{U}$ is generated by a collection of bounded real-valued continuous functions. The result now follows. □

19.4 PROPOSITION *If S is a c-embedded subspace of a completely regular space X, then S is C-embedded in X. If S is a c^*-embedded subspace of a completely regular space X, then S is C^*-embedded in X.*

Proof. These are immediate from 17.3. □

Neither converse in the above proposition is true in general. The example after 18.2 provides an example of a space X that is both C-embedded and C^*-embedded in βX but is neither c-embedded nor c^*-embedded in βX.

19.5 THEOREM *Suppose that S is a subset of a completely regular topological space $(X, \mathscr{T})$ and that $\mathscr{U}$ is an admissible precompact uniformity on S. If $\mathscr{U}$ can be extended to an admissible (respectively continuous) uniformity on X then $\mathscr{U}$ can be extended to an admissible (respectively continuous) precompact uniformity on X.*

Proof. Let $\mathscr{A}$ be the collection of all $f \in C^*(S)$ that are uniformly continuous with respect to $\mathscr{U}$ and observe that $\mathscr{U}$ is generated by $\mathscr{A}$. Let $\mathscr{V}$ be an admissible (respectively continuous) uniformity on X such that $\mathscr{V}|S \times S = \mathscr{U}$. Let $\mathscr{A}^*$ be the set of all $f \in C^*(X)$ such that $f|S \in \mathscr{A}$ and let $\mathscr{U}(\mathscr{A}^*)$ be the uniformity on X generated by $\mathscr{A}^*$. Clearly $\mathscr{A}^*|S \subset \mathscr{A}$ and therefore

$$\mathscr{U}(\mathscr{A}^*)|S \times S \subset \mathscr{U}.$$

To show that $\mathscr{U} \subset \mathscr{U}(\mathscr{A}^*)|S \times S$, let $f \in \mathscr{A}$. Since $(S, \mathscr{U})$ is a uniform subspace of $(X, \mathscr{V})$ and since f is uniformly continuous on $(S, \mathscr{U})$ there exists by 17.5, 17.2(3), and 2.2 $g \in C^*(X)$ such that $g|S = f$. But then $g \in \mathscr{A}^*$ and therefore $f \in \mathscr{A}^*|S$. It follows that $\mathscr{U}(\mathscr{A}^*)|S \times S \subset \mathscr{U}$.

Now let us show that $\mathscr{U}(\mathscr{A}^*)$ is continuous. Since

$$\mathscr{U}(\mathscr{A}^*) \subset \mathscr{C}^*(X), \quad \mathscr{T}(\mathscr{U}(\mathscr{A}^*)) \subset \mathscr{T}(\mathscr{C}^*(X)) = \mathscr{T}.$$

Moreover if $G \in \mathscr{T}$ and if $x \in G$ then by 13.13, there exists an $f \in C(X, E)$ uniformly continuous on $(X, \mathscr{V})$ such that $f(x) = 0$

and $f(y) = 1$ for $y \in X - G$. But then $W = \{y \in X : f(y) < 1\}$ is an element of $\mathcal{T}(\mathcal{U}(\mathcal{A}^*))$ and $x \in W \subset G$. Hence $G \in \mathcal{T}(\mathcal{U}(\mathcal{A}^*))$ and so $\mathcal{T} \subset \mathcal{T}(\mathcal{U}(\mathcal{A}^*))$. Therefore $\mathcal{T}(\mathcal{U}(\mathcal{A}^*))$ is an admissible precompact extension of $\mathcal{U}$ to X. □

By simple modification of the previous proof we may obtain

19.6 THEOREM *Suppose that S is a subset of a completely regular topological space $(X, \mathcal{T})$ and that $\mathcal{U}$ is an admissible prerealcomplete uniformity on S. Then there exists an admissible (respectively continuous) uniformity $\mathcal{U}^*$ on X that is generated by a collection of continuous real-valued functions on X such that $\mathcal{U}^* | S \times S = \mathcal{U}$.*

The concepts of 'precompact uniformly embedded' and 'prerealcomplete uniformly embedded' are the proper uniform analogues of 'C^*-embedding' and 'C-embedding'. This is not difficult to demonstrate.

19.7 THEOREM *If S is a subset of a completely regular space X, then the following are equivalent.*

(1) *The subset S is C-embedded in X.*
(2) *The subset S is $\aleph_0$-uniformly embedded in X.*
(3) *The subset S is prerealcompact uniformly embedded in X.*

Proof. The equivalence of (1) and (2) follows from 16.4 and 18.3. In 16.4, it is shown that S is C-embedded in X if and only if it is $P^{\aleph_0}$-embedded in X. In 18.3, it is shown that a subset S is C-embedded in X if and only if it is $\aleph_0$-uniformly embedded in X.

In showing that (2) *implies* (3) let $\mathcal{U}$ be an admissible uniformity on S generated by a collection $\mathcal{G}$ of continuous real-valued functions on S. By 13.31(6), the pseudometric ψ_f is $\aleph_0$-separable for $f \in \mathcal{G}$. Hence (2) applies to $\mathcal{U}$ and gives (3).

Finally in showing (3) *implies* (1) consider $\mathcal{C}(S)$. Since this prerealcompact uniformity is admissible there is a continuous uniformity $\mathcal{V}$ on X such that $\mathcal{V} | S \times S = \mathcal{C}(S)$. Thus $(S, \mathcal{C}(S))$ is a uniform subspace of $(X, \mathcal{V})$ and therefore by 17.5, $(S, \mathcal{C}(S))$ is almost P-embedded in X. By 17.6 it now follows that S is C-embedded in X. □

Using arguments similar to the above together with 17.9 and 17.10 we can give the following analogue.

19.8 THEOREM *If S is a subset of a completely regular space $(X,\mathcal{T})$ then the following are equivalent.*

(1) *The subset S is C^*-embedded in X.*

(2) *Every admissible uniformity on S generated by a collection of totally bounded continuous pseudometrics has a continuous extension to X.*

(3) *The subset S is precompact uniformly embedded in X.*

Let us now investigate when C or C^*-embedded subsets are respectively c- or c^*-embedded.

19.9 LEMMA *If S is a C-embedded (respectively C^*-embedded) subset of a topological space X then* cl S *is C-embedded (respectively, C^*-embedded) in X.*

Proof. Any continuous function f on the cl S restricts to a continuous function on the dense subset S. The latter function must then extend to a continuous function f^* on X. Since these functions map into a Hausdorff space, $f^*|\text{cl } S = f$. □

Now the following results are immediate.

19.10 THEOREM *If S is a closed C-embedded (respectively, closed C^*-embedded) subset of a completely regular topological space X, then S is c-embedded (respectively, c^*-embedded) in X.*

19.11 COROLLARY *If S is a C-embedded (respectively, C^*-embedded) subset of a completely regular topological space X, then* cl S *is c-embedded (respectively, c^*-embedded) in X.*

Let us now clarify the relationship between the concepts of c-embedding, $u_{\aleph_0}$-embedding and c^*-embedding.

19.12 THEOREM *If $(X,\mathcal{T})$ is a topological space and if $S \subset X$ then the following statements are equivalent.*

(1) *The subset S is $u_{\aleph_0}$-embedded in X.*

(2) *The subset S is c-embedded in X.*

(3) *The subset S is C-embedded and c^*-embedded in X.*

Proof. The implication, (1) *implies* (2), follows immediately from the fact that ψ_f is $\aleph_0$-separable for all $f \in C(X)$. By 19.2 every c-embedded subset is c^*-embedded and by 19.4 a c-embedded subset is C-embedded. Thus (2) *implies* (3).

To show (3) *implies* (1) assume that $\mathscr{U}$ is an admissible uniformity on S generated by a collection of continuous $\aleph_0$-separable pseudometrics on S. Let $\mathscr{P}$ be the collection of all uniformly continuous $\aleph_0$-separable pseudometrics on $(S, \mathscr{U})$. Note that $\mathscr{U}$ is $\mathscr{U}(\mathscr{P})$ by 13.8. Let $\mathscr{P}^*$ be the set of all continuous $\aleph_0$-separable pseudometrics d on X such that $d|S \times S \in \mathscr{P}$ and let $\mathscr{U}^* = \mathscr{U}(\mathscr{P}^*)$. Clearly $\mathscr{P}^*|S \times S \subset \mathscr{P}$. On the other hand if $d \in \mathscr{P}$ then d is a uniformly continuous $\aleph_0$-separable pseudometric on $(S, \mathscr{U})$. By hypothesis, S is C-embedded in X. Hence S is $P^{\aleph_0}$-embedded in X (16.4). Therefore d can be extended to a continuous $\aleph_0$-separable pseudometric e on X. By definition $e \in \mathscr{P}^*$ and

$$e|S \times S = d \in \mathscr{P}^*|S \times S.$$

It follows that $\mathscr{P} = \mathscr{P}^*|S \times S$ and therefore $\mathscr{U} = \mathscr{U}^*|S \times S$.

It remains to show that $\mathscr{U}^*$ is admissible. Now $\mathscr{U}^* \subset \mathscr{C}$ so $\mathscr{T}(\mathscr{U}^*) \subset \mathscr{T}(\mathscr{C}(X)) = \mathscr{T}$. To prove $\mathscr{T} \subset \mathscr{T}(\mathscr{U}^*)$, let $\mathscr{A}$ be the collection of all bounded uniformly continuous real-valued functions on $(S, \mathscr{U})$. Then $\{\psi_f : f \in \mathscr{A}\} \subset \mathscr{P}$. Let $\mathscr{V}$ be the uniformity on S generated by $\mathscr{A}$, Clearly $\mathscr{T}(\mathscr{V}) \subset \mathscr{T}(\mathscr{U}) = \mathscr{T}|S$. If $G \in \mathscr{T}|S$ and if $x \in G$ then by 13.12, there exists an $f \in C(X, E)$ uniformly continuous on $(S, \mathscr{V})$ such that $f(x) = 0$ and $f(y) = 1$ for $y \in S - G$. Let $W = \{y \in S : \psi_f(x, y) < 1\}$. Since $f \in \mathscr{A}$, $W \in \mathscr{T}(\mathscr{V})$ and $x \in W \subset G$. Thus $G \in \mathscr{T}(\mathscr{V})$. Therefore $\mathscr{V}$ is an admissible uniformity on S. Since $\mathscr{V}$ is generated by bounded functions it is precompact. Since S is c^*-embedded in X, there exists an admissible precompact uniformity $\mathscr{V}^*$ on X such that

$$\mathscr{V}^*|S \times S = \mathscr{V}.$$

Let $\mathscr{B}$ be the set of all bounded uniformly continuous real-valued functions on $(X, \mathscr{V}^*)$. If $f \in \mathscr{B}$ then $f|S$ is uniformly continuous with respect to $\mathscr{V}$ and hence with respect to $\mathscr{U}(\mathscr{V} \subset \mathscr{U})$. Therefore $\psi_{f|S} \in \mathscr{P}$ and so $\psi_f \in \mathscr{P}^*$. Therefore $\mathscr{V}^* \subset \mathscr{U}^*$ whence

$$\mathscr{T} = \mathscr{T}(\mathscr{V}^*) \subset \mathscr{T}(\mathscr{U}^*). \square$$

19.13 Theorem *If X is a completely regular topological space then the following statements are equivalent.*

(1) *The space X is normal.*

(2) *Every closed subset of X is $\aleph_0$-uniformly embedded in X.*

(3) *Every closed subset of X is precompact uniformly embedded in X.*

(4) *Every closed subset of X is prerealcompact uniformly embedded in X.*

(5) *Every closed subset of X is $u_{\aleph_0}$-embedded in X.*

(6) *Every closed subset of X is c-embedded in X.*

(7) *Every closed subset of X is c*-embedded in X.*

(8) *For every closed subset F of X, $\mathscr{U}_{\aleph_0}(F) = \mathscr{U}_{\aleph_0}(X)|F \times F$.*

(9) *For every closed subset F of X, $\mathscr{U}_{\aleph_0}(F)$ has an admissible extension to X.*

(10) *For every closed subset F of X, $\mathscr{C}(F)$ has an admissible extension to X.*

(11) *For every closed subset F of X, $\mathscr{C}^*(F)$ has an admissible extension to X.*

(12) *Every countable discrete family of closed subsets of X is uniformly discrete in $(X, \mathscr{U}_{\aleph_0}(X))$.*

(13) *Every countable discrete family of closed subsets of X is uniformly discrete in $(X, \mathscr{C}(X))$.*

(14) *Every finite family of pairwise disjoint closed subsets of X is uniformly discrete in $(X, \mathscr{C}^*(X))$.*

Proof. By 19.7 and 19.8 (1), (2), (3) and (4) are all equivalent. By 19.12 and 19.10 (1), (5), (6) and (7) are all equivalent. By 18.8 and the fact that X is normal if and only if X is $\aleph_0$-collectionwise normal, (1), (8), (9), and (12) are all equivalent. By 17.7, (1), (10), and (11) are equivalent. Thus it remains to prove that (1), (13), and (14) are equivalent.

To show (1) *implies* (13), let us suppose $(F_n)_{n \in \mathbf{N}}$ is a discrete family of closed subsets of X. Let $F = \bigcup_{n \in \mathbf{N}} F_n$ and note that F is a closed subset of the normal space X. Therefore F is C-embedded in X. Let f be a real-valued function on F defined by

$$f(x) = n \quad \text{if and only if} \quad x \in F_n.$$

Then $f \in C(S)$ and therefore there exists $g \in C(X)$ such that

$g|F = f$. Note that $U = U(\psi_f, \frac{1}{2}) \in \mathscr{C}(X)$ and $F_n \subset U(F_n)$. It is a standard argument to show that $(U(F_n))_{n \in \mathbf{N}}$ is pairwise disjoint and so (13) holds.

To prove (1) *implies* (14) we proceed as above after observing that $f \in C^*(F)$ since the indexing set is finite. Clearly (13) *implies* (12).

Finally for (14) *implies* (1) suppose that F_1 and F_2 are disjoint closed subsets of X. By (14) there exists $U \in \mathscr{C}^*(X)$ such that $U(F_1) \cap U(F_2) = \varnothing$. Let $V = \operatorname{int} U$ and observe that $V(F_1)$ and $V(F_2)$ are disjoint open sets containing F_1 and F_2 respectively. Therefore X is normal. □

19.14 Exercises

(1) The Stone–Čech compactification βX of X is that unique compact Hausdorff space containing X densely such that every admissible precompact uniformity on X has a continuous extension.

(2) The Hewitt–Nachbin realcompletion υX of X is that unique realcomplete Hausdorff space containing X densely such that every admissible prerealcomplete uniformity on X has a continuous extension.

(3) If S is a realcomplete subset of a completely regular space then S is c-embedded in X if and only if S is closed and C-embedded in X.

(4) If S is a compact subset of a completely regular space then S is c^*-embedded in X if and only if S is closed and C^*-embedded in X.

20 Some applications

In the exercises of Chapter I, we defined the concepts pertinent to locally convex linear spaces, which are by far the most useful and richest of the linear spaces. It was interesting to see how the Minkowski functional and pseudonorms (which are therefore pseudometrics) were relevant. We now consider how our previous results are related to the extensions of continuous functions on a space. That is, we are interested in some 'Tietze–Urysohn type' extension theorems as was given in Chapter II.

In order to prove 14.2 we considered continuous functions defined on a closed subset of a metric space with values in a convex subset of a locally convex linear topological space. We now consider a similar kind of result for closed subsets of spaces more general than metric spaces. To do this, one of our basic techniques will be to map the subspace S into the space $C^*(S)$ as was suggested by the proof in 14.3 and as we explicated immediately after that proof. We will also make use of the functions d^x in $C^*(X)$ as defined in the proof of 14.3. More explicitly,

20.1 DEFINITION If d is a bounded continuous pseudometric on the space X, then the family $(d^x)_{x \in X}$ is called the family of *partial functions associated with d.*

For our work in this section (see [12], [396]), we need the following definition of extensions of functions.

20.2 DEFINITION Let S be a subset of X and Y a subset of Z, where X and Z are topological spaces. If f is a continuous function from S into Z such that $f(S) \subset Y$, then f *extends continuously to X relative to Y* if there is a continuous function f^* from X into Z such that $f^*|S = f$ and $f^*(X) \subset Y$.

We emphasize the phrase 'relative to Y' to stress that in the extension process the space Y must not be enlarged.

20.3 DEFINITION Let X be a topological space, let γ be an infinite cardinal, and let f be a function from X to a locally convex topological vector space (abbreviated LCTV space). We call f an *M-valued function* if the image of X under f is contained in a complete convex metrizable subset M of L (abbreviated CCM subspace). The function f is a *(γ, M)-valued function* if it is an M-valued function and if the image of X under f is a γ-separable subset of M.

In Exercise 12 of 13.31, a Fréchet space was defined as a complete, metrizable LCTV space. Also in the first part of the same exercise, it was asked to show that $C^*(X)$ is a Banach space under the norm ρ defined by

$$\rho(g) = \sup_{x \in X} |g(x)|.$$

The following theorem characterizes the extendability of M-valued functions and (γ, M)-valued functions in the same spirit as previous results. The first two equivalences were originally shown by Arens in [28] for closed subsets of a topological space. The rest of the results may be found in [12].

20.4 THEOREM *Let S be a non-empty subspace of X, let γ be an infinite cardinal, and let A be a discrete space such that $|A| \geqslant |S|$. The following statements are equivalent.*

(1) *The subspace S is P^{γ}-embedded in X.*

(2) *Given a* CCM *subspace M of a* LCTV *space L, every continuous (γ, M)-valued function on S extends continuously to X relative to M.*

(3) *Given a* CCM *subspace M of a* LCTV *space L, every continuous (γ, M)-valued function on S extends to a continuous function from X to L.*

(4) *Every continuous function from S to a Fréchet space, such that the image of S is γ-separable, extends to a continuous function on X.*

(5) *Every continuous function from S into $C^*(S)$, such that the image of S is γ-separable, extends to a continuous function on X.*

(6) *Every continuous function from S into $C^*(A)$, such that the image of S is γ-separable, extends to a continuous function on X.*

Furthermore, the above conditions are also equivalent to the conditions obtained from (2) *through* (6) *by requiring the image of S to be a bounded subset of the locally convex space in question.*

Proof. We first show that (1) *implies* (2). Let f be a continuous (γ, M)-valued function from S to the LCTV space L, and let m be a complete metric for M. One easily verifies (see 2.14) that $d = m \circ (f \times f)$ is a γ-separable continuous pseudometric on S. Hence by (1) it has an extension to a continuous pseudometric d^* on X. The space (S, d) is a subspace of the pseudometric space (X, d^*). The function f is uniformly continuous as a mapping from (S, d) into (M, m) since $d(x, y) \leqslant \epsilon$ implies $m(f(x), f(y)) \leqslant \epsilon$ for all x, y in S and for all $\epsilon > 0$. The space (M, m) is a complete Hausdorff uniform space. Hence f extends to a uniformly continuous function f^* from the closure of (S, d) in (X, d^*) to (M, m).

Now we have a continuous function from a closed subset of a pseudometric space into a LCTV space L.

By Lemma 2 in Section 14 the function f^* extends to a continuous function g from (X, d^*) into M, since M is convex. The function g is continuous with respect to the topology generated by d^*. Since this topology is contained in the original topology of X, the mapping g is the continuous extension of f relative to M that we seek.

The implications (2) *implies* (3) *implies* (4) *implies* (5) are immediate.

We now show that (5) *implies* (1). By 14.5 it is sufficient to show that every bounded γ-separable continuous pseudometric on S extends to a continuous pseudometric on X. Let d be a bounded γ-separable continuous pseudometric on S. To define the function f from S into $C^*(S)$, let $f(x)$ be the partial function d^x for each x in S. The function f is continuous for if $\epsilon > 0$ and if $d(x, y) \leqslant \epsilon$ for x, y in S, then

$$\begin{aligned}\|f(x)-f(y)\| &= \sup_{z\in S}|f(x)(z)-f(y)(z)| \\ &= \sup_{z\in S}|d(x,z)-d(y,z)| \\ &= d(x,y) \leqslant \epsilon.\end{aligned}$$

Let $(x_\alpha)_{\alpha\in I}$ be a dense subset of power at most γ. It is easily verified that $(f(x_\alpha))_{\alpha\in I}$ is a dense subset of $f(S)$. Therefore, by (5), the function f extends to a continuous function f^* mapping X into $C^*(S)$. Define a function d^* on $X \times X$ by

$$d^*(x,y) = \|f^*(x)-f^*(y)\|$$

for all x, y in X. It is readily seen that d^* is a continuous pseudometric on X and it extends the pseudometric d. In fact, if x and y are in S, then

$$\begin{aligned}d^*(x,y) = \|f(x)-f(y)\| &= \sup_{z\in S}|f(x)(z)-f(y)(z)| \\ = \sup_{z\in S}|d(x,z)-d(y,z)| &= d(x,y).\end{aligned}$$

Thus S is P^γ-embedded in X.

The implication (1) *implies* (6) is now clear from the equivalence of (1) and (4).

To show that (6) *implies* (1), let d be a bounded continuous γ-separable pseudometric on S. Let G denote S with the discrete topology. Since the cardinality of the discrete space A is large enough, we may identify G with some copy of it in A. Hence in the sequel we will consider G as a subset of the space A. Define a function f from S into $C^*(A)$ by setting for each x in S,

$$f(x)(a) = d(x, a)$$

if a is in G and 0 otherwise. Since A is discrete and d is bounded, $f(x)$ is an element of $C^*(A)$ for all x in F. Thus the map f is continuous. In fact if $\epsilon > 0$ and $d(x, y) \leqslant \epsilon$, then

$$\begin{aligned} \|f(x) - f(y)\| &= \sup_{a \in S} |f(x)(a) - f(y)(a)| \\ &= \sup_{a \in G} |f(x)(a) - f(y)(a)| \\ &= \sup_{a \in S} |d(x, a) - d(y, a)| \\ &= d(x, y) \leqslant \epsilon. \end{aligned}$$

Again it is easy to check that the γ-separability of d implies the γ-separability of $f(S)$. Hence, by (6), the function f extends to a continuous function f^* on X. Defining d^* on $X \times X$ by

$$d^*(x, y) = \|f^*(x) - f^*(y)\|$$

for all x, y in X, we see that d^* is a continuous pseudometric on X. By similar computations as those above, it is easily verified that d^* extends the pseudometric d. Therefore S is P^{γ}-embedded in X.

To prove the last statement, call (2*) through (6*) the new conditions resulting from requiring the image of S to be bounded in (2) through (6), respectively. The implications (1) *implies* (2*) *implies* (3*) *implies* (4*) *implies* (5*) hold. The proof of (5*) *implies* (1) is as that of (5) *implies* (1) after one notes the following. By 14.5, it is sufficient to show that every bounded, γ-separable, continuous pseudometric on S extends to a continuous pseudometric on X. If d is bounded, then $f(S)$ is a bounded subset of $C^*(S)$. The implication (1) *implies* (6*) is clear. The implication (6*) *implies* (1) is like (6) *implies* (1), noting that $f(S)$ is a bounded subset of $C^*(A)$. □

Since a subspace S of a topological space X is P-embedded in X if and only if it is P^γ-embedded in X for all infinite cardinal numbers γ, it is clear from 20.4 that we obtain characterizations of P-embedding by removing all mention of cardinality. In particular, we obtain the following

20.5 COROLLARY *Let S be a subspace of a topological space X. Then S is P-embedded in X if and only if every continuous function from S into a bounded closed convex subset of a Banach space extends to a continuous function on X.*

Thus 20.5 gives an interesting characterization of collectionwise normal spaces when considered in conjunction with 15.7. In particular, we can now characterize collectionwise normal spaces in terms of the extension from closed subsets of a particular class of continuous functions just as in the case of the Tietze extension theorem for normal spaces.

20.6 COROLLARY *A topological space X is collectionwise normal if and only if for every closed subset F of X, every continuous function from F into a bounded closed convex subset B of a Banach space can be extended continuously to X relative to B.*

Let us consider the result of 20.6 in conjunction with Bing's example (see Section 9) of a normal Hausdorff, non-collectionwise normal space. Together these show that a continuous function defined on a closed subset of a normal space with values in a bounded closed convex subset of a Banach space, need not necessarily extend to a continuous function on the whole space.

Let us now consider 20.4 in conjunction with the result that C-embedding of a subset is equivalent to its $P^{\aleph_0}$-embedding. That is, let $\gamma = \aleph_0$ in 20.4 and let us improve the Tietze–Urysohn extension theorem by giving the following corollary.

20.7 COROLLARY *For any non-empty topological space X the following are equivalent.*

(1) *The space X is normal.*

(2) *For every closed subset F of X, any continuous $(\aleph_0, M)$-valued*

function from F into a CCM *subset M of an* LCTV *space L can be continuously extended to X relative to M.*

The following result about the Hewitt–Nachbin realcompletion υX of a completely regular T_1-space X follows in a similar way.

20.8 COROLLARY *The Hewitt–Nachbin realcompletion υX of a Tychonoff space X is that unique realcompletion of X for which every continuous function f from X to a Fréchet space, such that $f(X)$ is separable, can be extended to a continuous function on υX.*

Moreover, if X has non-measurable cardinality, then υX is that unique realcompletion of X such that every continuous function from X into a Fréchet space extends to a continuous function on υX.

Proof. The result follows from characterizations of υX given in 5.20 and from 20.4, 16.4, and 15.3. □

The following theorem characterizes T- or C^*-embedding in terms of the extension of totally bounded linear space-valued functions in a spirit analogous to 20.4.

20.9 THEOREM *Let S be a non-empty subspace of X and let A be a discrete space such that* card $A \geqslant$ card S. *The following statements are equivalent.*

(1) *The subspace S is C^*-embedded in X.*

(2) *Given a* CCM *subspace M of a* LCTV *space L, every continuous M-valued function f on S, such that $f(S)$ is totally bounded, extends continuously to X relative to M.*

(3) *Given a* CCM *subspace M of a* LCTV *space L, every continuous M-valued function f on S, such that $f(S)$ is totally bounded, extends to a continuous function from X to L.*

(4) *Every continuous function f from S to a Fréchet space, such that $f(S)$ is totally bounded, extends to a continuous function on X.*

(5) *Every continuous function f from S into $C^*(S)$, such that $f(S)$ is totally bounded, extends to a continuous function on X.*

(6) *Every continuous function f from S into $C^*(A)$, such that $f(S)$ is totally bounded, extends to a continuous function on X.*

Proof. We first show that (1) *implies* (2). Let f be a continuous M-valued function on S such that $f(S)$ is totally bounded. Let m be a complete metric for M. Then $d = m\circ(f\times f)$ is a continuous pseudometric on S which is also totally bounded (see Exercise 1, Section 17). By 17.10 and (1) there exists a continuous pseudometric d^* on X that extends d. The proof now proceeds exactly as the proof of (1) *implies* (2) of 20.4.

The implications (2) *implies* (3) *implies* (4) *implies* (5) are immediate. To show that (5) *implies* (1) it is sufficient to show, by 17.10 and 17.9, that every totally bounded continuous pseudometric on S extends to a continuous pseudometric on X. Let d be a totally bounded continuous pseudometric on S and define a function f from S into $C^*(S)$ by sending x into the partial function d^x for each x in S. As we saw in the proof of (5) *implies* (1) of 20.4, the function f is continuous. The image of S under f is totally bounded. In fact if S is the union of the d-spheres of radius ϵ centered about the points of the finite set $F_\epsilon \subset X$, then

$$f(S) \subset \bigcup_{x\in F_\epsilon} S(f(x)),$$

where

$$S(f(x)) = \{z \in C^*(S): \|z-f(x)\| < \epsilon\}.$$

The proof now proceeds in a fashion similar to (5) *implies* (1) of 20.4.

If S is C^*-embedded in X, it is now clear that (6) holds. Conversely, if (6) holds, the proof of (6) *implies* (1) of 20.4 and the remarks above pertaining to $C^*(S)$ applied to $C^*(A)$, give the result that S is C^*-embedded in X. □

Since any completely regular T_1-space is C^*-embedded in its Stone–Čech compactification we obtain the following corollary to 20.9. (Recall 5.4.)

20.10 Corollary *The Stone–Čech compactification of a completely regular T_1-space X is that unique compactification βX of X for which every continuous function f from X into a Fréchet space, where the range of f is totally bounded, can be extended to a continuous function on βX.*

We now obtain easily the following result on the extension of uniformly continuous functions with values in a Fréchet space.

20.11 THEOREM *If S is a non-empty subspace of a uniform space X and if L is any Fréchet space, then every uniformly continuous function from S into L can be extended to a continuous function on X.*

Proof. Let f be a uniformly continuous function from S to a Fréchet space L with complete metric m. Then $d = m \circ (f \times f)$ is a uniformly continuous pseudometric on S. Hence by 17.5, d extends to a continuous pseudometric d^* on X. The proof now proceeds as in (1) *implies* (2) of 20.4. □

In [141], an example is given which shows that we cannot request that the extended function also be uniformly continuous. In fact for the case in which the range space is the real line, the boundedness of the function is needed to insure uniform continuity of the extended function.

Now we consider another direction for applying our results on the extension of pseudometrics. Since a pseudometric on a space X is a function on the product set $X \times X$, it is of interest to relate the extension of pseudometrics to the extension of functions (without the triangle inequality) on $X \times X$.

By utilizing the results we have just seen, we can show the following theorem.

20.12 THEOREM *If X is a Tychonoff space then the following statements are equivalent.*

(1) *S is P-embedded in X.*

(2) *The product set $S \times \beta S$ is C^*-embedded in the product space $X \times \beta S$.*

(3) *For all compact Hausdorff spaces A, $S \times A$ is C^*-embedded in $X \times A$.*

For Tychonoff spaces X, this implies that X is P-embedded in υX if and only if $\upsilon(X \times \beta X) = \upsilon X \times \beta X$. Moreover for spaces X of non-measurable cardinality, this also implies that

$$\upsilon(X \times A) = \upsilon X \times A$$

for all compact Hausdorff spaces A.

To give the results alluded to above, the following concept is necessary.

20.13 DEFINITION A topological space X is *hemicompact* if it is the union of a countable collection of compact subsets $(K_i)_{i \in \mathbf{N}}$ such that every compact subset of the space is contained in some K_i.

Compact spaces, Euclidean spaces $\mathbf{R}^n$, and countable direct sums of compact spaces are hemicompact spaces. Our interest in hemicompact spaces can be gleaned from the exercises. We want to consider the set $C(A, B)$ of continuous functions from a topological space A into a topological space B. This set is a vector space if B is a vector space. What is interesting is that if we consider the compact-open topology on $C(A, B)$ (see Exercise 1) and if B is a locally convex linear topological space, then $C(A, B)$ is a locally convex linear topological space. Moreover, if A is a hemicompact, Hausdorff space and if B is a Fréchet space, then $C(A, B)$ is also a Fréchet space (see Exercise 2).

We can now prove the essential result in this line of thinking which is found in [396].

20.14 THEOREM *Let S be a subspace of a completely regular T_1-space X. The following are equivalent.*

(1) *The subspace S is P-embedded in X.*

(2) *For all locally compact, hemicompact Hausdorff spaces A, the product set $S \times A$ is P-embedded in the product space $X \times A$.*

(3) *For all locally compact, hemicompact Hausdorff spaces A, the product set $S \times A$ is C-embedded in the product space $X \times A$.*

(4) *For all locally compact, hemicompact Hausdorff spaces A, the product set $S \times A$ is C^*-embedded in the product space $X \times A$.*

(5) *The product set $S \times \beta S$ is P-embedded in the product space $X \times \beta S$.*

(6) *The product set $S \times \beta S$ is C-embedded in the product space $X \times \beta S$.*

(7) *The product set $S \times \beta S$ is C^*-embedded in the product space $X \times \beta S$.*

Proof. To show that (1) *implies* (2), let A be a locally compact, hemicompact Hausdorff space. By the equivalence of (1) and (4) in 20.4 and the remark after 20.4 relating to P-embedding, it is sufficient to prove that if f is a continuous function from the product set $S \times A$ into a Fréchet space B, then f extends to a continuous function on $X \times A$. Let f be a continuous function from $S \times A$ into a Fréchet space B, and define a map ϕ from S into $C(A, B)$ by $(\phi(x))(a) = f(x, a)$ for all a in A and all x in S. Then ϕ is continuous by Exercise 3 and $C(A, B)$ is a Fréchet space by Exercise 2. By assumption S is P-embedded in X. Hence by 20.4 the map ϕ extends to a continuous function ϕ^* from X into $C(A, B)$. Define a map f^* from $X \times A$ into B by

$$f^*(x, a) = (\phi^*(x))(a)$$

for all (x, a) in $X \times A$. The function f^* is an extension of f and is continuous by Exercise 3.

The implications (2) *implies* (3) *implies* (4) *implies* (7), and (2) *implies* (5) *implies* (6) *implies* (7) are clear. It remains to show that (7) *implies* (1). By 20.4 it is sufficient to prove that every bounded continuous function from S into $C^*(S)$ extends to a continuous function on X. Let ϕ be such a function. The Banach space $C^*(S)$ is isomorphic to $C^*(\beta S)$ by the mapping $f \to f^\beta$ which assigns to every bounded real-valued continuous function f on S its unique extension f^β to βS. Hence we may think of ϕ as mapping S into $C^*(\beta S)$.

Define f from $S \times \beta S$ into $\mathbf{R}$ by $f(x, z) = (\phi(x))(z)$ for all (x, z) in $S \times \beta S$. The map f is continuous by Exercise 3. Since ϕ is bounded, there is a constant K such that $\|\phi(x)\| \leqslant K$ for all x in S. Hence

$$\sup_{z \in \beta S} |\phi(x)(z)| = \sup_{z \in \beta S} |f(x, z)| \leqslant K$$

for all x in S. Therefore, f is a bounded continuous function on $S \times \beta S$. By hypothesis f extends to a continuous real-valued function f^* on $X \times \beta S$. Defining a function ϕ^* from X into $C^*(\beta S)$ by $(\phi^*(x))(z) = f^*(x, z)$ for all x in X and z in βS, we see that ϕ^* is continuous by Exercise 3 and is therefore the desired extension of the map ϕ. □

As a result of this theorem extending pseudometrics from a subspace S to the space X is the same as extending bounded continuous real-valued functions from $S \times A$ to $X \times A$ where A is a compact Hausdorff space.

20.15 Corollary *Let S be a subspace of a completely regular T_1-space X. Then S is P-embedded in X if and only if $S \times A$ is C^*-embedded in $X \times A$ for all compact Hausdorff spaces A.*

Now 20.14 gives some additional characterizations of Tychonoff spaces which are collectionwise normal. Recall that a space is collectionwise normal if and only if every closed subset is P-embedded. We state here two characterizations for these spaces which result from 20.14.

20.16 Corollary *Let X be a completely regular T_1-space. The following are equivalent.*

(1) *The space X is collectionwise normal.*

(2) *For all locally compact hemicompact Hausdorff spaces A and for all closed subsets F of X, the product set $F \times A$ is C^*-embedded in $X \times A$.*

(3) *For all closed subsets F of X the product set $F \times \beta F$ is C^*-embedded in $X \times \beta F$.*

The other characterizations can be obtained in a likewise fashion. For the Hewitt–Nachbin realcompletion of a product the following result is obtained. For further results in this area, the interested reader is referred to [75] as well as the more recent papers [241], [396], [399], and [402].

20.17 Corollary *If X is a completely regular T_1-space, then X is P-embedded in υX if and only if $\upsilon(X \times \beta X) = \upsilon X \times \beta X$. Moreover, if X has non-measurable cardinality, then*

$$\upsilon(X \times A) = \upsilon X \times A$$

for all compact Hausdorff spaces A.

Proof. If X is P-embedded in υX, then by 20.14 the space $X \times \beta X$ is C-embedded in $\upsilon X \times \beta X$. The product of a compact Hausdorff space and a realcomplete space is realcomplete. Therefore, $\upsilon X \times \beta X$ is a realcomplete space in which $X \times \beta X$ is dense and C-embedded. By the uniqueness of the Hewitt–

Nachbin realcompletion, it follows that $\upsilon(X \times \beta X) = \upsilon X \times \beta X$. Conversely, if $\upsilon(X \times \beta X) = \upsilon X \times \beta X$, then $X \times \beta X$ is C-embedded in $\upsilon X \times \beta X$. Therefore, by 20.14 it follows that X is P-embedded in υX.

To prove the second statement we recall 15.3. Consequently if X has non-measurable cardinality then X is P-embedded in υX. Thus the statement follows from (1) *implies* (2) of 20.14. This completes the proof. □

Now recall that every C^*-embedded subset is z-embedded but the converse is not true. In 20.14, statement (7) cannot be improved by stating 'the product set $S \times \beta S$ is z-embedded in the product space $X \times \beta S$'. In fact an equivalence to (1) cannot be obtained even if S is required to be closed. The following examples are instructive.

Let $\mathbf{R}$ be the real line and let S be the open interval $(0, 1)$. Since S is a cozero-set of $\mathbf{R}$ and since $S \times \beta S$ is a cozero-set of $\mathbf{R} \times \beta S$, then S is z-embedded in $\mathbf{R}$ and $S \times \beta S$ is z-embedded in $\mathbf{R} \times \beta S$ (see 7.9). However S is not P-embedded in $\mathbf{R}$ since it is not C-embedded.

Now let X be the Tychonoff plank, that is

$$X = [0, \Omega] \times [0, \omega] - (\Omega, \omega)$$

and let F be the closed subset $\{(\Omega, \alpha) \colon \alpha < \omega\}$. Since F is a Lindelöf subset of X and since $F \times \beta F$ is a Lindelöf subset of $X \times \beta F$, it follows that $F \times \beta F$ is z-embedded in $X \times \beta F$. However, F is not P-embedded in X since it is not C^*-embedded in X.

In [366], Tamano proved that if BX is any Hausdorff compactification of a completely regular T_1-space X, then X is paracompact if and only if $X \times BX$ is normal. Since $X \times BX$ is paracompact if X is paracompact, this can be restated as: a Tychonoff space X is paracompact if and only if $X \times BX$ is normal if and only if $X \times BX$ is paracompact. The next theorem (see [396]) is a parallel result for collectionwise normality. If A is a compact Hausdorff space in which the space X is C^*-embedded, this result will show that the collectionwise normality of X is equivalent to the following conditions: (1) For all closed subsets F of X, the product $F \times A$ is C^*-embedded in $X \times A$, and (2) for all closed subsets F of X, the product $F \times A$ is P-embedded in $X \times A$. We know that every closed subset of a normal space is C^*-embedded and every closed subset of a paracompact Hausdorff space is P-embedded. Hence it is clear that these conditions are weaker than the normality or paracompactness of $X \times A$.

20.18 THEOREM *Let A be a compact Hausdorff space in which X is C^*-embedded. The following statements are equivalent.*

(1) *The space X is collectionwise normal.*

(2) *For all closed subsets F of X, the product set $F \times A$ is P-embedded in the product space $X \times A$.*

(3) *For all closed subsets F of X, the product set $F \times A$ is C^*-embedded in the product space $X \times A$.*

Proof. The implication (1) *implies* (2) follows from (1) *implies* (2) of 20.14 and the fact that every closed subset of a collectionwise normal space is P-embedded in it. The implication (2) *implies* (3) is immediate. Hence it suffices to show that (3) *implies* (1).

It is easy to show that (3) *implies* that every closed subset of X is C^*-embedded in X; hence X is normal. Let $(F_\alpha)_{\alpha \in I}$ be a discrete family of closed subsets of X. For each α in I, the sets F_α and $H_\alpha = \bigcup_{\beta \neq \alpha} F_\beta$ are disjoint closed subsets of X. For each α in I, by the normality of X, there is an $f_\alpha \in C(X, E)$ such that $f_\alpha(x) = 0$ if $x \in F_\alpha$, and $f_\alpha(y) = 1$ if $y \in H_\alpha$. Since X is C^*-embedded in A, let $f_\alpha^* \in C(A, E)$ such that $f_\alpha^* | \{x\} = f_\alpha$. Let F be the union of F_α for α in I. Let us note that F is a closed subset of X.

Define a real-valued function f on $F \times A$ by $f(x, a) = f_\alpha^*(a)$ for the unique α in I such that $f_\alpha(x) = 0$. The function f is clearly bounded and continuous. To see this, let (x_0, a_0) be in $F \times A$, let $\epsilon > 0$, and suppose that x_0 is in F_α. There is a neighborhood U of a_0 such that if a is in U, then $|f_\alpha^*(a) - f_\alpha^*(a_0)| < \epsilon$. If (x, a) is in $F_\alpha \times U$, then $|f(x, a) - f(x_0, a_0)| < \epsilon$. Therefore by assumption f extends to a continuous real-valued function f^* on $X \times A$.

Define g from X into $C(A)$ by $(g(x))(a) = f^*(x, a)$ for all a in A and x in X. By Exercise 3 the function g is continuous. Hence the pseudometric d^* defined on X by $d^*(x, y) = \|g(x) - g(y)\|$ for all x, y in X is continuous. Let $G_\alpha = \bigcup_{x \in F_\alpha} S(x; d^*, \frac{1}{4})$ for all α in I. Then G_α is an open subset of X containing F_α. It remains to show that $(G_\alpha)_{\alpha \in I}$ is a pairwise disjoint family.

Suppose t is in G_α and G_β where $\alpha \neq \beta$. Then there exist x in F_α and y in F_β such that $d^*(t, x) < \frac{1}{4}$ and $d^*(t, y) < \frac{1}{4}$. Hence $\|g(t) - g(x)\| < \frac{1}{4}$ which means that

$$\sup_{a \in A} |f^*(t, a) - f^*(x, a)| < \tfrac{1}{4}.$$

Similarly,

$$\sup_{a \in A} |f^*(t, a) - f^*(y, a)| < \tfrac{1}{4}.$$

Let $a = x$. Then

$$|f^*(t, x) - f^*(x, x)| = |f^*(t, x) - f(x, x)| = |f^*(t, x)| < \tfrac{1}{4}$$

and

$$|f^*(t, x) - f^*(y, x)| = |f^*(t, x) - f(y, x)| = |f^*(t, x) - 1| < \tfrac{1}{4},$$

which is a contradiction. This completes our proof. □

Now as a corollary we have Tamano's original result (see [366]).

20.19 Corollary *A completely regular T_1-space X is collectionwise normal if and only if $F \times \beta X$ is C^*-embedded in $X \times \beta X$ for all closed subsets F of X.*

20.20 Exercises

(1) Let A and B be topological spaces and let $C(A, B)$ denote the set of continuous functions from A to B equipped with the *compact-open* topology which we now define. A subbase for this topology is the collection $\mathscr{U}$ of sets

$$(K, W) = \{f \in C(A, B) : f(K) \subset W\},$$

where K is a compact subset of A and W is an open subset of B. If A is a Hausdorff space and if B is a locally convex linear topological space, then $C(A, B)$ is also a locally convex linear topological space. (For each compact set K and each f in the linear space $C(A, B)$, $f(K)$ is bounded in B. Thus for $f \in C(A, B)$, K a compact subset of A, and W an open subset of B, there is a $\lambda > 0$ such that $f(K) < \lambda W$. Hence $f \in (K, \lambda W) = \lambda(K, W)$. The collection $\mathscr{U}$ of all subsets (K, W) in $C(A, B)$ for K a compact subset of A and W an open subset of B, is a local neighborhood base of the zero vector 0 in $C(A, B)$. See the exercises in Section 1. See also pages 257 ff. of [IX].)

(2) Let A be a hemicompact Hausdorff space and let B be a Fréchet space. The space $C(A, B)$ of Exercise 1 with the compact open topology is a Fréchet space. (From (1) it remains to show that $C(A, B)$ is complete and metrizable. Since A is hemicompact there is a countable local base for the zero vector and therefore for the topology of the Tychonoff space $C(A, B)$.)

(3) Let X, A and B be topological spaces and let $C(A, B)$ be as in Exercises 1 and 2. Let f be a continuous function from $X \times A$ into B. The function ϕ defined from X into $C(A, B)$ by

$$(\phi(x))\,(a) = f(x, a)$$

for all x in X and a in A is continuous. Conversely, if A is regular and locally compact, and if ϕ is a continuous map from X into $C(A, B)$, then the map f from $X \times A$ into B defined by

$$f(x, a) = (\phi(x))\,(a)$$

for all (x, a) in $X \times A$ is continuous (see Fox [121]).

Appendix

Introduction

In both the past and present development of set theoretical topology the allusiveness of the solution of several problems has encouraged new developments to arise in both the orientation of this branch of topology and the application of its results to other areas. To establish the themes of this appendix, let us look at some of these areas.

In applications to analysis, metric spaces and their investigation are a central part of the study. There are also many reasons for wanting to generalize metric spaces, that is, to have a topological structure that will permit applications to more structures of analysis. Also the topologist, whose interests may lie wholly within his subject or who is motivated by applications has many reasons for wanting to look to more general topological structures. We will allude to some of these in our discussion.

Many classes of topological spaces are hereditary (that is, subspaces of a member of the class belong to the class), and either finitely or arbitrarily productive (that is, the product of either a finite or arbitrary subcollection of members of the class belong to the class). However, the class of normal topological spaces is neither hereditary nor finitely productive (see Section 9, Examples 2 and 4). Such properties (and a few others that will be mentioned below) are significant and desirable in the development of the theory.

On the other hand, as a generalization of metric spaces, normal spaces have many significant properties. Every continuous real-valued function on a closed subspace of such a space is the restriction of such a function (with no expansion of the range) on the entire space (see 6.8). For any open point-finite cover $\mathscr{U}$ there is an open cover $\mathscr{V}$ such that closures of members of $\mathscr{V}$ refine $\mathscr{U}$ (see 10.1). Thus for normal spaces, closed subsets are

C- (C^*-) embedded and open covers are strongly screened. If one considers proximity spaces (see [xxv]), normal topological spaces are those spaces for which disjoint subsets are 'distant' with respect to the Čech proximity of the space. Spaces which are generalizations of pseudometric spaces would be worthwhile if these properties were also satisfied.

Let us return then to the theme of the penultimate paragraph. If one requires that our generalization be compact then one may obtain finite productivity since the product of a finite number of compact Hausdorff spaces (and therefore normal spaces) is a compact Hausdorff space (and again, is normal). However, generalizing the compactness to paracompactness deprives us of the productivity just obtained, for E. Michael has shown that not even the product of a paracompact Hausdorff (and therefore normal) space and a metric space (which is always paracompact) need be normal and thus need not be paracompact (see [268] and 1.26(9)). Both compact spaces and metric spaces are paracompact. As one observes very early in elementary topology and/or analysis, compact spaces and metric spaces possess very significant properties, some of which carry over very nicely to the more general class of paracompact spaces. The example of Michael demonstrates its unsuitability for productivity, yet the topologist is motivated to find a 'reasonable' class of spaces which will include the compact spaces and the metric spaces.

The image of a metric space under a closed continuous map need not be a metric space (see the w-star space defined in 22.8). Similarly the sum of a countable number of closed metric subspaces need not be metrizable. In topology and analysis both of these operations are significant and spaces which are obtained as a consequence of these operations also have important properties. However metric spaces are not general enough to include these.

In the realm of dimension theory there is a satisfactory theory for general metric spaces. In the case of spaces that are not metrizable, or even in the case of paracompact spaces, the theory is not satisfactory. In such spaces there are several concepts for the notion of dimension. An important theorem that one would like to obtain is the validity of the product theorem, that is,

$$\text{(P)} \qquad \dim (X \times Y) \leqslant \dim X + \dim Y.$$

The conditions yielding this result appear to be attached to the normality of the product $X \times Y$. Thus the search for a decent generalization of metric space for which the topologist will have a meaningful theory of dimension should yield this result and the inherent normality. One reasonable class of spaces that have evolved are the paracompact M-spaces. It is unresolved as to whether or not (P) will be true for this class of spaces.

With regard to the question of productivity let us mention one very nice result. In 1948, Stone (see [359]) showed that for an arbitrary product of metric spaces, the normality of the Tychonoff topology is equivalent to anyone of the following: collectionwise normality, paracompactness, the product has at most a countable number of non-compact spaces as factors (that is, the product is homeomorphic to the product of a metric space and a compact space if the cardinality of the index set is greater than $\aleph_0$). Hence in the search for a reasonable generalization of metric spaces and the relation of it to the question of productivity, one should be mainly concerned with the question of productivity for a countable collection of factor spaces.

For those who are interested in homotopy theory and its applications, the question of productivity for normality is pertinent for a finite number of factor spaces. Here the problem was to determine the class of Hausdorff topological spaces X for which the Tychonoff product space $X \times E$ is a normal space. In 1951, Dowker (see [99]) resolved this question: that is for $X \times E$ to be a normal space it is necessary and sufficient that X be a countably paracompact normal space. As a further weakening (over the concept of paracompactness) of compactness it is interesting to see that countable paracompactness is sufficient for the normality of $X \times E$.

On the other hand Dowker's solution promoted a question which was for a long period a hard unsolved problem in set theoretical topology. Dowker asked if there existed a normal Hausdorff space that was not countably paracompact. Recently this problem was resolved by Rudin (see [333] and Example 8, Section 9). Prior to this solution, the problem generated much interest and led to various new classes of spaces and deeper insights into the concepts generalizing compactness (e.g., paracompactness, countable paracompactness, and the concepts generalizing metric spaces).

Here then the discussion will begin in leading us to the ideas mentioned above. In a book such as this, practicality inhibits one from including many of the ideas that seem worthwhile. Consequently selections must be made and because of this, many concepts will be omitted however small the intention to slight them. We have had to make some choices and so a partial defense has been established to placate the objectors.

We will begin then with Section 21, which will concentrate on countable paracompactness and its characterizations. We will relate it to the Dowker problem. We will also consider some of the classes of spaces whose definitions are motivated by it and their relationship to generalized metric spaces. In Section 22 we will concentrate on generalized metric spaces, giving definitions and results of some of the more interesting concepts. It is hoped that the reader will be motivated, by reading these sections, for further research and study in these areas. Consequently we feel it wise to leave the proofs, for the most part, to the reader (where some hints have been provided) or to state the appropriate references where the proofs may be found. We hope that in this way much will be gained.

21 Countably paracompact spaces and related questions

One of the nicest characterizations of paracompactness is that which deals with partitions of unity, that is, suitable collections of continuous real-valued functions. Let us keep this in mind as we move forward in our considerations for weakened versions of paracompactness.

The concept of countable paracompactness generalizes paracompactness by restricting the class of open covers of the space to countable covers. That is, a space is *countably paracompact* if every countable open cover has a locally finite open refinement.

This restriction, however, was potent enough to permit Dowker in [99] to prove the following result: a Hausdorff topological space X is countably paracompact and normal if and only if the product space $X \times E$ is normal if and only if $X \times Y$ is normal for any compact metric space Y. Later, in [101] he showed that normality in this theorem could be replaced by collectionwise normality.

This result brought Dowker to make an interesting conjecture: *is every normal Hausdorff space countably paracompact?* This conjecture had been one of the most difficult problems facing set theoretical topology for almost twenty years. It has recently been resolved by Rudin in [333] where she exhibits a collectionwise normal (and therefore normal) Hausdorff space that is not countably paracompact. We have exhibited this example in Section 9.

Thus we see how countably paracompactness relates to the normality (collectionwise normality) of a very important product space – especially of importance to those who are interested in homotopy theory. Furthermore, there is much evidence that one cannot omit collectionwise normality when discussing metrization theory or extension theory of mappings. Hence the definition of Katětov (see [225]) that a topological space X is *strongly normal* if it is countably paracompact and collectionwise normal.

In considering the class of strongly normal spaces, the following concept, stronger than P-embedding, will characterize this class of spaces in a manner similar to a characterization for the class of collectionwise normal spaces (see [20]).

21.1 Definition A non-empty subset S of a topological space X is *strongly P-embedded in X* if every σ-locally finite open cover of S has a refinement that can be extended to a locally finite cozero-set cover of X.

It is clear that a strongly P-embedded subset is always P-embedded. It follows also that compact subsets of Tychonoff spaces are strongly P-embedded since in such spaces compact sets and closed sets are completely separated.

At first glance it appears that the concept 'strongly P-embedded' is 'too close' to the concept of 'P-embedded' to be useful. However, the facts are that 'strongly P-embedding' is to 'strongly normal spaces' as 'P-embedding' is to 'collectionwise normal spaces' as 'C- (or C^*-) embedding' is to 'normal spaces'. Moreover using some previous results it will follow as a corollary of 21.3 that the concept of 'strongly P-embedded' is definitely stronger than P-embedding.

Motivated by the definition of strongly P-embedded subspaces we define the notion of a strongly C-embedded subspace.

21.2 DEFINITION Let S be a non-empty subset of a topological space X. The subset S is *strongly C-embedded in X* if every countable open cover of S has a refinement that can be extended to a locally finite cozero-set cover of X.

Note that $\mathscr{U}$ is a countable σ-locally finite open cover if and only if $\mathscr{U}$ is a countable open cover. Thus S is strongly C-embedded in X if and only if every countable σ-locally finite open cover has a refinement that can be extended to a locally finite cozero-set cover of X. Consequently every strongly P-embedded subset is also strongly C-embedded. A strongly C-embedded subset is also C-embedded.

Let us now consider strongly normal spaces and normal countably paracompact spaces in the following way. (Note that we are not assuming that X is a Hausdorff space.) The concepts of strongly P-embedding and strongly C-embedding imply P-embedding and C-embedding, respectively. Some consideration of the converse implications may be obtained.

21.3 THEOREM *If S is a non-empty subspace of a topological space X then the following statements are equivalent.*

(1) *The subspace S is countably paracompact, normal, and C-embedded in X.*

(2) *The subspace S is strongly C-embedded in X.*

Moreover, if in (1) *it is assumed that S is also P-embedded in X, then this is equivalent to*:

(2′) *The subspace S is strongly P-embedded in X.*

Proof. The statement (1) *implies* (2) follows from 11.7, 10.10, and 16.1, and the fact that refinements of C-embedded subsets extend to locally finite cozero-set covers on the whole space.

Moreover, a subset S of a topological space X is P-embedded in X if and only if every locally finite cozero-set cover of S has a refinement that can be extended to a locally finite cozero-set cover of X.

The converse is easily shown using 11.7. □

21.4 Corollary *A topological space is countably paracompact and normal if and only if every closed subspace is strongly C-embedded in the space.*

Now we may characterize strongly normal spaces as was intended.

21.5 Corollary *A topological space is strongly normal if and only if every closed subset is strongly P-embedded.*

Proof. Closed subsets of collectionwise normal spaces are always P-embedded (see 15.7). □

21.6 Corollary *A T_1-space X is countably paracompact and normal if and only if it is strongly C-embedded in its Hewitt–Nachbin realcompletion.*

Moreover, if the cardinality of X is non-measurable then X is countably paracompact and normal if and only if it is strongly P-embedded in X.

Proof. Dense C-embedded subsets of non-measurable cardinality are also P-embedded (see 15.3). □

Thus restating Dowker's result, one has that the product of a compact metric space with any strongly C-embedded subset is a normal space, whereas the product of a compact metric space with a strongly P-embedded subset is collectionwise normal.

Now the Tychonoff plank, T, is a countably paracompact P-embedded subset in $\upsilon T = \beta T$. Since T is not normal, T is neither strongly C-embedded nor strongly P-embedded in βT. Another example is a collectionwise normal Hausdorff space Y (see 9.7) that is not countably paracompact. Since Y is collectionwise normal every closed subset is P-embedded (and therefore C-embedded). Consequently, there must be some closed subsets of Y which are not strongly C-embedded in Y and some which are not strongly P-embedded in Y. Thus a C-embedded subset need not be strongly C-embedded nor need a P-embedded subset be strongly P-embedded. On the other hand an example of a countably paracompact normal space that is not collectionwise normal will show that there are strongly C-embedded subsets that are not strongly P-embedded and some which are not P-embedded. Such an example of a Hausdorff space which is countably paracompact, normal but not collectionwise normal was discussed in Section 9.

We have now set the theme for considering a group of characterizations of spaces that are countably paracompact and normal, of normal spaces that are countably paracompact and of spaces that are solely countably paracompact. First, we need to give the definitions that pervade the literature. In the characterizations we will see how the concepts play their role.

We have emphasized the role of continuous functions in the realm of paracompactness. The following classes of functions will become important in the realm of countable paracompactness.

21.7 DEFINITION A real-valued function f on a topological space X is *locally bounded* if each point of X has a neighborhood on which the function is bounded. The space X is called a cb-*space* if for each locally bounded function h, there is an $f \in C(X)$ for which $|h| \leqslant f$.

J. G. Horne initiated the study of cb-spaces (see [186]) and J. E. Mack extended their study (see [246]). They are a class of countably paracompact spaces that emphasize the role of real-valued functions. In fact when normality is added to the hypothesis that the space be countably paracompact, then the space must be a cb-space. Before we see how this is possible, let's consider a weakening of the concept considered by Mack and Johnson in [250].

21.8 DEFINITION A topological space X is a *weak* cb-*space* if every locally bounded lower semi-continuous function h on X is bounded above by a continuous function, that is, there is an $f \in C(X)$ such that $|h| \leqslant f$.

It is clear that every cb-space is a weak cb-space. In fact a space is a cb-space if and only if it is both a countably paracompact space and a weak cb-space. Since normality and countable paracompactness also imply that the space is a cb-space, one may take the example of an uncountable product of real lines which is a non-normal weak cb-space. The Tychonoff plank is a weak cb-space that is not countably paracompact and in Exercise **3** we will state an example of a countably paracompact non-weak cb-space.

Furthermore the importance of weak cb-spaces is demonstrated by the fact that for a Tychonoff space X, the Dedekind completion of $C(X)$ is isomorphic to some $C(Y)$ if and only if υX is a weak cb-space (see [250]). In [250] there is also constructed a realcomplete space that is not a weak cb-space. It is not known whether every normal space is a weak cb-space.

The closeness of the concept of a cb-space to that of countably paracompactness as stated above is even more evidenced by the following characterizations of cb-spaces and weak cb-spaces.

21.9 THEOREM *If X is a topological space then the following statements are equivalent.*

(1) *The space X is a weak* cb-*space.*

(2) *Every increasing regular open cover of X has a locally finite partition of unity subordinate to it.*

(3) *Every increasing regular open cover of X has a countable partition of unity subordinate to it.*

(4) *Every increasing regular open cover of X has a locally finite cozero-set refinement.*

(5) *Each increasing regular open cover of X has a countable cozero-set refinement.*

(6) *Given a decreasing sequence $(F_n)_{n\in\mathbf{N}}$ of regular closed sets with empty intersection, there exists a sequence $(Z_n)_{n\in\mathbf{N}}$ of zero-sets with empty intersection such that $F_n \subset Z_n$ for all $n \in \mathbf{N}$.*

The results in 21.9 are found in [250]. Considering these characterizations, it is interesting to see as in the next theorem, how the countability of the open cover changes the concept from that of a weak cb-space to that of a cb-space.

21.10 THEOREM *For any topological space X, the following statements are equivalent.*

(1) *The space X is a* cb-*space.*

(2) *Given an upper semi-continuous function h on X there exists $f \in C(X)$ such that $h < f$.*

(3) *Given a positive (non-vanishing) lower semi-continuous function g on X there exists $f \in C(X)$ such that $\mathbf{0} < f \leqslant g$.*

(4) *Every countable increasing open cover of X has a locally finite partition of unity subordinate to it.*

(5) *Every countable increasing open cover of X has a (countable) partition of unity subordinate to it.*

(6) *Each countable increasing open cover of X has a locally finite cozero-set refinement.*

(7) *Each countable increasing open cover of X has a σ-locally finite cozero-set refinement.*

(8) *Each countable increasing open cover of X has a countable cozero-set refinement.*

(9) *Given a decreasing sequence $(F_n)_{n \in \mathbf{N}}$ of closed sets in X with empty intersection, there exists a sequence $(Z_n)_{n \in \mathbf{N}}$ of zero-sets with empty intersection such that $F_n \subset Z_n$ for each $n \in \mathbf{N}$.*

In the above characterizations for a cb-space, the deleting of the word 'increasing' in statements (4) through (8) and the deleting of the word 'decreasing' in statement (9) provides a characterization of normal and countably paracompact spaces. The results of 21.10 may be found in [246]. Of course in normal spaces, locally finite open refinements have locally finite cozero-set refinements (see 11.7). Thus 21.10 shows readily that countably paracompact normal spaces must be cb-spaces. It is easy to show that

$$W(w_1) \times W(w_1 + 1)$$

is a cb-space that is not normal.

In normal spaces zero-sets are equivalent to closed G_δ-sets. Consequently one would expect that the role of the zero-sets in 21.10 might be replaced by closed G_δ-sets in normal countably paracompact spaces.

21.11 THEOREM *If X is a Hausdorff topological space then the following statements are equivalent.*

(1) *X is countably paracompact and normal.*

(2) (*Mansfield* [258]) *Every countable open cover of X has a countable open star-refinement.*

(3) (*Shapiro* [337]) *Every countable open cover has a countable star-finite normal cozero-set star-refinement.*

(4) (*Morita* [279]) *Every countable open cover of X is normal.*

(5) (*Ishii* [204]) *Every countable open cover of X has a (locally finite) partition of unity subordinate to it.*

(6) (*Alò and Shapiro* [20]) *Every closed subset of X is strongly C-embedded in X.*

(7) (*Dowker* [99]) *Every countable open cover* $(U_n)_{n\in\mathbf{N}}$ *of X admits an open refinement* $(V_n)_{n\in\mathbf{N}}$ *such that* $\operatorname{cl} V_n \subset U_n$.

(8) (*Dowker* [99]) *The product space* $X \times E$ *is normal.*

(9) (*Tamano* [367]) *The product space* $X \times M$ *is normal for some compact metric space M containing infinitely many points.*

(10) (*Dowker* [99]) *If g is a lower semi-continuous function and if h is an upper semi-continuous function, and* $h < g$, *there exists an* $f \in C(X)$ *such that* $h < f < g$.

This theorem can be shown readily by the reader using 11.7 and 10.10.

Now we turn our attention to the class of normal spaces and we characterize those that are countably paracompact. However, there are results obtainable under somewhat weaker separation axioms. For example in [345], Singal and Arya use Frolík's definition of a weakly regular space to show that if a space is either regular or weakly regular than X is countably paracompact if and only if every proper regular closed subset is countably paracompact.

The following weakening of the concept of closure preserving refinements has had considerable interest. We give its definition and the desired characterization.

21.12 Definition Suppose that

$$\mathscr{U} = (U_\alpha)_{\alpha\in I} \quad \text{and} \quad \mathscr{V} = (V_\beta)_{\beta\in J}$$

are two families of subsets of a space X. We say that $\mathscr{V}$ is a *cushioned refinement of* $\mathscr{U}$ if there exists a function $\pi: J \to I$ such that for all $K \subset J$

$$\text{cl}\,(\bigcup_{\beta \in K} V_\beta) \subset \bigcup_{\beta \in K} U_{\pi(\beta)}.$$

The family $\mathscr{V}$ is a *σ-cushioned refinement of* $\mathscr{U}$ in case $\mathscr{V} = \bigcup_{n \in \mathbf{N}} \mathscr{V}_n$ and each $\mathscr{V}_n$ is cushioned in $\mathscr{U}$.

21.13 THEOREM *If X is a normal topological space then the following statements are equivalent.*

(1) *The space X is countably paracompact.*

(2) (*Mansfield* [258]) *Every countable open cover of X has a countable locally finite closed refinement.*

(3) (*Mansfield* [258]) *Every countable open cover of X has a countable closure preserving closed refinement.*

(4) (*Mansfield* [258]) *Every countable open cover of X has a σ-discrete closed refinement.*

(5) (*Mansfield* [258]) *Every countable open cover of X has a σ-locally finite closed refinement.*

(6) (*Mansfield* [258]) *Every countable open cover of X has a σ-closure preserving closed refinement.*

(7) (*Iseki* [198]) *Every countable open cover of X has a countable star-finite open star-refinement.*

(8) (*Mansfield* [258]) *Every countable open cover of X is even.*

(9) (*Alò and Shapiro* [20]) *Every σ-star-finite open cover of X has a closure preserving open refinement.*

(10) (*Alò and Shapiro* [20]) *Every σ-discrete open cover of X has a closure preserving open refinement.*

(11) (*Alò and Shapiro* [20]) *Every σ-discrete open cover of X has a cushioned open refinement.*

(12) (*Swaminathan* [365]) *Every countable open cover of X has a cushioned open refinement.*

(13) (*Swaminathan* [365]) *Every countable open cover of X has a σ-cushioned open refinement.*

(14) (*Dowker* [99]) *Every countable open cover of X has a point finite open refinement.*

(15) (*Ishii* [205]) *Every countable filter base* $\mathscr{F}$ *in X whose image has a cluster point in any metric space into which X is continuously mapped, has a cluster point in X.*

Proof. The proofs of (2) *implies* (3) *implies* (4) *implies* (5) *implies* (6); (9) *implies* (10) *implies* (11) *implies* (12) *implies* (13) are immediate. Theorem 11.7 assists in the proof of (1) *implies* (2) and (1) *implies* (14). Corollary 11.5 assists in proving that (6) *implies* (7), (14) *implies* (1), and (8) *implies* (1). The reader can readily supply the proof of (7) *implies* (8). We refer the reader to [365] for the proof of (13) *implies* (1) and to [205] for the equivalence of (1) and (15). □

The word 'closed' in conditions (4), (5), and (6) cannot be replaced by 'open'. A countable open cover of an arbitrary topological space is a σ-discrete refinement of itself. Thus if we could replace 'closed' with 'open' we would have that every normal space is countably paracompact.

The word 'closed' in (3) cannot be replaced by 'open'. The real line with base $\{G_a = \{x \in \mathbf{R}: x < a\}: a \in \mathbf{R}\}$ has the property that every collection of open subsets is closure preserving. *It may be that* a normal Hausdorff space is countably paracompact if and only if each countable open cover has a countable closure preserving open refinement. Conditions (1), (12), and (13) might be pertinent here.

Note also that it is not true that a normal space X is countably paracompact if and only if each countable open cover of X has a locally finite refinement since each countable open cover of an arbitrary space has a locally finite refinement.

Finally the reader is asked to consider the necessity of the normality condition in 21.13. It would be good to compare the results of 21.13 with those in 21.11.

In 21.11, normal countably paracompact spaces were characterized in terms of the Dowker result that $X \times E$ must be normal. Considering this result in conjunction with cb-spaces and weak cb-spaces we have several interesting concepts.

21.14 Definition Suppose that X is a topological space, that S is a subspace of X, and that γ is an infinite cardinal number. We say that S is an *S_γ-set* (respectively, *regular S_γ-set*) in case S is the intersection of at most γ open sets (respectively, at most γ closed sets whose interiors contain S). If $\gamma = \aleph_0$, we then have the more familar terms *G_δ-set* and *regular G_δ-set*. The space X is *γ-normal* if for every pair of disjoint closed sets one of which is a regular G_γ-set, there are disjoint neighborhoods containing them. For $\aleph_0$-normal we shall use the more suggestive term *δ-normal*

It is clear that a zero-set of any continuous real-valued function is a regular G_δ-set and that the intersection of no more than γ such zero-sets is a regular G_γ-set. Moreover, a normal space is clearly γ-normal and a regular space is normal if and only if it is γ-normal for all infinite cardinal numbers γ. On the other hand, a compact T_1-space that is not Hausdorff is γ-normal for all infinite cardinal numbers γ but fails to be normal. It can be shown that a γ-paracompact space is γ-normal and thus, in particular, every countably paracompact space is δ-normal.

Now considering Dowker's result on the normality of the product space $X \times E$, the following theorem is most interesting (see [249]).

21.15 THEOREM *A topological space is countably paracompact if and only if its product with the closed unit interval is δ-normal.*

21.16 DEFINITION A subset F of a space X is *δ-normally separable* if each zero-set disjoint from it is completely separated from it. The space X is *δ-normally separated* (respectively, *weakly* (or *wδ-*) *δ-normally separated*) if each closed set (respectively, regular closed set) is δ-normally separable.

For a topological space in which every regular G_δ-set is a zero-set, δ-normally separated implies δ-normal but not conversely. On the other hand Hewitt's example [181] of an infinite regular Hausdorff space on which every continuous real-valued function is constant is a δ-normally separated space that is not δ-normal.

QUESTION For the class of completely regular spaces does δ-normally separated imply δ-normal?

Normality implies δ-normally separated which implies weakly δ-normally separated and the converse for the latter is true for

δ-normal spaces. The reader is referred to [249] where some of the above concepts were introduced.

The concepts δ-normally separated and weakly δ-normally separated are related to cb-spaces and weak cb-spaces in the following interesting way (see [249]).

21.17 THEOREM *If X is a topological space then the following statements hold.*

(1) *The space X is a* cb-*space if and only if $X \times E$ is δ-normally separated.*

(2) *The space X is a weak* cb-*space if and only if $X \times E$ is weakly δ-normally separated.*

If X is normal then $X \times E$ must be normal provided X is δ-normal. This suggests that if X is a regular δ-normally separated space then $X \times E$ might be δ-normal. In [249], Mack observed that if the hypothesis that X be regular is dropped then $X \times E$ need not be δ-normal. However, M. E. Rudin's example of a collectionwise normal Hausdorff space (that is not countably paracompact) is δ-normally separated. If its product with E were δ-normal then by 21.15, this space would have to be countably paracompact.

QUESTION What condition on X is necessary and sufficient for $X \times E^{\gamma}$ to be γ-normal for γ an infinite cardinal number?

The fact that every γ-paracompact space is γ-normal implies that γ-paracompactness of X is sufficient. However it remains unknown whether it is necessary.

A weakening of the notion of paracompactness has recently been defined by Burke in [61]. In our discussion on generalized metric spaces it will be pertinent in the consideration of developable spaces.

21.18 DEFINITION A topological space X is *subparacompact* if every open cover of X has a σ-locally finite closed refinement.

The following characterization places it in the framework of our present endeavors.

21.19 THEOREM *For a topological space X the following statements are equivalent.*

(1) *The space X is subparacompact.*

(2) *For any open cover $\mathscr{U}$ of X there exists a sequence $(\mathscr{U}_n)_{n \in \mathbf{N}}$ of open covers of X such that if $x \in X$, there is an $m(x) \in \mathbf{N}$ and $U \in \mathscr{U}$ with* $\operatorname{st}(x, \mathscr{U}_{m(x)}) \subset U$.

(3) *Every open cover of X has a σ-discrete closed refinement.*

(4) *Every open cover of X has a σ-closure preserving closed refinement.*

Condition (2) above was originally investigated by Arhangelskii in [30] where he called it *σ-paracompact*. Condition (3) was originally investigated by McAuley in [243] where he called a space having this property *F_σ-screenable*. It is clear that paracompact regular spaces are subparacompact.

Some of the important results concerning subparacompact spaces are the following.

21.20 THEOREM *A collectionwise normal subparacompact space is paracompact.*

21.21 THEOREM *A countably compact subparacompact space is compact.*

Proof. This is immediate since a σ-discrete collection in a countably compact space must be countable. □

An example of a normal countably paracompact subparacompact space that is not paracompact was given by Bing and is discussed in [50]. This example is a modification of 9.6.

The concept of an expandable space has recently been introduced by Krajewski (see [231]). These spaces are of special interest when considering the ideas presented here.

DEFINITION Suppose that X is a topological space and that γ is an infinite cardinal number. We say that X is *γ-expandable* if for every locally finite collection $\mathscr{F}$ of closed subsets of X of cardinality at most γ there exists a locally finite collection $\mathscr{G}$ of open subsets of X such that $\mathscr{F}$ is a screen of $\mathscr{G}$. We say that X is *expandable* if X is γ-expandable for all infinite cardinal numbers γ.

First let us observe that a countably compact space is expandable and then that a paracompact space is expandable.

21.22 PROPOSITION *A countably compact topological space is expandable.*

Proof. This follows from the fact that a space X is countably compact if and only if every locally finite collection of subsets is finite (see [231, Theorem 2.7]). □

21.23 THEOREM *If γ is an infinite cardinal number and if X is a γ-paracompact topological space then X is a γ-expandable space* (see [231]).

21.24 COROLLARY *If X is a paracompact topological space, then X is expandable* (see [231]).

The converse to 21.23 is false. The space $W(\omega_1)$ is countably compact (see Section 9) and hence expandable (21.22). However, the open cover $\{[0, \alpha): \alpha < \omega_1\}$ has no open locally finite refinement. Thus $W(\omega_1)$ is an expandable normal space that is not $\aleph_1$-paracompact. However, for countably paracompact spaces we have the following.

21.25 THEOREM *A topological space X is $\aleph_0$-expandable if and only if it is countably paracompact.*

Proof. Sufficiency follows from 21.23. The necessity is left to the reader. □

The importance of expandable spaces lies in the following result (see [231]).

21.26 THEOREM *A topological space X is normal and expandable if and only if it is collectionwise normal and countably paracompact.*

Modifying Bing's example (see Section 9) one can construct a normal countably paracompact space that is not $\aleph_1$-expandable (see [231]). For a more detailed study of expandable spaces the reader is referred to [231].

Let us complete, now, our considerations for countably paracompact spaces. Putting the results of the previous elaborations together we have characterizations of countably paracompact spaces. We first need the following definition of a directed cover.

21.27 DEFINITION A cover $\mathscr{U}$ of a space X is called a *directed cover* if it is directed by set inclusion, that is, if U and V are in $\mathscr{U}$ then there exists a W in $\mathscr{U}$ such that $U \cup V \subset W$.

21.28 THEOREM *If X is a topological space then the following statements are equivalent.*

(1) *The space X is countably paracompact.*

(2) (*Mack* [246]) *Every countable increasing open cover of X has a locally finite open refinement.*

(3) (*Ishikawa* [208]) *If $(F_n)_{n\in\mathbf{N}}$ is a decreasing sequence of closed sets such that $\bigcap_{n\in\mathbf{N}} F_n = \varnothing$ then there exists a decreasing sequence $(G_n)_{n\in\mathbf{N}}$ of open sets such that $\bigcap_{n\in\mathbf{N}} G_n = \varnothing$ and $F_n \subset G_n$ for all $n \in \mathbf{N}$.*

(4) (*Ishikawa* [208]) *If $(F_n)_{n\in\mathbf{N}}$ is a decreasing sequence of closed subsets of X such that $\bigcap_{n\in\mathbf{N}} F_n = \varnothing$ then there exists a sequence $(G_n)_{n\in\mathbf{N}}$ of open subsets of X such that $\bigcap_{n\in\mathbf{N}} \operatorname{cl} G_n = \varnothing$ and $F_n \subset G_n$ for all $n \in \mathbf{N}$.*

(5) (*Mack* [246]) *If $(U_n)_{n\in\mathbf{N}}$ is an increasing sequence of open subsets of X then there exists an open refinement $(V_n)_{n\in\mathbf{N}}$ such that $\operatorname{cl} V_n \subset U_n$ for all $n \in \mathbf{N}$.*

(6) (*Krajewski* [231]) *The space X is $\aleph_0$-expandable.*

(7) (*Alò and Shapiro* [20]) *Every σ-locally finite open cover of X has a locally finite open refinement.*

(8) (*Alò and Shapiro* [20]) *Every σ-star finite open cover of X has a locally finite open refinement.*

(9) (*Alò and Shapiro* [20]) *Every σ-discrete open cover of X has a locally finite open refinement.*

(10) (*Mack* [249]) *The space $X \times E$ is δ-normal.*

(11) (*Mack* [247]) *If $(U_n)_{n\in\mathbf{N}}$ is a countable directed cover of X then there is a σ-closure preserving open refinement $(V_n)_{n\in\mathbf{N}}$ such that $\operatorname{cl} V_n \subset U_n$.*

(12) (*Tamano* [367]) *If A is closed in $X \times E$ and if K is closed in E such that A and $X \times K$ are disjoint, then A and $X \times K$ have disjoint neighborhoods.*

(13) (*Mack* [246]) *For each locally bounded function h defined on X there exists a locally bounded lower semicontinuous function g such that $|h| \leqslant g$.*

(14) (*Mack* [249]) *If g is a strictly positive lower semicontinuous function then there exist real-valued functions u and v where u is lower semicontinuous and v is upper semicontinuous and such that $\mathbf{0} < u \leqslant v \leqslant g$.*

The reader may try proving these results using the tools developed here. A scheme for the proof might be (1) *implies* (7) *implies* (8) *implies* (9) *implies* (1); (1) *is equivalent to* (6); (1) *implies* (5) *implies* (13) *implies* (2) *implies* (1); (5) *is equivalent to* (4); (1) *is equivalent to* (11); (1) *implies* (10) *implies* (12) *implies* (3) *implies* (1) *and* (13) *implies* (14) *implies* (1).

The implications (7) *implies* (8) *implies* (9) *implies* (1) *implies* (5) are either obvious or almost obvious after some slight thought. Theorem 12.2 is helpful in proving (1) *implies* (6). Slight modification of a proof for (5) *implies* (4) proves the converse. In (1) *implies* (10) we know that X is δ-normal and that $X \times E$ is countably paracompact. In (10) *implies* (12) any closed subset K of E is a regular G_δ-set and thus $X \times K$ is a regular G_δ-set in the δ-normal space $X \times E$. For the consideration of statements (11) and (14), the reader is referred to [247] and [249]. The other implications should be in the grasp of the reader.

We wish to state one final characterization of countably paracompact spaces that involves uniformities.

21.29 Definition A filter $\mathscr{F}$ in a uniform space $(X, \mathscr{U})$ is *weakly Cauchy* if for each $U \in \mathscr{U}$ there exists a filter $\mathscr{H}_U$ such that $\mathscr{F} \subset \mathscr{H}_U$ and $H \times H \subset U$ for some $H \in \mathscr{H}_U$.

21.30 Theorem (*Morita* [270]) *A normal Hausdorff space X is countably paracompact if and only if there exists a uniformity of X such that every countable weakly Cauchy filter has a cluster point.*

Weakly Cauchy filters are also pertinent to the study of paracompact spaces (see Corson [82]).

21.31 Exercises

(1) Show by using results in the text the following theorem of Morita (see [281]). Every σ-locally finite open cover of a countably paracompact normal space is normal.

(2) In [19] a subspace S was said to be *strongly countably paracompact* in the space $(X, \mathcal{T})$ if every countable cover of S which is an open cover in X has a refinement which is locally finite in X. It was shown there that for closed subsets of a normal space the notions of countably paracompactness and strong countably paracompactness are equivalent. Now show that for closed subsets of normal spaces these notions are equivalent to the subset being strongly C-embedded. If in addition the space is collectionwise normal then all of the above notions are equivalent to closed subsets being strongly P-embedded in the topological space.

(3) Let us look at an example interesting in its own right, of a countably paracompact space (that is locally compact) but not a weak cb-space. Let T be a completely regular space and let A and B be closed subsets of T such that $A \cap B$ is compact. In $T \times \mathrm{N}$, identify $A \times \{2n-1\}$ with $A \times \{2n\}$ and $B \times \{2n\}$ with $B \times \{2n+1\}$. The resulting topological space X inherits any of the following properties that T may possess: normality, σ-compactness, realcompactness, paracompactness, and countable paracompactness. Moreover, if A and B are disjoint subsets, then local compactness of T will imply local compactness of X. If T is countably paracompact but non-normal and if A and B are disjoint closed sets that are not contained in disjoint open sets, then X is not a weak cb-space.

In particular, let W and W^* be the spaces of ordinals $W(\omega_1)$ and $W(\omega_1 + 1)$ respectively. Set

$$T = \{(\sigma, \tau) \in W \times W^* : \sigma \leqslant \tau\}.$$

This space is locally compact and countably compact but not normal (the diagonal A and the upper edge B are disjoint closed sets that cannot be separated by open sets). If we set $W = (\omega_1, \omega_1)$ then $\beta T = T^* = T \cup \{w\} (\subset W^* \times W^*)$.

Let X and X^* be the spaces obtained from T and T^* by identifying images of A and B, and those of $A \cup \{w\}$ and $B \cup \{w\}$, respectively, as in the above construction. Then X is locally compact and countably paracompact but not a weak cb-space, while X^* is σ-compact (hence realcompact and a weak cb-space). Note that X^* is not locally compact. Since each continuous function on T is constant on a deleted neighborhood of w, it follows that X is C-embedded in X^*, whence $X^* = \upsilon X$. Thus X is a locally compact non-weak cb-space such that υX is a weak cb-space. This example is found in [250].

(4) A topological space is a cb-space if and only if it is both a countably paracompact and a weak cb-space (see [246]).

22 Generalized metric spaces and related results

In the introduction to this appendix, we have mentioned several properties we feel a new class of spaces should satisfy. There are several others that are also of interest. For the purposes of our discussion, let us formulate some of these, adding more to the list as necessary. Thus when we discuss a new class of spaces we will be interested in checking this list as to the appropriateness of any one of its items. First, the following definition is needed.

22.1 DEFINITION If $\mathscr{F}$ is a collection of closed subsets of the space X, then X is said to be *dominated by* $\mathscr{F}$ if a subset K of X is closed if and only if there is a subcollection $\mathscr{F}'$ of $\mathscr{F}$ which covers K and such that $F \cap K$ is closed for all $F \in \mathscr{F}'$.

Let $\mathscr{C}$ be a class of topological spaces.

($\mathscr{C}$.H) If whenever $X \in \mathscr{C}$ and $X' \subset X$ it follows that $X' \in \mathscr{C}$, then $\mathscr{C}$ is called a *hereditary class for arbitrary subsets* (*or just a hereditary class*).

($\mathscr{C}$.CP) If $X_i \in \mathscr{C}$, $i \in \mathbf{N}$ implies that the countable product $\prod_{i \in \mathbf{N}} X_i \in \mathscr{C}$ then $\mathscr{C}$ is called a *countably productive class.*

($\mathscr{C}$.CI) If whenever $X \in \mathscr{C}$ and Y is the image of X under a closed continuous map it follows that $Y \in \mathscr{C}$, then $\mathscr{C}$ is called a *class closed under closed continuous images.*

($\mathscr{C}.\sigma$) If whenever X is the countable union of a class of closed subsets $X_i \in \mathscr{C}$ implies $X \in \mathscr{C}$ then $\mathscr{C}$ is called a *σ-closed class.*

($\mathscr{C}$.D) If whenever X is dominated by a closed cover $\{X_\alpha : \alpha \in I\}$ of elements of $\mathscr{C}$ implies that $X \in \mathscr{C}$ then $\mathscr{C}$ is called a *class closed under domination.*

Some classes of spaces which have recently had considerable interest are the M-spaces of Morita (see [280]), the p-spaces of Arhangelskii (see [30]), the σ-spaces of Okuyama (see [312]) and the M_i-spaces ($i = 1, 2, 3$) of Ceder (see [66]). So let us present these concepts first.

22.2 Definition A space X is called an *M-space* if there is a normal sequence $(\mathscr{U}_i)_{i \in \mathbf{N}}$ of open covers of X satisfying the following condition (M).

(M) If $(K_i)_{i \in \mathbf{N}}$ is a decreasing sequence of non-empty closed subsets of X such that $K_i \subset \operatorname{st}(x, \mathscr{U}_i)$ for a fixed $x \in X$, then $\bigcap_{i \in \mathbf{N}} K_i \neq \varnothing$.

22.3 Definition A Tychonoff space is called a *p-space* if there is a sequence $(\mathscr{U}_i)_{i \in \mathbf{N}}$ of covers of X by open sets in βX such that for each $x \in X$, the intersection of the star sets $\operatorname{st}(x, \mathscr{U}_i)$ (in βX), is contained in X.

22.4 Definition Let $\mathscr{B}$ be a collection of subsets (not necessarily open) of a regular Hausdorff space X. If, for each point $x \in X$ and for each open subset G of X, $x \in G$, there is a $B \in \mathscr{B}$ such that $x \in B \subset G$, then $\mathscr{B}$ is called a *network* of X. If the network is a σ-locally finite (respectively, σ-discrete) collection of subsets of X, then $\mathscr{B}$ is called a *σ-locally finite network* (respectively, *σ-discrete network*). The space X is called a *σ-space* if it has a σ-locally finite network. If the network $\mathscr{B}$ is a countable collection of subsets of X, then X is called a *cosmic space.*

From 21.12, we see that the concept of a cushioned refinement and hence a cushioned collection of subsets may be stated as follows. If $\mathscr{U}$ is a collection of pairs (U_α, U'_α), $\alpha \in I$, of subsets U_α and U'_α of X then $\mathscr{U}$ is called *cushioned* if

$$\mathrm{cl} \cup \{U_\alpha : \alpha \in I'\} \subset \cup \{U'_\alpha : \alpha \in I'\}$$

for any subset I' of I. Thus we may give the following definition.

22.5 DEFINITION If U_α is an open set and if the collection $\mathscr{U}$ of such pairs (U_α, U'_α) is the union of countably many cushioned pair collections such that for any x in X and for any neighborhood V of x, there is an $\alpha \in I$ such that $x \in U_\alpha \subset U'_\alpha \subset V$ then $\mathscr{U}$ is called a *σ-cushioned pair base of X.*

22.6 DEFINITION A regular T_1-space X is called an *M_1-space* if it has a *σ-closure preserving base* (that is, a base which is the union of countably many closure preserving collections of open sets). A regular T_1-space X is called an *M_3-space* or a *stratifiable space* if it has a σ-cushioned pair base.

Let us further examine these definitions. Every locally finite collection is closure preserving. Any closure preserving family forms a cushioned collection of pairs (where $\mathscr{U}'_\alpha = \mathrm{cl}\,\mathscr{V}_\alpha$). Since metric spaces are paracompact, it is clear that the definitions of M_1-space and M_3 or stratifiable spaces are true generalizations of metric spaces. Also it is understandable that the latter is more general than the former.

The Nagata–Smirnov metrization theorem states that a necessary and sufficient condition for a regular T_1-space X to be metrizable is that it has a σ-locally finite base of open sets. The definition of a σ-space weakens the concept of base by dropping the requirement that its members be open subsets. Thus we see from the definitions that all metric spaces, or, more generally, all spaces that are expressed as the countable union of closed metrizable spaces and all cosmic spaces are σ-spaces.

We will consider p-spaces in more detail in a short while after we have considered some other characterizations of the spaces above.

The Alexander–Urysohn metrization theorem states that a

T_1-space X is metrizable if and only if it has a sequence $(\mathscr{U}_n)_{n\in\mathbf{N}}$ of open covers of X such that

(i) $\mathscr{U}_{n+1} <^* \mathscr{U}_n$ for each $n \in \mathrm{N}$,

(ii) $\{\mathrm{st}(x, \mathscr{U}_n): n \in \mathrm{N}\}$ is a neighborhood base for each point x in X.

22.7 Definition A T_1-space X is *developable* if there is a sequence of covers $(\mathscr{U}_n)_{n\in\mathbf{N}}$ which satisfy condition (ii) above.

Thus we see that a developable space is a generalization of metric space. Moreover we see quickly how uniform spaces also are developed by dropping the countability of the collection of covers (see [213]). However our main reason for considering the above metrization theorem was to analyse M-spaces somewhat. But we see readily that the definition of an M-space is just *another* modification of statement (ii) above. Every metric space is thus an M-space. For a countably compact space X, the covers $\mathscr{U}_i = \{X\}$, $i \in \mathrm{N}$, trivially satisfy the conditions, and so X is also an M-space. It can also be easily seen that metric spaces and locally compact spaces are p-spaces.

The spaces defined above may be classified (for reasons to be discussed below) as two types. Later we will define some other spaces which will belong to one of these classes. So we consider:

Type I: The spaces M, p and related spaces.

Type II: The spaces M_i, σ and related spaces.

The spaces of Type I are grouped as such for they are general enough to generalize both metric spaces and compact spaces, yet they are still concrete enough to allow extensions of theorems for some of the more classical spaces. On the other hand, spaces of Type II, in several aspects, have nicer properties than metric spaces. For example the class of σ-spaces is a σ-closed class of spaces, that is it satisfies $(\mathscr{C}.\sigma)$. This is not true for metrizable spaces as the following example will show.

Let I be a given infinite indexing set and let $\{E_\alpha: \alpha \in I\}$ be a family of real unit intervals, that is $E_\alpha = [0, 1]$ for each $\alpha \in I$. Let S be the discrete sum $\Sigma\{E_\alpha: \alpha \in I\}$ of the metric spaces E_α. By identifying the zero elements in each E_α, $\alpha \in I$, we have a map of

S onto the 'star shaped' set $S(I)$. A topology on $S(I)$ is defined as follows: the subset U of $S(I)$ is *open* if and only if for every $\alpha \in I$, $U \cap E_\alpha$ is open in E_α where E_α has the usual relative topology. With this topology $S(I)$ is a *paracompact Hausdorff space which is not metrizable.* Essentially the zero in $S(I)$ has no countable local neighborhood base. However $S(I)$ is the *closed continuous image* of the discrete sum $\Sigma\{E_\alpha: \alpha \in I\}$, of metric spaces, it is *dominated* by the closed covering $\{E_\alpha: \alpha \in I\}$, and it is also the *union of countably many metric spaces* if I is countable.

22.8 Definition The space $S(I)$ described above is called a *w-star space.*

With this discussion of a w-star space we see a concrete motive for wanting to generalize metric spaces. The class of metric spaces does not satisfy ($\mathscr{C}$.CI), ($\mathscr{C}.\sigma$), or ($\mathscr{C}$.D). Yet these properties are quite useful in topology.

With regard to our classifying scheme given after 22.1, the spaces of Type I usually satisfy very few of them. For example, M-spaces satisfy none of the statements. Moreover Isiwata in [210] has shown that the product of Tychonoff M-spaces need not be an M-space. The image of an M-space by a perfect map is not necessarily an M-space (see [282]). For spaces of Type II, however, the situation is remarkably different. The class of σ-spaces satisfies all of the conditions in 22.1 except that in ($\mathscr{C}$.I) the regularity (which is built into the definition of a σ-space) is essential and for ($\mathscr{C}$.D) the normality of X is needed (see [314]). Stratifiable spaces statisfy all the statements except ($\mathscr{C}.\sigma$) (see [59]). We will say more about these conditions and their relation to the spaces at hand as we go along. We will also in due time give some relationships between the spaces. In the meantime, let us consider some important results.

First let us go back to the w-star spaces. A w-star space $S(I)$ is non-metrizable as we have seen. But it is an M_1-space. Each E_α, $\alpha \in I$, used in constructing the space $S(I)$ has a sequence $\mathscr{U}_{\alpha,n}$, $n \in \mathbf{N}$, of locally finite open covers for which the (often called the *mesh* of $\mathscr{U}_{\alpha,n}$)

$$\inf\{\operatorname{diam} U: U \in \mathscr{U}_{\alpha,n}\}$$

converges to zero as n becomes arbitrarily large. Let $\mathscr{U}'_{\alpha,n}$ be the restriction of $\mathscr{U}_{\alpha,n}$ to the interval $(1/n, 1]$ in E_α. Then

$$B_n = \cup \{\mathscr{U}'_{\alpha,n} : \alpha \in I\}$$

is a locally finite open collection in $S(I)$. For each $\alpha \in I$, assign a natural number n_α. The union of the intervals $[0, 1/n_\alpha)$, $\alpha \in I$, is an open subset in $S(I)$. Let $\mathscr{V}$ be the collection of all such open sets. Then $\mathscr{V}$ is closure preserving and $S(I)$ has a σ-closure preserving base $\mathscr{B} = \bigcup_{n=0}^{\infty} B_n$, where $B_0 = \mathscr{V}$.

QUESTION 1 Are M_1- and M_3-spaces equivalent?

Quite a bit is known about M_3-spaces which is mainly due to Borges in [56] and Ceder in [66]. The following characterization is very interesting (see [66] for the proof).

22.9 THEOREM *A T_1-space X is an M_3-space if and only if to each open set V of X there is a sequence $(V_n)_{n\in\mathbf{N}}$ of open subsets of X such that*

(i) $\operatorname{cl} V_n \subset V$,

(ii) $V = \bigcup_{n\in\mathbf{N}} V_n$, *and*

(iii) $V_n \subset V'_n$ *whenever* $V \subset V'$.

In defining M_3-spaces we mentioned that they are also called stratifiable spaces. To this end, let us then consider the following definitions.

22.10 DEFINITION A T_1 topological space $(X, \mathscr{T})$ is called *semistratifiable* if there is a function S mapping $\mathbf{N} \times \mathscr{T}$ into the closed subsets of X such that

(*a*) if $U \in \mathscr{T}$ then $U = \bigcup_{n\in\mathbf{N}} S(n, U)$,

(*b*) if U and V are in $\mathscr{T}$ and if $U \subset V$ then $S(n, U) \subset S(n, V)$ for each $n \in \mathbf{N}$.

The function S is called a *semistratification of* X (see [88], [89]). If the function S also satisfies

(*c*) the set $U = \bigcup_{n\in\mathbf{N}} \operatorname{int} S(n, U)$,

then S is called a *stratification* of X.

Thus we see that if a T_1-space X has a stratification S, then the interiors of the sets $S(n, U)$ for $U \in \mathscr{T}$ satisfy the conditions of 22.9. Conversely from 22.9 we see that every M_3-space has a stratification S defined by

$$S(n, U) = \operatorname{cl} U_n,$$

where U and U_n, $n \in \mathbf{N}$ are as in the theorem. Thus a *stratifiable space* is often defined as a T_1-space X that has a stratification.

Now stratifiable spaces (or M_3-spaces) satisfy ($\mathscr{C}$.CP), ($\mathscr{C}$.H), ($\mathscr{C}$.D), and ($\mathscr{C}$.CI) (see [59]). As for ($\mathscr{C}.\sigma$), Heath (see [176]) has shown that there is a countable regular space that is not stratifiable.

Closely related to the above is the concept of a semimetric.

22.11 Definition A *semimetric* on a set X is a function d from $X \times X$ into $\mathbf{R}$ such that for x and y in X, $d(x, y) \geqslant 0$, $d(x, y) = d(y, x)$, and $d(x, y) = 0$ if and only if $x = y$. A semimetric d is said to be *compatible with a topology* $\mathscr{T}$ on X provided that for any subset A of X, the closure of A with respect to $\mathscr{T}$ is the set $\{y \in X: \inf\{d(x, y): x \in A\} = 0\}$. A space X is *semimetrizable* if there is a semimetric on X which is compatible with the topology of X.

In [88], Creede has shown:

22.12 Theorem *A T_1-space is semimetrizable if and only if it is first countable and semistratifiable.*

In [174], Heath gave an example of a paracompact semimetrizable space that was not stratifiable. When is a semimetrizable space stratifiable? (see Heath [174] where it was originally posed). Or now more generally

Question 2 When is a semistratifiable space stratifiable?

This last question was posed by Lutzer in [240]. He gave there a partial answer to it as well as an answer to Heath's question. To do so he made the following definition.

22.13 DEFINITION A *k-semistratification* of a T_1-space X is a semistratification S of X with the property that whenever $C \subset U$ with C compact and U open in X, there is an $n \in \mathrm{N}$ with $C \subset S(n, U)$. A space that admits a k-semistratification is said to be *k-semistratifiable*.

Then it is clear that any stratifiable space is k-stratifiable (for any stratifiable space X, the stratification S has the property that, for any open set U, $S(n, U) \subset S(n+1, U)$ for $n \in \mathrm{N}$). Every k-stratifiable space is semistratifiable. These implications are not reversible however (see [174] and [240]).

The following theorem of [240] settles one of the questions above.

22.14 THEOREM *A T_1-semimetrizable space is stratifiable if and only if it is k-semistratifiable.*

What properties do k-semistratifiable spaces have? Let's look at the following proposition (see [240] for a sketch of a proof of (*c*)).

22.15 PROPOSITION *The following properties hold.*

(*a*) *Any subspace of a k-semistratifiable space is k-semistratifiable.*

(*b*) *Any countable product of Hausdorff k-semistratifiable spaces is k-semistratifiable.*

(*c*) *Any closed image of a paracompact k-semistratifiable space is k-semistratifiable.*

As for semistratifiable spaces, we have the following.

22.16 THEOREM *Let $\mathscr{C}$ be the class of regular T_1-semistratifiable spaces. Then the class $\mathscr{C}$ satisfies ($\mathscr{C}$.H), ($\mathscr{C}$.CP), and ($\mathscr{C}.\sigma$). Moreover, if $\mathscr{C}$ is restricted to be the class of normal T_1-spaces then the class $\mathscr{C}$ also satisfies ($\mathscr{C}$.D) and ($\mathscr{C}$.CI).*

The proof of the ($\mathscr{C}$.H) and ($\mathscr{C}.\sigma$) properties follow from the definition, whereas ($\mathscr{C}$.CP) may be found in [88], ($\mathscr{C}$.D) may be found in [314], and ($\mathscr{C}$.CI) is in [358].

Some further remarks are necessary regarding the ($\mathscr{C}$.CP) property of 22.16. It says that the closed continuous image Y of a normal T_1-semistratifiable space X is a normal T_1-semistratifiable space. If f is the map of interest, then the set

$$\{y \in Y : f^{-1}(y) \text{ is not compact}\}$$

is σ-discrete in Y (see [358]).

Let us turn our attention now to σ-spaces. The intention with the definition of a σ-space was to modify the concept of base to the concept of a network. The following theorem due to Nagata and Siwiec in [306] indicates that the nature of these concepts are essentially different despite the similarity in their definitions. Essentially this fact contributes to the advantages σ-spaces have over metric spaces in the way of properties.

22.17 THEOREM *For a regular T_1-space X the following statements are equivalent.*

(1) *The space X is a σ-space.*
(2) *The space X has a σ-closure preserving network.*
(3) *The space X has a σ-discrete network.*

In [312], Okuyama has shown that σ-spaces do enjoy many properties similar to those of metric spaces.

22.18 THEOREM *Let $\mathscr{C}$ be the class of regular T_1 σ-spaces. Then the class $\mathscr{C}$ satisfies ($\mathscr{C}$.H), ($\mathscr{C}$.CP), and ($\mathscr{C}.\sigma$). Moreover, if $\mathscr{C}$ is restricted to the class of regular T_1-spaces then $\mathscr{C}$ satisfies ($\mathscr{C}$.CI) and if $\mathscr{C}$ is restricted to the class of normal T_1-spaces, then $\mathscr{C}$ also satisfies ($\mathscr{C}$.D).*

The proof of the first three properties may be found in [312], the proof of property ($\mathscr{C}$.D) may be found in [314], while the proof of property ($\mathscr{C}$.CI) is in [305] and [313].

If $\mathscr{C}$ is taken to be either the class of all collectionwise normal T_1 σ-spaces or the class of all collectionwise normal T_1-semistratifiable spaces then separability and Lindelöf are equivalent. In fact, we have even more as the following theorem states. The proof for semistratifiable spaces is found in [358] (see also [88]). Also all σ-spaces are semistratifiable.

22.19 **Theorem** *If X is a collectionwise normal T_1 σ-space then the following statements are equivalent.*

(1) *The space X is separable.*

(2) *The space X is Lindelöf.*

(3) *The space X satisfies the countable chain condition (that is, X does not contain uncountably many pairwise disjoint non-empty subsets).*

Recall that the class $\mathscr{C}$ of metric spaces satisfies ($\mathscr{C}$.H), ($\mathscr{C}$.CP), but not ($\mathscr{C}$.CI),† ($\mathscr{C}.\sigma$), nor ($\mathscr{C}$.D). Thus we have given certain classes of σ-spaces and semistratifiable spaces, properties which the class of metric spaces enjoy and does not enjoy. So what now is the relationship between σ-spaces and metric spaces? A definition is in order to state the theorem.

22.20 **Definition** A collection $\mathscr{U}$ of open subsets of X is called a *point countable base* if it is a base such that each x in X is contained in at most countably many members of $\mathscr{U}$.

This concept has been very useful in metrization theorems as we will see in the results to follow, especially 22.31. However, for now we may just state that if *X is a collectionwise normal T_1 σ-space which is also an M-space then X is metrizable.* Then we may show that if *X is a collectionwise normal σ-space with a point countable base then X is metrizable.* However, these results will soon be superseded.

To better understand p-spaces let us consider them together with M-spaces in the light of some characterizations.

A normal sequence was used in the definition of an M-space. We know that every normal sequence determines a pseudometric. From what we have seen of normal sequences and pseudometrics, then it is not surprising that an M-space is the preimage of a metric space by a closed continuous map whose inverse image of points are countably compact.‡

† The class $\mathscr{C}$ of metrizable M_1-spaces has just been shown to satisfy ($\mathscr{C}$.CI) (see [415]).

‡ If the inverse image of points are metrizable and if the preimage is just a normal space X, then X is metrizable if and only if X has a G_δ-diagonal (see [404] and [416]).

22.21 DEFINITION Let f be a closed continuous map of the topological space X onto the topological space Y. The map f is said to be *perfect* (respectively, *quasi-perfect*) if $f^{-1}(y)$ is compact (respectively, countably compact) for each $y \in Y$.

We state the above mentioned result as a theorem due to Morita in [281].

22.22 THEOREM *The T_1-space X is an M-space if and only if there is a quasi-perfect map from X onto a metric space Y.*

Thus we have another interesting way for considering M-spaces. We have mentioned previously how countably compact T_1-spaces are M-spaces.

Now we can present the relationship that exists between paracompact T_1 M-spaces, the concept of p-spaces, metric spaces, and compact spaces. In the next theorem the equivalence of (1), (2), and (3) is due to Morita in [281] and Arhangelskii in [29], whereas the equivalence of (1) and (4) is due to Nagata in [304].

22.23 THEOREM *The following conditions on a Hausdorff space X are equivalent.*

(1) *The space X is a paracompact M-space.*

(2) *The space X is a paracompact p-space.*

(3) *There is a metric space Y and a perfect map of X onto Y.*

(4) *The space X is homeomorphic to a closed subset of the product of a metric space and a compact Hausdorff space.*

Not only are metric spaces and countably compact T_1-spaces M-spaces but Frolík in [131] has shown that any paracompact T_1-space X which is a G_δ-set in βX is also an M-space and therefore a p-space. Of course every metric space is a p-space and not every metric space is a G_δ set in βX.

Condition (4) of 22.23 raises the following questions due to Nagata.

QUESTION 3 Does there exist a unique space Z such that a space X is a paracompact M-space with weight α if and only if it is homeomorphic to a closed subset of Z? (Similar questions can be asked for other classes of generalized metric spaces.)

Question 4 Is every M-space homeomorphic to a closed subset of the product of a metric space and a countably compact space?

A special case of the last question is given by Nagata in [304].

In comparing p-spaces X with M-spaces, the fact that they are defined in terms of open collections in βX prevents a somewhat facile means of handling this problem. In [62], Burke has given an interesting internal characterization which assists in the comparison.

22.24 Theorem *A Tychonoff space X is a p-space if and only if there is a sequence $(\mathscr{U}_n)_{n\in\mathbf{N}}$ of open covers of X satisfying the following. If x is a point in X and if $V_n \in \mathscr{U}_n$ such that $x \in V_n$, $n \in \mathbf{N}$, then*

(1) *the $\bigcap_{n\in\mathbf{N}} \mathrm{cl}\, V_n$ is compact, and*

(2) *if $x_n \in \bigcap_{n\in\mathbf{N}} \mathrm{cl}\, V_n$ for each n in $\mathbf{N}$ then the sequence $(x_n)_{n\in\mathbf{N}}$ has a cluster point.*

We have already mentioned that the class $\mathscr{C}$ of M-spaces satisfies none of the five conditions. However the class of paracompact T_1 M-spaces does satisfy ($\mathscr{C}$.CP). Utilizing statement (3) of 22.23 one can show this result. The importance of the class of M-spaces however stems from the fact that they include both the metric spaces and the compact spaces. Also they were originally defined in [281] in connection with the problem of characterizing a space whose product with any metric space is normal.

Before leaving 22.24 however let us consider two other classes of spaces which are modifications of the definition of an M-space and are somewhat motivated by 22.24.

22.25 Definition A space X is a *$w\Delta$-space* (respectively, *M^*-space*) if it has a sequence $(\mathscr{U}_n)_{n\in\mathbf{N}}$ of open covers (respectively, locally finite closed covers) such that whenever $x_n \in \mathrm{st}\,(x, \mathscr{U}_n)$, $n \in \mathbf{N}$, for a fixed $x \in X$, then the sequence $(x_n)_{n\in\mathbf{N}}$ has a cluster point.

The $w\Delta$-spaces were introduced by Borges in [58] and the M^*-spaces were introduced by Ishii in [207]. Every M-space is an M^*-space and every M^*-space is a $w\Delta$-space (see [304]). The M^*-spaces are an interesting supplement to M-spaces for M^*-spaces are often easier to use than M-spaces. For normal spaces the two concepts coincide (see [207]). In [62], $w\Delta$-spaces and p-spaces are considered along with their relations. The following theorem of Nagata and Siwiec in [306] is of interest.

22.26 THEOREM *If X is a T_1 collectionwise normal σ-space and a $w\Delta$-space then it is metrizable.*

Another class of spaces has recently been defined by Nagami in [298]. They are of interest because they extract from both Type I and Type II spaces, that is they include M-spaces and σ-spaces.

22.27 DEFINITION A T_1-space X is called a *Σ-space* if there exists a sequence $(\mathscr{U}_n)_{n\in\mathbf{N}}$ of locally finite closed covers of X such that for each x in X

(i) the intersection $C(x) = \cap\{F \in \bigcup_{n\in\mathbf{N}} \mathscr{U}_n : x \in F\}$ is countably compact,

(ii) whenever $C(x)$ is contained in an open subset U of X there is an $F \in \bigcup_{n\in\mathbf{N}} \mathscr{U}_n$ for which $C(x) \subset F \subset U$ (that is, $\bigcup_{n\in\mathbf{N}} \mathscr{U}_n$ forms a network for $C(x)$).

Now, as we have mentioned, every M-space and every σ-space is a Σ-space (see [298]). Moreover the class of perfectly normal Σ-spaces satisfies ($\mathscr{C}$.D) (see [314]). Moreover a form of 22.26 due to Michael (see [269]) may also be given.

22.28 THEOREM *A paracompact regular Σ-space with a point countable base is metrizable.*

In [269], it is pointed out that the image of a Σ-space under a closed continuous map need not be a Σ-space (that is, the class of Σ-spaces does not satisfy ($\mathscr{C}$.I)). In fact the image of a locally finite closed cover of the domain via a closed continuous onto map is a hereditarily closure preserving closed cover of the range. Thus the closed continuous image of a Σ-space is a Σ^*-space which we now define.

22.29 Definition A space X is a Σ^*-*space* if there exists a sequence $(\mathscr{U}_n)_{n\in\mathbf{N}}$ of hereditarily closure preserving closed covers of X satisfying statement (ii) in 22.27. A space X is a $\Sigma^\#$-*space* if it satisfies statements (i) and (ii) of 22.27 with 'locally finite' replaced by 'closure preserving'.

As pointed out in [269], the class of $\Sigma^\#$-spaces is strictly larger than the class of Σ-spaces. However, it is clear that every Σ-space is a Σ^*-space and every Σ^*-space is a $\Sigma^\#$-space. The following results found in [315] point out readily the interest in these spaces.

22.30 Theorem

(1) *Every Lindelöf Hausdorff Σ^*-space is a Σ-space.*

(2) *A Σ^*-space X is a Σ-space if every open subset of X is an F_σ-set.*

(3) *For paracompact Hausdorff spaces X the following statements are equivalent.*

(*a*) *The space X is a Σ-space.*

(*b*) *The product space $X \times E$ is a Σ-space (E is the closed unit interval).*

(*c*) *The product space $X \times E$ is a Σ^*-space.*

In [269], it is shown that the product of a paracompact Hausdorff $\Sigma^\#$-space with E is a $\Sigma^\#$-space. Utilizing (3) of 22.30 it follows that the product of a paracompact Hausdorff Σ^* non Σ-space with E is a $\Sigma^\#$ non Σ^*-space. Thus the class of $\Sigma^\#$-spaces is strictly larger than the class of Σ^*-spaces which we have already seen to be strictly larger than the class of Σ-spaces.

In [315], it is shown that in (1) of 22.30 one cannot replace Σ^* by $\Sigma^\#$. It is not known whether Σ^* may be replaced by $\Sigma^\#$ in (2) of 22.30.

We have mentioned some metrization theorems in our discussion. For metrizability of a generalized metric space, a general rule has been observed in [59] and [312]. The rule is that metrizability is equivalent to a condition of Type I plus a condition of Type II and some additional conditions. For example, in [312], Okuyama observed that metrizability was equivalent to the space being a paracompact space which is both an M-space and a σ-space.

Slaughter in [346] much improved this result by showing metrizability to be equivalent to the space being Hausdorff and both an M-space and a σ-space. With respect to the concept of point countable base, Heath showed in [409], that every *pointwise paracompact* (every open cover has a point-finite open refinement) p-space with a point-countable base is developable. Hence a paracompact M-space (or p-space) with a point-countable base is metrizable. This latter result was also shown by Fillipov in [116] and was later generalized by Nagata (see [305]) and Michael and Slaughter (see [346]).

22.31 THEOREM *A Haudorff space X is metrizable if and only if it is an M^*-space and has a point-countable base.*

Shiraki in [341] improved 22.28 of Michael. However it comes as a corollary to another theorem of Michael (see [269]) which is of interest in its own right.

22.32 THEOREM† *Every Σ-space X with a point-countable base is a σ-space.*

22.33 COROLLARY *A space X is metrizable if and only if it is a collectionwise normal Σ-space with a point-countable base.*

In [408], Heath showed the following which generalizes the Nagata–Smirnov metrization theorem. Recall that in a *semimetric* space X there is a real-valued function d defined on $X \times X$ such that $x \in \operatorname{cl} A$, $A \subset X$, if and only if $\inf\{d(x, y): y \in A\} = 0$.

22.34 THEOREM *A T_1-semimetric space with a point-countable base is developable.*

Thus a paracompact σ-space with a point-countable base is metrizable.

† F. Slaughter has communicated to us that it is possible to prove that a Σ^*-space with point countable 'open ct-net' is a σ-space. This has also been shown by Shiraki (see [341]). In addition Slaughter and Shiraki have shown that a space is a σ-space if and only if it is both a Σ^*-space and a σ^*-space when σ^* is analogously defined (see [341] and [346]).

22.35 COROLLARY *A T_1 regular space is metrizable if it has a base that is point-countable and a base (possibly the same) that is σ-closure preserving.*

We note here that both types of bases mentioned in the above corollary, even if taken together are strictly weaker than a base being σ-locally finite. Many of the metrization theorems given do not have implications from one to the other. An intesting area for future research would be to develop a more unified theory for this aspect.

So many generalized metric spaces have been defined it would be worthwhile to have some means of characterizing them in a unified manner. Such an attempt has been made by Heath, Hodel and others using open neighborhoods. In this direction let us consider a sequence $\mathscr{U}_x = \{U(n,x): n = \mathrm{N}\}$ of open neighborhoods of a point x in the space X. Let us then consider the following conditions.

($\mathscr{U}$.NB) The sequence $\mathscr{U}_x$ is a neighborhood base for x.

($\mathscr{U}$.T) If $y \in U(n,x)$ then $U(n,y) \subset U(n,x)$.

($\mathscr{U}$.S) If x is not a point in a closed subset F of X then x is not an element of the set $\cup\{U(n,y): y \in F\} = U_F(n)$ for some n in N or equivalently if $x \in U(n,x_n)$ for $n \in \mathrm{N}$ then x is a cluster point of the sequence $(x_n)_{n\in\mathrm{N}}$.

($\mathscr{U}$.CS) If x is not a point in the closed set F then x is not in the closure of $U_F(n)$ for some $n \in \mathrm{N}$.

($\mathscr{U}$.S1) If the pair $\{x, x_n\}$ is in $U(n,y_n)$ for n in N then x is a cluster point of $(x_n)_{n\in\mathrm{N}}$.

($\mathscr{U}$.S2) If the pair $\{x, x_n\}$ is in $U(n,y_n)$ for n in N then $(x_n)_{n\in\mathrm{N}}$ has a cluster point.

($\mathscr{U}$.S3) If the intersection of $U(n,x)$ with $U(n,x_n)$ is nonempty for n in N then x is a cluster point of $(x_n)_{n\in\mathrm{N}}$.

($\mathscr{U}$.Sm) If x_n is in $U(n,x)$ for each n in N then x is a cluster point of the sequence $(x_n)_{n\in\mathrm{N}}$.

22.36 THEOREM *For each x in the T_1-space X, let $\mathscr{U}_x$ be a sequence $\{U(n,x): n \in \mathrm{N}\}$ of open neighborhoods of x. Then the following statements hold.*

(1) (*Creede* [189]) *The space X is semistratifiable if and only if $\mathscr{U}_x$ satisfies ($\mathscr{U}$.S) for each x in X.*

(2) (*Heath* [172]) *The space X is a semimetric space if and only if $\mathscr{U}_x$ satisfies* ($\mathscr{U}$.NB) *and* ($\mathscr{U}$.S) *for each x in X.*

(3) (*Heath–Hodel* [177]) *The space X is a σ-space if and only if $\mathscr{U}_x$ satisfies* ($\mathscr{U}.T$) *and* ($\mathscr{U}$.S) *for each x in X.*

(4) (*Heath* [173]) *The space X is stratifiable if and only if $\mathscr{U}_x$ satisfies* ($\mathscr{U}$.CS) *for each x in X.*

(5) (*Heath* [173]) *The space X is stratifiable and first countable (sometimes called a Nagata space) if and only if $\mathscr{U}_x$ satisfies* ($\mathscr{U}$.S3) *for each x in X.*

(6) (*Heath* [172]) *The space X is a developable space if and only if $\mathscr{U}_x$ satisfies* ($\mathscr{U}$.S1) *for each x in X.*

(7) (*Heath* [172]) *The space X is a $w\Delta$-space if and only if $\mathscr{U}_x$ satisfies* ($\mathscr{U}$.S2) *for each x in X.*

(8) *The space X is a first countable space if and only if $\mathscr{U}_x$ satisfies* ($\mathscr{U}$.Sm) *for each x in X.*

Unifying the approach to generalized metric spaces in this way does clarify the relationship between the various classes of spaces, suggests new classes of spaces, and gives rise to many natural questions and observations.

Another concept which is apparently interesting is the concept of a space being monotonically normal. The class of monotonically normal spaces generalizes the class of stratifiable spaces. However, like the class of Σ-spaces, they differ distinctly from the other spaces of Type II. The following definition and results may be found in [410].

22.37 Definition Let $\mathscr{F}$ be the collection of all ordered pairs of disjoint closed subsets of a topological space $(X, \mathscr{T})$. Then the space X is said to be *monotonically* (*monotone*) *normal* if there is a function G from $\mathscr{F} \times \mathscr{F}$ into $\mathscr{T}$ such that

(1) whenever $(F, K) \in \mathscr{F} \times \mathscr{F}$, $F \cap K = \varnothing$, then

$$F \subset G(F, K) \quad \text{and} \quad G(F, K) \cap G(K, F) = \varnothing,$$

and

(2) whenever $F \subset F_1$, $K_1 \subset K$, and $(F_1, K) \in \mathscr{F} \times \mathscr{F}$, $F_1 \cap K_1 = \varnothing$ then

$$G(F, K) \subset G(F_1, K_1).$$

With this definition it is easy to see that every monotonically normal space is collectionwise normal. In fact every subspace of a monotonically normal space is monotonically normal and therefore collectionwise normal.

Moreover every linearly ordered space (in fact generalized linear ordered space) is monotonically normal (see also [395]). Consequently monotone normality is the strongest separation axiom known to hold in such spaces. The Sorgenfrey line is also hereditarily monotone normal.

In respect to the classes of spaces discussed previously, monotone normality serves nicely in determining the stratifiable spaces among the semistratifiable ones.

22.38 THEOREM *A space X is stratifiable if and only if it is both semistratifiable and monotone normal.*

In respect to the extension theorems that we have discussed, monotone normal spaces not only have all their closed subsets C^*-embedded but in addition the extended functions preserve the order of comparable functions.

22.39 THEOREM *If F is a closed subset of a monotone normal T_1-space X then every $f \in C(F, E)$ has a continuous extension $\bar{f} \in C(X, E)$ such that whenever $f(x) \leqslant g(x)$ for all $x \in F$, $f, g \in C(F, E)$ then $\bar{f}(x) \leqslant \bar{g}(x)$ for all $x \in X$.*

In stratifiable spaces, the Dugundji extension theorem (see [104]) holds. An interesting question would be to determine if such a theorem holds for monotone normal spaces.

As for the converse of the above theorem, is it true as it stands or must one have a stronger statement such as requiring that if F_1 and F_2 are closed subsets of X, $F_1 \subset F_2$ and if $f_1 \in C^*(F_1, E)$, $f_2 \in C^*(F_2, E)$, $\bar{f}_2 = \bar{f}_1 | F_2$ then $\bar{f}_1 = \bar{f}_2$ (this conjecture is due to Heath).

Let us look at a diagram of implications suggested by our work in Sections 21 and 22 of this appendix. Note that developable spaces are related to spaces of Type I and Type II.

Now the space $W(\omega_1)$ discussed in Section 9 is a hereditarily montone normal (so is $W(\omega_1 + 1)$), countably paracompact, first

All spaces are assumed to be at least regular Hausdorff

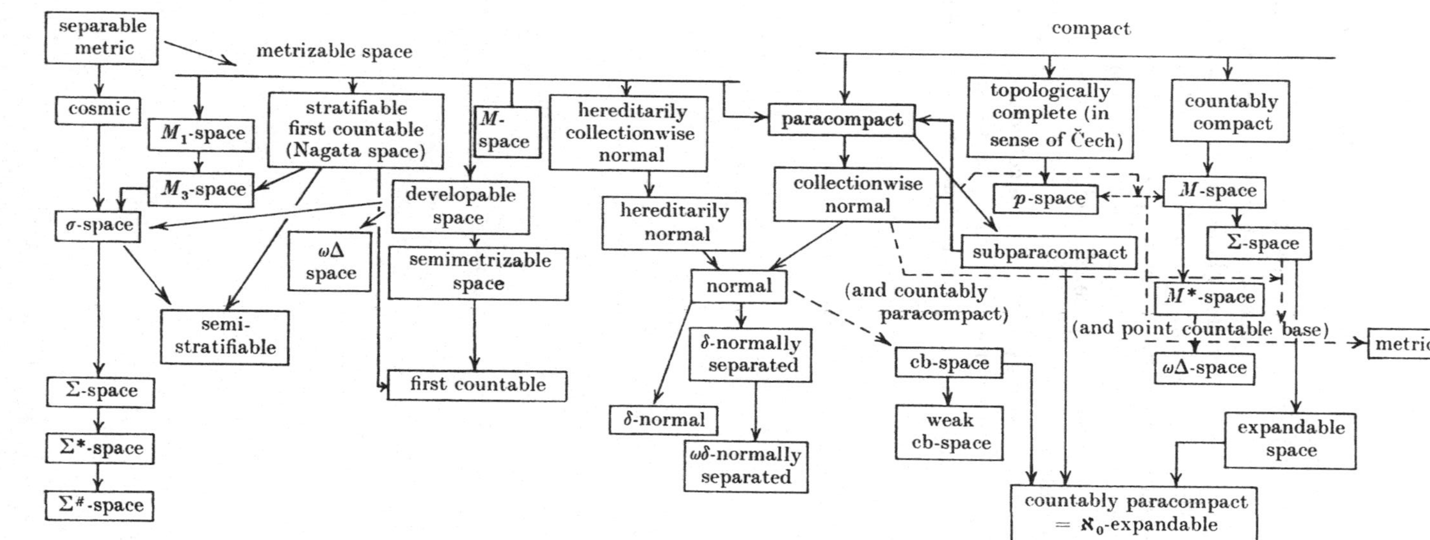

$a \longrightarrow b \equiv$ every a space is a b space

countable space that is neither paracompact nor semimetrizable. The space $W(\omega_1+1)$ is compact but not first countable. The Sorgenfry line is a hereditarily monotone normal, paracompact, first countable space that is neither semimetrizable nor compact. The Niemytzki plane is a Tychonoff (non-normal), developable, separable space that is neither countably paracompact nor metrizable. Bing's space is a normal (non-collectionwise normal) space that is not paracompact.

Let us now consider the space E^E which is the uncountable Cartesian product of the closed unit interval $E = [0, 1]$ with the Tychonoff product topology. Then E^E is a collectionwise normal (non-hereditarily collectionwise normal), separable, Lindelöf (and therefore paracompact) space that is not a first countable space.

References

Books

I. Alexandroff, P. and Hopf, H., *Topologie I*. Berlin: Springer-Verlag, 1935.

II. Berge, C., *Topological Spaces*. New York: Macmillan, 1963.

III. Bourbaki, N., *General Topology, Part 1*. Reading: Addison-Wesley, 1966.

IV. Bourbaki, N., *General Topology, Part 2*. Reading: Addison-Wesley, 1966.

V. Burgess, D. C. J., *Analytical Topology*. Princeton: Van Nostrand, 1966.

VI. Bushaw, D., *Elements of General Topology*. New York: John Wiley and Sons, 1963.

VII. Choquet, G., *Topology*. New York: Academic Press, 1966.

VIII. Cullen, H. F., *Introduction to General Topology*. Boston: D. C. Heath, 1968.

IX. Dugundji, J., *Topology*. Boston: Allyn and Bacon, 1966.

X. Engelking, R., *Outline of General Topology*. Amsterdam: North Holland, 1968.

XI. Gillman, L. and Jerison, M., *Rings of Continuous Functions*. Princeton: Van Nostrand, 1960.

XII. Greever, J., *Theory and Examples of Point-set Topology*. Belmont, California: Brooks/Cole, 1967.

XIII. Halmos, P. R., *Naïve Set Theory*. Princeton: Van Nostrand, 1960.

XIV. Hausdorff, F., *Grundzüge der Mengenlehre*. Leipzig: 1914. (Reprinted New York: Chelsea, 1949.)

XV. Hu, S. T., *Elements of General Topology*. San Francisco: Holden-Day, 1964.

XVI. Hu, S. T., *Introduction to General Topology*. San Francisco, Holden-Day, 1966.

XVII. Isbell, J. R., *Uniform Spaces*. Providence: American Mathematical Society, 1964.

XVIII. Kelley, J. L., *General Topology*. Princeton: Van Nostrand, 1955.

XIX. Kowalsky, H. J., *Topological Spaces*. New York: Academic Press, 1965.

XX. Kuratowski, C., *Topology I*. New York: Academic Press, 1966.

XXI. McCarty, G., *Topology*. New York: McGraw-Hill, 1967.

XXII. Moore, T. O., *Elementary General Topology*. Englewood Cliffs: Prentice-Hall, 1964.

XXIII. Nagami, K., *Dimension Theory*. New York: Academic Press, 1970.

XXIV. Nagata, J., *Modern Dimension Theory*. New York: Interscience, 1965.

xxv. Naimpally, S. A. and Warrack, D. B., *Proximity Spaces*. London: Cambridge University Press, 1969.
xxvi. Schaeffer, H., *Topological Vector Spaces*. New York: Macmillan, 1966.
xxvii. Thron, W. J., *Topological Structures*. New York: Holt, Rinehart and Winston, 1966.
xxviii. Tukey, J. W., *Convergence and Uniformity in Topology*. Princeton: Princeton University Press, 1940.
xxix. Weil, A., *Sur les Espaces à Structure Uniforme et sur la Topologie Générale*. Paris: Hermann, 1937.
xxx. Willard, S., *General Topology*. Reading: Addison-Wesley, 1970.

Articles

1. Aarts, J. M., Every metric compactification is a Wallman-type compactification, *Proc. of Intrl. Conf. on Topology, Herzeg-Novi* (ed. Kurepa, D. R.) (1968), 29–34.
2. Aarts, J. M. and DeGroot, J., Complete regularity as a separation axiom, *Canad. J. Math.* **21** (1969), 96–105.
3. Aarts, J. M., DeGroot, J. and McDowell, R. H., Cotopology for metrizable spaces, *Duke Math. J.* **37** (1970), 291–5.
4. Alexandroff, P., Some results in the theory of topological spaces obtained within the last twenty-five years, *Russian Math. Surveys*, **15** (1960), 23–83.
5. Alexandroff, P. and Urysohn, P., Une condition necessaire et suffisante pour qu'une class ($\mathscr{L}$) soit une class ($\mathscr{D}$), *C.R. Acad. Sci. Paris* **177** (1923), 1274–7.
6. Alfsen, E. M. and Fenstad, J. E., On the equivalence between proximity structures and totally bounded uniform structures, *Math. Scand.* **7** (1959), 353–60.
7. Alfsen, E. M. and Fenstad, J. E., Correction to a paper on proximity and totally bounded uniform structures, *Math. Scand.* **9** (1961), 258.
8. Alfsen, E. M. and Njastad, O., Proximity and generalized uniformity, *Fund. Math.* **52** (1963), 235–52.
9. Alò, R. A. and Frink, O., Topologies of lattice products, *Canad. J. Math.* **18** (1966), 1004–14.
10. Alò, R. A. and Frink, O., Topologies of chains, *Math. Ann.* **171** (1967), 239–46.
11. Alò, R. A., Imler, L. and Shapiro, H. L., *P*- and *z*-embedded subspaces, *Math. Ann.* **188** (1970), 13–22.
12. Alò, R. A. and Sennott, L. I., Extending linear space-valued functions, *Math. Ann.* **191** (1971), 79–86.
13. Alò, R. A. and Shapiro, H. L., A note on compactifications and seminormal bases, *J. Australian Math. Soc.* **8** (1968), 102–8.
14. Alò, R. A. and Shapiro, H. L., Normal base compactifications, *Math. Ann.* **175** (1968), 337–40.
15. Alò, R. A. and Shapiro, H. L., Extensions of totally bounded pseudometrics, *Proc. Amer. Math. Soc.* **19** (1968), 877–84.
16. Alò, R. A. and Shapiro, H. L., Wallman-compact and realcompact spaces, *Proc. of Intrl. Conf. on Topology, Berlin, Contributions to Extension Theory of Topological Structures*, Veb Deutscher, Verlag der Wissenschaften, Berlin (1969), 9–14.
17. Alò, R. A. and Shapiro, H. L., $\mathscr{Z}$-realcompactifications, *J. Australian Math. Soc.* **9** (1969), 489–95.

18. Alò, R. A. and Shapiro, H. L., Continuous uniformities, *Math. Ann.* **185** (1970), 322–8.
19. Alò, R. A. and Shapiro, H. L., Paracompact subspaces, *Acta Math.* **21** (1970), 115–19.
20. Alò, R. A. and Shapiro, H. L., Countably paracompact, normal and collectionwise noımal spaces, *Indag. Math.* (to appear).
21. Anderson, F. W., Approximation in systems of real-valued continuous functions, *Trans. Amer. Math. Soc.* **103** (1962), 249–71.
22. Anderson, F. W. and Blair, R. L., Characterizations of the algebra of all real-valued continuous functions on a completely regular space, *Illinois J. Math.* **3** (1959), 121–33.
23. Aquaro, G. R., Aperti e strutture uniformi sopra uno spazio topologico, *Ann. Mat. Pura Appl.* (4) **47** (1959), 319–89.
24. Aquaro, G. R., Spazii collettivamente normali ed estensione di applicazioni continue, *Riv. Mat. Univ. Parma* (2) **2** (1961), 77–90.
25. Arens, R. F., A topology for spaces of transformations, *Ann. Math.* **47** (1946), 480–95.
26. Arens, R. F. and Dugundji, J., Topologies for function spaces, *Pacific J. Math.* **1** (1951), 5–31.
27. Arens, R. F., Extension of functions on fully normal spaces, *Pacific J. Math.* **2** (1952), 11–22.
28. Arens, R. F., Extensions of coverings, of pseudometrics, and of linear space valued mappings, *Canad. J. Math.* **5** (1953), 211–15.
29. Arhangelskii, A. V., On a class of spaces containing all metric and all locally bicompact spaces, *Soviet Math.* **4** (1963), 751–4.
30. Arhangelskii, A. V., Mappings and spaces, *Russian Math. Surveys* **21** (1966), 115–62.
31. Arhangelskii, A. V., On closed mappings, bicompact spaces, and a problem of P. Aleksandrov, *Pacific J. Math.* **18** (1966), 201–8.
32. Atsuji, M., Necessary and sufficient conditions for the normality of the product of two spaces, *Proc. Japan Acad.* **45** (1969), 894–8.
33. Aull, C. E., A note on countably paracompact spaces and metrization, *Proc. Amer. Math. Soc.* **16** (1965), 1316–17.
34. Aull, C. E., Collectionwise normal subsets, *J. London Math. Soc.* **4** (2) (1969), 155–62.
35. Bagley, R. W., On Àscoli theorems, *Proc. Japan Acad.* **37** (1961), 444–6.
36. Bagley, R. W., Connell, E. H. and McKnight, J. D., On properties characterizing pseudometric spaces, *Proc. Amer. Math. Soc.* **9** (1958), 500 –6.
37. Bagley, R. W. and Weddington, D. D., Products of k'-spaces, *Proc. Amer. Math. Soc.* **22** (1969), 392–4.
38. Bagley, R. W. and Yang, J. S., On k-spaces and function spaces, *Proc. Amer. Math. Soc.* **17** (1966), 703–5.
39. Ball, E., Weak normality and related properties, *Canad. J. Math.* **22** (1970), 997–1001.
40. Banaschewski, B., Spaces of dimension zero, *Canad. J. Math.* **9** (1957), 38–46.
41. Banaschewski, B., On homeomorphisms between extension spaces, *Canad. J. Math.* **12** (1960), 252–62.
42. Banaschewski, B., Orderable spaces, *Fund. Math.* **50** (1961), 21–34.
43. Banaschewski, B., Normal systems of sets, *Math. Nachr.* **24** (1962), 53–75.

44. Banaschewski, B., On Wallman's method of compactification, *Math. Nachr.* 27 (1963), 105–14.
45. Banaschewski, B., Extensions of topological spaces, *Canad. Math. Bull.* **7** (1964), 1–22.
46. Bartle, R. G., Nets and filters in topology, *Amer. Math. Monthly* **62** (1955), 551–7.
47. Biesterfeldt, H. J., Uniformization of convergence spaces, *Math. Ann.* **177** (1968), 31–42, 43–48.
48. Biles, C., Wallman type compactifications, *Proc. Amer. Math. Soc.* **25** (1970), 363–8.
49. Bing, R. H., Extending a metric, *Duke Math. J.* **14** (1947), 511–19.
50. Bing, R. H., Metrization of topological spaces, *Canad. J. Math.* **3** (1951), 175–86.
51. Blair, R. L., Direct decomposition of lattices of continuous functions, *Proc. Amer. Math. Soc.* **13** (1962), 631–4.
52. Blair, R. L., Mappings that preserve realcompactness, *Pre-print*, Ohio University.
53. Blefko, R., On E-compact spaces, *Thesis*, Pennsylvania State University, 1965.
54. Blefko, R. and Mrowka, S., On the extensions of continuous functions from dense subspaces, *Proc. Amer. Math. Soc.* **17** (1966), 1396–1400.
55. Bockstein, M., Un théorème de séparabilité pour les produits topologiques, *Fund. Math.* **35** (1948), 242–6.
56. Borges, C. J. R., On stratifiable spaces, *Pacific J. Math.* **17** (1966), 1–16.
57. Borges, C. J. R., On function spaces of stratifiable spaces and compact spaces, *Proc. Amer. Math. Soc.* **17** (1966), 1074–8.
58. Borges, C. J. R., On metrizability of topological spaces, *Canad. J. Math.* **22** (1968), 795–804.
59. Borges, C. J. R., A survey of M-spaces, open questions and partial results, *Proc. Top. Symposium* 70 (1971), 79–84.
60. Borsuk, K., Sur les retracts, *Fund. Math.* **17** (1931), 152–70.
61. Burke, D. K., On subparacompact spaces, *Proc. Amer. Math. Soc.* **23** (1969), 655–63.
62. Burke, D. K., On p-spaces and $w\Delta$-spaces, *Pacific J. Math.* **35** (1970), 285–90.
63. Cain, G. L., Compact and related mappings, *Duke Math. J.* **33** (1966), 639–46.
64. Cain, G. L., Compactification of mappings, *Proc. Amer. Math. Soc.* **23** (1969), 298–303.
65. Čech, E., On bicompact spaces, *Ann. Math*, (4) **38** (1937), 823–44.
66. Ceder, J., Some generalizations of metric spaces, *Pacific J. Math.* **11** (1961), 105–25.
67. Chevalley, C. and Frink, O., Bicompactness of cartesian products, *Bull. Amer. Math. Soc.* **47** (1941), 612–14.
68. Ciampa, S., Successioni di Cauchy e completamento degli spazi uniformi, *Rend. Sem. Mat. Univ. Padova*, **34** (1964), 427–33. (Italian.)
69. Cohen, H. B., The k-externally disconnected spaces as projectives, *Canad. J. Math.* **16** (1964), 253–60.
70. Cohen, H. J., Sur un problème de M. Dieudonné, *C.R. Acad. Sci. Paris*, **234** (1952), 290–2.
71. Cohen, P. J., The independence of the continuum hypothesis, *Proc. Nat. Acad. Sci.* (I) **50** (1963), 1143–8; (II) **51** (1964), 105–10.

72. Comfort, W. W., Retractions and other continuous maps from βX onto $\beta X - X$, *Trans. Amer. Math. Soc.* **114** (1964), 1–9.
73. Comfort, W. W., Locally compact realcompactifications, General Topology and its Relations to Modern Analysis and Algebra, II, *Proc. Second Prague Top. Symp.* (1966), 95–100.
74. Comfort, W. W., A theorem of Stone–Čech type, and a theorem of Tychonoff type, without the axiom of choice; and their realcompact analogues, *Fund. Math.* **63** (1968), 97–109.
75. Comfort, W. W., On the Hewitt realcompactification of a product space, *Trans. Amer. Math. Soc.* **131** (1968), 107–18.
76. Comfort, W. W., Closed Baire sets are (sometimes) zero-sets, *Proc. Amer. Math. Soc.* **25** (1970), 870–5.
77. Comfort, W. W. and Gordon, H., Disjoint open subsets of $\beta X - X$, *Trans. Amer. Math. Soc.* **111** (1964), 513–20.
78. Comfort, W. W. and Negrepontis, S., Extending continuous functions on $X \times Y$ to subsets of $\beta X \times \beta Y$, *Fund. Math.* **54** (1966), 1–12.
79. Comfort, W. W. and Negrepontis, S., Homeomorphs of three subspaces of $\beta N - N$, *Math. Z.* **107** (1968), 53–8.
80. Comfort, W. W. and Ross, K. A., On the infinite product of topological spaces, *Arch. der Math.* **14** (1963), 62–4.
81. Comfort, W. W. and Ross, K. A., Pseudocompactness and uniform continuity in topological groups, *Pacific J. Math.* **16** (1966), 483–96.
82. Corson, H. H., The determination of paracompactness by uniformities, *Amer. J. Math.* **80** (1958), 185–90.
83. Corson, H. H., Normality in subsets of product spaces, *Amer. J. Math.* **81** (1959), 785–96.
84. Corson, H. H., Examples relating to normality in topological spaces, *Trans. Amer. Math. Soc.* **99** (1961), 205–11.
85. Corson, H. H. and Isbell, J. R., Some properties of strong uniformities, *Quart. J. Math.* **11** (1960), 17–33.
86. Corson, H. H. and Isbell, J. R., Euclidean covers of topological spaces, *Quart. J. Math.* **11** (1960), 34–42.
87. Corson, H. H. and Michael, E., Metrizability of certain countable unions, *Illinois J. Math.* **8** (1964), 351–60.
88. Creede, G., Semistratifiable spaces, *Topology Conference, Arizona State University* (ed. Grace, E. E.) (1968), 318–23.
89. Creede, G., Concerning semistratifiable spaces, *Pacific J. Math.* **32** (1970), 47–54.
90. Davis, A. S., The uniform continuity of continuous functions on a topological space, *Proc. Amer. Math. Soc.* **14** (1963), 977–80.
91. DeGroot, J. and McDowell, R. H., Extension of mappings on metric spaces, *Fund. Math.* **48** (1960), 251–63.
92. Derwent, J., A note on numerable covers, *Proc. Amer. Math. Soc.* **19** (1968), 1130–2.
93. Dickinson, A., Compactness conditions and uniform structures, *Amer. J. Math.* **75** (1953), 224–7.
94. Dickman, R. F. and Zame, A., Functionally compact spaces, *Pacific J. Math.* **31** (1969), 303–11.
95. Dieudonné, J. A., Une généralisation des espaces compacts, *J. Math. Pures Appl.* **23** (1944), 65–76.
96. Dieudonné, J. A., Un critère de normalité pour les espaces produits, *Colloq. Math.* **6** (1958), 29–32.

97. Doicinov, D., A unified theory of topological spaces, proximity spaces and uniform spaces, *Soviet Math. Dokl.* **5** (1964), 595–8.
98. Doss, R., On uniform spaces with a unique structure, *Amer. J. Math.* **71** (1949), 19–23.
99. Dowker, C. H., On countably paracompact spaces, *Canad. J. Math.* **3** (1951), 219–24.
100. Dowker, C. H., On a theorem of Hanner, *Ark. Mat.* **2** (1952), 307–13.
101. Dowker, C. H., On homotopy extension theorems, *Proc. London Math. Soc.* **6** (1956), 110–16.
102. Duda, E., Compactness of mappings, *Pacific J. Math.* **29** (1969), 259–66.
103. Duda, R. and Telgarsky, R., On some covering properties of metric spaces, *Czech. Math. J.* (93) **18** (1968), 66–80.
104. Dugundji, J., A generalization of Tietze's theorem, *Pacific J. Math.* **1** (1951), 353–66.
105. Dykes, N., Mappings and realcompact spaces, *Pacific J. Math.* **32** (1969), 347–58.
106. Dykes, N., Generalizations of realcompact spaces, *Pacific J. Math.* **33** (1970), 571–81.
107. Dyter, J. A. and Johnson, W. B., Some properties of spaces of uniformly quasi continuous functions, *Amer. Math. Monthly* **76** (1969), 489–94.
108. Eilenberg, S., Ordered topological spaces, *Amer. J. Math.* **63** (1941), 39–45.
109. Elliott, R. J., Inductive limits of uniform spaces, *J. London Math. Soc.* **42** (1967), 93–100.
110. Ellis, R. L., Extending continuous functions on zero dimensional spaces, *Math. Ann.* **186** (1970), 114–22.
111. Engelking, R., Remarks on realcompact spaces, *Fund. Math.* **55** (1964), 303–8.
112. Engelking, R. and Pelczynski, A., Remarks on dyadic spaces, *Colloq. Math.* **11** (1963), 55-63.
113. Engelking, R. and Sklyarenko, E. G., On compactifications allowing extensions of mappings, *Fund. Math.* **53** (1953), 65–79.
114. Fan, K. and Gottesman, N., On compactifications of Freudenthal and Wallman, *Indag. Math.* **55** (1952), 504–10.
115. Ferrier, J. P., Paracompacité dans les espaces uniformes: applications, *C.R. Acad. Sci. Paris* **260** (1965), 2672–4.
116. Filippov, V., On feathered paracompacta, *Soviet Math.* **9** (1968), 161–4.
117. Fine, N. J. and Gillman, L., Extension of continuous functions in βN, *Bull. Amer. Math. Soc.* **66** (1960), 376–81.
118. Fine, N. J. and Gillman, L., Remote points in β**R**, *Proc. Amer. Math. Soc.* **13** (1962), 29–36.
119. Fletcher, P., McCoy, R. A. and Slover, R., On boundedly metacompact and boundedly paracompact spaces, *Proc. Amer. Math. Soc.* **25** (1970), 335–42.
120. Formin, S., Extensions of topological spaces, *Ann. Math.* **44** (1943), 471–80.
121. Fox, R. H., On topologies for function spaces, *Bull. Amer. Math. Soc.* **51** (1945), 429–32.
122. Freudenthal, H., Entwicklungen von Räumen und ihren Gruppen, *Compositio Math.* **4** (1937), 145–234.
123. Freudenthal, H., Kompaktisierungen und bikompaktisierungen, *Indag. Math.* **13** (1951), 184–92.

124. Freudenthal, H., Neuaufbau der Endentheorie, *Ann. Math.* **43** (1942), 261–79.
125. Freudenthal, H., Enden und Priminden, *Fund. Math.* **39** (1953), 189–210.
126. Fréchet, M., Sur quelques points du calcul fonctionnel, *Rendiconti di Palermo*, **22** (1906), 1–74.
127. Frink, A. H., Distance functions and the metrization problem, *Bull. Amer. Math. Soc.* **43** (1937), 133–42.
128. Frink, O., Topology in lattices, *Trans. Amer. Math. Soc.* **51** (1942), 567–82.
129. Frink, O., Compactifications and semi-normal spaces, *Amer. J. Math.* **86** (1964), 602–7.
130. Frolík, Z., Generalizations of compact and Lindelöf spaces, *Czech Math. J.* **9** (84) (1959), 212–17.
131. Frolík, Z., On the topological product of paracompact spaces, *Bull. Acad. Pol. Sci.* **8** (1960), 747–50.
132. Frolík, Z., The topological product of two pseudocompact spaces, *Czech. Math. J.* **10** (85) (1960), 339–49.
133. Frolík, Z., Applications of complete families of continuous functions to the theory of Q spaces, *Czech. Math. J.* **11** (1961), 115–31.
134. Frolík, Z., On almost realcompact spaces, *Bull. Acad. Pol. Des Sciences Math. Astr. et Phys.* **9** (1961), 247–50.
135. Frolík, Z., Locally topologically complete spaces, *Soviet Math. Dokl.* **2** (1961), 355–7.
136. Frolík, Z., A generalization of realcompact spaces, *Czech. Math. J.* **13** (88) (1963), 127–37.
137. Gal, I. S., On a generalized notion of compactness, I, *Indag. Math.* **19** (60) (1957), 421–35.
138. Gal, I. S., On the theory of (m, n)-compact topological spaces, *Pacific J. Math.* **8** (1958), 721–34.
139. Gal, I. S., Uniformizable spaces with a unique structure, *Pacific J. Math.* **9** (1959), 1053–60.
140. Gal, I. S., Proximity relations and precompact structures, I, II, *Indag. Math.* **21** (1959), 304–14, 315–26.
141. Gantner, T. E., Extensions of uniformly continuous pseudometrics, *Trans. Amer. Math. Soc.* **132** (1968), 147–57.
142. Gantner, T. E., Some corollaries to the metrization lemma, *Amer. Math. Monthly* **76** (1969), 45–8.
143. Gantner, T. E., Extensions of uniformities, *Fund. Math.* **66** (1970), 263–81.
144. Gillman, L., A P-space and an extremally disconnected space whose product is not an F-space, *Arch. Math.* **11** (1960), 53–5.
145. Gillman, L., A note on F-spaces, *Arch. Math.* **12** (1961), 67–8.
146. Gillman, L. and Henriksen, M., Concerning rings of continuous functions, *Trans. Amer. Math. Soc.* **77** (1954), 340–62.
147. Gillman, L. and Jerison, M., Stone–Čech compactification of a product, *Arch. Math.* **10** (1959), 443–6.
148. Ginsburg, S. and Isbell, J., Some operators on uniform spaces, *Trans. Amer. Math. Soc.* **93** (1959), 145–68.
149. Gleason, A., Projective topological spaces, *Illinois J. Math.* **2** (1958), 482–9.
150. Glicksberg, I., Stone–Čech compactifications of products, *Trans. Amer. Math. Soc.* **90** (1959), 369–82.

151. Gould, G. G., A Stone–Čech Alexandroff type compactification and its application to measure theory, *Proc. London Math. Soc.* **14** (3) (1964), 221–44.

152. Grunbaum, F. and Zarantonello, E. H., On the extension of uniformly continuous mappings, *Mich. J. Math.* **15** (1968), 65–74.

153. Hager, A. W., On the maximal ring of quotients of $C(X)$, *Bull. Amer. Math. Soc.* **72** (1966), 850–2.

154. Hager, A. W., Projections of zero sets (and the fine uniformity on a product), *Trans. Amer. Math. Soc.* **140** (1969), 87–94.

155. Hager, A. W., Approximation of real continuous functions on Lindelöf spaces, *Proc. Amer. Math. Soc.* **22** (1969), 156–63.

156. Hager, A. W. and Johnson, D. G., A note on certain subalgebras of $C(X)$, *Canad. J. Math.* **20** (1968), 389–93.

157. Hammer, P. C., Extended topology: continuity I, *Port. Math.* **25** (1964), 77–93.

158. Hammer, P. C., Extended topology: connected sets and Wallace separations, *Port. Math.* **22** (1965), 167–87.

159. Hanai, S., On closed mappings, *Proc. Japan Acad.* **30** (1954), 285–8.

160. Hanai, S., On closed mappings, II, *Proc. Japan Acad.* **32** (1956), 388–91.

161. Hanai, S., On open mappings, *Proc. Japan Acad.* **33** (1957), 177–80.

162. Hanai, S., Inverse images of closed mappings, I, II, III, *Proc. Japan Acad.* **37** (1961), 298–301, 302–4, 457–8.

163. Hanai, S. On open mappings, II, *Proc. Japan Acad.* **37** (1961), 233–8.

164. Hanai, S., Open bases and continuous mappings, II, *Proc. Japan Acad.* **38** (1962), 448–51.

165. Hanai, S., Open mappings and metrization theorems, *Proc. Japan Acad.* **39** (1963), 450–4.

166. Hanai, S. and Morita, K., Closed mappings and metric spaces, *Proc. Japan Acad.* **32** (1956), 9–14.

167. Hanai, S. and Okuyama, A., On paracompactness of topological spaces, *Proc. Japan Acad.* **36** (1960), 466–9.

168. Hanai, S. and Okuyama, A., On pseudocompactness and continuous mappings, *Proc. Japan Acad.* **38** (1962), 444–7.

169. Hanner, O., Solid spaces and absolute retracts, *Ark. Math.* **1** (1951), 375–82.

170. Hanner, O., Retraction and extension of mappings of metric and non-metric spaces, *Ark. Math.* **2** (1952), 315–60.

171. Hausdorff, F., Erweiterung einer stetigen abbildung, *Fund. Math.* **30** (1938), 40–7.

172. Heath, R. W., Arcwise connectedness in semi-metric spaces, *Pacific J. Math.* **12** (1962), 1301–19.

173. Heath, R. W., On open mappings and certain spaces satisfying the first countability axiom, *Fund. Math.* **62** (1965), 91–6.

174. Heath, R. W., On certain first countable spaces, *Topology Seminar Wisconsin* (1965), 103–13.

175. Heath, R. W., A paracompact semi-metric space which is not an M_3-space, *Proc. Amer. Math. Soc.* **17** (1966), 868–70.

176. Heath, R. W., An easier proof that a certain countable space is not stratifiable, *Proc. Washington State Conference on General Topology* Mar. 1970, 56–9.

177. Heath, R. W., and Hodel, R. E., Characterization of σ-spaces, *Fund. Math.* (to appear).

178. Henriksen, M. and Isbell, J. R., Local connectedness in the Stone–Čech compactification, *Illinois J. Math* **1** (1957), 574–82.
179. Henriksen, M. and Isbell, J. R., Some properties of compactifications, *Duke Math. J.* **25** (1958), 83–105.
180. Herrlich, H., *E*-kompakte Räume, *Math. Z.* **96** (1967), 228–55.
181. Hewitt, E., On two problems of Urysohn, *Ann. Math.* (2) **47** (1946), 503–9.
182. Hewitt, E., Rings of real-valued continuous functions, I, *Trans. Amer. Math. Soc.* **64** (1948), 45–99.
183. Hewitt, E., Linear functions on spaces of continuous functions, *Fund. Math.* **37** (1950), 161–89.
184. Himmelberg, C. J., Preservation of pseudo-metrizability by quotient maps, *Proc. Amer. Math. Soc.* **17** (1966), 1378–84.
185. Hodel, R. E., Total normality and the hereditary property, *Proc. Amer. Math. Soc.* **17** (1966), 462–5.
186. Horne, J. G., Countable paracompactness and cb-spaces, *Notices Amer, Math. Soc.* **6** (1959), 629–30.
187. Husek, M., The Hewitt realcompactification of a product, *Comment. Math. Univ. Carolinae* **11** (1970), 393–5.
188. Husek, M., Pseudo-*m* compactness and $\upsilon(P \times Q)$, *Indag. Math.* **33** (1971), 320–6.
189. Husek, M., Realcompactness of function spaces on $\upsilon(P \times Q)$, *J. General Topology* (to appear).
190. Imler, L., Extensions of pseudometrics and linear space-valued functions, *Thesis*, Carnegie–Mellon University (1969).
191. Insel, A. J., A note on the Hausdorff separation property in first countable spaces, *Amer. Math. Monthly* **72** (1965), 289–90.
192. Isbell, J. R., On finite-dimensional uniform spaces, *Pacific J. Math.* **9** (1957), 107–21.
193. Isbell, J. R., Algebras of uniformly continuous functions, *Ann. Math.* **68** (1958), 96–125.
194. Isbell, J. R., Euclidean and weak uniformities, *Pacific J. Math.* **8** (1958), 67–86.
195. Isbell, J. R., Structure of categories, *Bull. Amer. Math. Soc.* **72** (1966), 619–55.
196. Iseki, K. I., A note on hypocompact spaces, *Math. Japan.* **3** (1953), 16–17.
197. Iseki, K. I., On extension of continuous mappings on countably paracompact normal spaces, *Proc. Japan Acad.* **30** (1954), 736–9.
198. Iseki, K. I., A note on countably paracompact spaces, *Proc. Japan Acad.* **30** (1954), 350–1.
199. Iseki, K. I., On the property of Lebesgue in uniform spaces, *Proc. Japan Acad.* **31** (1955), 220–1.
200. Iseki, K. I., A remark on countably compact normal spaces, *Proc. Japan Acad.* **33** (1957), 131–3.
201. Iseki, K. I., A characterization of pseudocompact spaces, *Proc. Japan Acad.* **33** (1957), 320–2.
202. Iseki, K. I., New characterizations of compact spaces, *Proc. Japan Acad.* **34** (1958), 144–5.
203. Iseki, K. I. and Kasahara, S., On pseudocompact and countably compact spaces, *Proc. Japan Acad.* **33** (1957), 100–2.
204. Ishii, T., Some characterizations of *M* paracompact spaces, I, II, *Proc. Japan Acad.* **38** (1962), 480–3, 651–4.

205. Ishii, T., On M- and M^*-spaces, *Proc. Japan Acad.* **44** (1968), 1028–30.
206. Ishii, T., On closed mappings and M-spaces, I, II, *Proc. Japan Acad.* **43** (1969), 752–6, 756–61.
207. Ishii, T., On wM-spaces, I, II, *Proc. Japan Acad.* **46** (1970), 5–10, 11–15.
208. Ishikawa, F., On countably paracompact spaces, *Proc. Japan Acad.* **44** (1968), 1028–30.
209. Isiwata, T., On locally Q complete spaces, I, II, III, *Proc. Japan Acad.* **35** (1959), 232–6, 263–7, 431–5.
210. Isiwata, T., Mappings and spaces, *Pacific J. Math.* **20** (1967), 455–80.
211. Isiwata, T., The product of M-spaces need not be an M-space, *Proc. Japan Acad.* **45** (1969), 154–6.
212. Jensen, G. A., A note on complete separation in the Stone topology, *Proc. Amer. Math. Soc.* **21** (1969), 113–16.
213. Jones, F. B., Moore spaces and uniform spaces, *Proc. Amer. Math. Soc.* **9** (1958), 483–5.
214. Jones, F. B., On the first countability axiom for locally compact Hausdorff spaces, *Coll. Math.* **7** (1959), 33–4.
215. Kakutani, S., Simultaneous extension of continuous functions considered as a positive linear operation, *Japan Math.* **17** (1940), 1–4.
216. Kasahara, S., On weakly compact regular spaces, II, *Proc. Japan Acad.* **33** (1957), 255–9.
217. Kasahara, S., A note on some topological spaces, *Proc. Japan Acad.* **33** (1957), 453–4.
218. Katětov, M., Complete normality of cartesian products, *Fund. Math.* **35** (1948), 271–4.
219. Katětov, M., On nearly discrete spaces, *Casopis Pro Pestovani Matematiky A Fysiky* **75** (1950), 69–78.
220. Katětov, M., On real-valued functions in topological spaces, *Fund. Math.* **38** (1951), 85–91.
221. Katětov, M., Measures in fully normal spaces, *Fund. Math.* **38** (1951), 73–84.
222. Katětov, M., On the dimension of non-separable spaces I, *Czech. Math. J.* **2** (1952), 333–68. (Russian: English summary.)
223. Katětov, M., Correction to 'On real-valued functions in topological spaces', *Fund, Math.* **40** (1953), 203–5.
224. Katětov, M., On the dimension of non-separable spaces II, *Czech. Math. J.* **6** (1956), 485–516. (Russian: English summary.)
225. Katětov, M., Extension of locally finite covers, *Colloq. Math.* **6** (1958), 145–51. (Russian.)
226. Kenderov, P., On Q-spaces, *Dokl. Akad. Nauk SSSR*, **175** (1967), 288–91.
227. Kennison, J. F., m-pseudocompactness, *Trans. Amer. Math. Soc.* **104** (1962), 436–42.
228. Kennison, J. F., Full reflective subcategories and generalized covering spaces, *Illinois J. Math.* **12** (1968), 353–65.
229. Kent, D. C., On the order topology in a lattice, *Illinois J. Math.* **10** (1966), 90–6.
230. Kister, J. M., Uniform continuity and compactness in topological groups, *Proc. Amer. Math. Soc.* **13** (1962), 37–40.
231. Krajewski, L. L., On expanding locally finite collections, *Canad. J. Math.* **23** (1971), 58–68.
232. Leader, S., On pseudometrics for generalized uniform structures, *Proc. Amer. Math. Soc.* **16** (1965), 493–5.

233. Lelek, A., On totally paracompact metric spaces, *Proc. Amer. Math. Soc.* **19** (1968), 168–70.
234. Lin, Y. F., Relations on topological spaces: Urysohn's lemma, *J. Australian Math. Soc.* **8** (1968), 37–42.
235. Liu, C. T., Absolutely closed spaces, *Trans. Amer. Math. Soc.* **130** (1968), 86–104.
236. Loeb, P. A., Compactifications of Hausdorff spaces, *Proc. Amer. Math. Soc.* **22** (1969), 627–34.
237. Lozier, F. W., A class of compact rigid zero dimensional spaces, *Canad. J. Math.* **21** (1969), 817–21.
238. Lubkin, S., Theory of covering spaces, *Trans. Amer. Math. Soc.* **104** (1962), 205–38.
239. Lutzer, D. J., A metrization theorem for linearly orderable spaces, *Proc. Amer. Math. Soc.* **22** (1969), 557–8.
240. Lutzer, D. J., Semimetrizable and stratifiable spaces, *General Top. and Applications* **1** (1971), 43–8.
241. McArthur, W. G., Hewitt realcompactifications of products, *Canad. J. Math.* **22** (1970), 645–56.
242. McAuley, L. F., A relation between perfect separability, completeness, and normality in semimetric spaces, *Pacific J. Math.* **6** (1956), 315–26.
243. McAuley, L. F., A note on complete collectionwise normality and paracompactness, *Proc. Amer. Math. Soc.* **9** (1958), 796–9.
244. McCandless, B. H., On order paracompact spaces, *Canad. J. Math.* **21** (1969), 400–5.
245. McDowell, R. H., Extension of functions from dense subspaces, *Duke Math. J.* **25** (1958), 297–304.
246. Mack, J. E., On a class of countably paracompact spaces, *Proc. Amer. Math. Soc.* **16** (1965), 467–72.
247. Mack, J. E., Directed covers and paracompact spaces, *Canad. J. Math.* **19** (1967), 649–54.
248. Mack, J. E., Product spaces and paracompactness, *J. London Math. Soc.* **1** (2) (1969), 90–4.
249. Mack, J. E., Countable paracompactness and weak normality properties, *Trans. Amer. Math. Soc.* **148** (1970), 265–72.
250. Mack, J. E. and Johnson, D., The Dedekind completion of $C(X)$, *Pacific J. Math.* **20** (1967), 231–43.
251. Magill, K. D., N-point compactifications, *Amer. Math. Monthly* **72** (1965), 1075–81.
252. Magill, K. D., Some embedding theorems, *Proc. Amer. Math. Soc.* **16** (1965), 126–30.
253. Magill, K. D., A note on compactifications, *Math. Z.* **94** (1966), 322–5.
254. Mandelker, M., Prime z-ideal structure of $C(R)$, *Fund. Math.* **63** (1968), 145–66.
255. Mandelker, M., Prime ideal structure of rings of bounded continuous functions, *Proc. Amer. Math. Soc.* **19** (1968), 1432–8.
256. Mandelker, M., F' spaces and z-embedded subspaces, *Pacific J. Math.* **28** (1969), 615–21.
257. Mansfield, M. J., Some generalizations of full normality, *Trans. Amer. Math. Soc.* **86** (1957), 489–505.
258. Mansfield, M. J., On countably paracompact normal spaces, *Canad. J. Math.* **9** (1957), 443–9.
259. Marik, J., On pseudocompact spaces, *Proc. Japan Acad.* **35** (1959), 120–1.

260. Mazur, S., On continuous mappings on cartesian products, *Fund. Math.* **39** (1952), 229–38.
261. Michael, E. A., Some extension theorems for continuous functions, *Pacific J. Math.* **3** (1953), 789–806.
262. Michael, E. A., A note on paracompact spaces, *Proc. Amer. Math. Soc.* **4** (1953), 831–8.
263. Michael, E. A., Review of 'On a theorem of Hanner', by C. H. Dowker, *Math. Reviews*, **14** (1953), 396.
264. Michael, E. A., Local properties of topological spaces, *Duke Math. J.* **21** (1954), 163–71.
265. Michael, E. A., Point finite and locally finite coverings, *Canad. J. Math.* **7** (1955), 275–9.
266. Michael, E. A., Another note on paracompact spaces, *Proc. Amer. Math. Soc.* **8** (1957), 822–8.
267. Michael, E. A., Yet another note on paracompact spaces, *Proc. Amer. Math. Soc.* **10** (1959), 309–14.
268. Michael, E. A., The product of a normal space and a metric space need not be normal, *Bull. Amer. Math. Soc.* **69** (1963), 375–6.
269. Michael, E. A., On Nagami's Σ spaces and related matters, *Proc. Wash. State University Conference on General Topology* (ed. Stoltenberg, R.), (1969), 13–19.
270. Morita, K., Star finite coverings and the star finite property, *Math. Japan* **1** (1948), 60–8.
271. Morita, K., On the dimension of normal spaces, I, *Japanese J. Math.* **20** (1950), 5–36.
272. Morita, K., On the dimension of normal spaces, II, *J. Math. Soc. Japan* **2** (1950), 16–33.
273. Morita, K., On the simple extension of a space with respect to a uniformity, I, II, III, *Proc. Japan Acad.* **27** (1951), 65–72, 130–7, 166–71.
274. Morita, K., Normal families and dimension theory for metric spaces, *Math. Ann.* **128** (1954), 350–62.
275. Morita, K., On closed mappings and dimension, *Proc. Japan Acad.* **32** (1956), 161–5.
276. Morita, K., On closed mappings, *Proc. Japan Acad.* **32** (1956), 539–43.
277. Morita, K., Note on mapping spaces, *Proc. Japan Acad.* **32** (1956), 671–5.
278. Morita, K., On closed mappings, II, *Proc. Japan Acad.* **33** (1957), 325–7.
279. Morita, K., Paracompactness and product spaces, *Fund. Math.* **50** (1962), 223–36.
280. Morita, K., On the product of a normal space with a metric space, *Proc. Japan Acad.* **39** (1963), 148–50.
281. Morita, K., Products of normal spaces with metric spaces, *Math. Ann.* **154** (1964), 365–82.
282. Morita, K., Some properties of *M* spaces, *Proc. Japan Acad.* **43** (1967), 869–72.
283. Morita, K., A survey of the theory of *M* spaces, *Proc. Topology Symposium* **70** (1971), 48–55.
284. Morita, K. and Hanai, S., Closed mappings and metric spaces, *Proc. Japan Acad.* **32** (1956), 10–14.
285. Mrowka, S., On completely regular spaces, *Fund. Math.* **41** (1954), 105–6.

286. Mrowka, S., Some properties of Q-spaces, *Bull. Acad. Pol. Sci. Ser. Sci. Math. Astronom. Phys.* **5** (10) (1957), 947–50.
287. Mrowka, S., Functionals on uniformly closed rings of continuous functions, *Fund. Math.* **46** (1958), 81–7.
288. Mrowka, S., An example of a non normal completely regular space, *Bull. Acad. Pol. Sci.* **6** (1958), 161–3.
289. Mrowka, S., On function spaces, *Fund. Math.* **45** (1958), 273–82.
290. Mrowka, S., Compactness and product spaces, *Coll. Math.* **7** (1959), 19–22.
290. Mrowka, S., On E-compact spaces, II, *Bull. Acad. Pol. Sci. Ser. Sci. Math. Astronom. Phys.* **14** (11) (1966), 597–605.
292. Mrowka, S., Further results on E-compact spaces, I, *Acta Math.* **120** (1968), 161–85.
293. Mrowka, S. and Blefko, R., On the extensions of continuous functions from dense subspaces, *Proc. Amer. Math. Soc.* **6** (1966), 1396–1400.
294. Myers, S. B., Equicontinuous sets of mappings, *Ann. Math.* (2), **47** (1946), 496–502.
295. Nachbin, L., Sur les espaces topologiques ordonnés, *C.R. Acad. Sci. Paris* **226** (1948), 381–2.
296. Nachbin, L., On the continuity of positive linear transformations, *Proc. Int. Congress Math. I*, (1950), 464.
297. Nagami, K., Paracompactness and strong screenability, *Nagoya Math. J.* **8** (1955), 83–8.
298. Nagami, K., Σ-spaces, *Fund. Math.* **65** (1969), 160–92.
299. Nagami, K., σ-spaces and product spaces, *Math. Ann.* **181** (1969), 109–18.
300. Nagata, J., On a necessary and sufficient condition of metrizability, *J. Inst. Polytech. Osaka City University*, **1** (1950), 93–100.
301. Nagata, J., A contribution to the theory of metrization, *J. Inst. Polytech., Osaka City University*, **8** (1957), 185–192.
302. Nagata, J., A theorem for metrizability of a topological space, *Proc. Japan Acad.* **33** (1957), 128–30.
303. Nagata, J., Mappings and M-spaces, *Proc. Japan Acad.* **45** (1969), 140–4.
304. Nagata, J., A note on M-spaces and topologically complete spaces, *Proc. Japan Acad.* **45** (1969), 541–3.
305. Nagata, J., A note on Filippov's theorem, *Proc. Japan Acad.* **45** (1969), 30–3.
306. Nagata, J. and Siwiec, F., A note on nets and metrization, *Proc. Japan Acad.* **44** (1968), 623–7.
307. Negrepontis, S., Baire sets in topological spaces, *Arch. Math.* **18** (1967), 603–8.
308. Negrepontis, S., On the product of F-spaces, *Trans. Amer. Math. Soc.* **136** (1969), 339–46.
309. Njastad, O., On uniform spaces where all uniformly continuous functions are bounded, *Monatsh Math.* **69** (1965), 167–76.
310. Njastad, O., A note on compactification by bounding systems, *J. London Math. Soc.* **40** (1965), 526–32.
311. Njastad, O., On Wallman-type compactifications, *Math. Z.* **91** (1966), 267–76.
312. Okuyama, A., Some generalizations of metric spaces, their metrization theorems and product spaces, *Sci. Rep. Tokyo Kyoiku Daigaku Sec. A*, **9** (1967), 236–54.

313. Okuyama, A., σ-spaces and closed mappings, I, *Proc. Japan Acad.* **44** (1968), 427–77.
314. Okuyama, A., A survey of the theory of σ-spaces, *Proc. Topo. Symposium* **70** (1971), 57–63.
315. Okuyama, A., On a generalization of Σ-spaces (to appear).
316. Onuchic, N., On the Nachbin uniform structure, *Proc. Amer. Math. Soc.* **11** (1960), 177–9.
317. Pervin, W. J., Uniformization of neighborhood axioms, *Math. Ann.* **147** (1962), 313–15.
318. Pervin, W. J., Quasi uniformization of topological spaces, *Math. Ann.* **147** (1962), 316–17.
319. Ponomarev, V., Open mappings of normal spaces, *Dokl. Akad. Nauk USSR* **126** (1959), 716–18. (Russian.)
320. Ponomarev, V., Axioms of countability and continuous mappings, *Bull. Acad. Polon. Sci. Ser. Sci. Math. Astr. Phys.* **8** (1960), 127–34. (Russian.)
321. Ponomarev, V., On paracompact and finally compact spaces, *Soviet Math. Dokl.* **2** (1961), 1510–12.
322. Porter, J. and Thomas, J., On *H* closed and minimal Hausdorff spaces, *Trans. Amer. Math. Soc.* **138** (1969), 159–70.
323. Ptak, V., Concerning spaces of continuous functions, *Czech. Math. J* **5** (80) (1955), 412–31.
324. Rajagopalan, M. and Wilansky, A., Reversible topological spaces, *Australian Math. Soc. J.* **6** (1966), 129–38.
325. Reichaw-Reichbach, M., On compactification of metric spaces, *Israel J. Math.* **1** (1963), 61–74.
326. Robinson, S., The intersection of the free maximal ideals in a complete space, *Proc. Amer. Math. Soc.* **17** (1966), 468–9.
327. Robinson, S., Some properties of $\beta X - X$ for complete spaces, *Fund. Math.* (to appear).
328. Rogers, C. A. and Willmott, R. C., On the uniformization of sets in topological spaces, *Acta Math.* **120** (1968), 1–52.
329. Ross, K. A. and Stone, A. H., Products of separable spaces, *Amer. Math. Monthly* **71** (1964), 398–403.
330. Rudin, M. E., Countable paracompactness and Souslin's problem, *Canad. J. Math.* **7** (1955), 543–7.
331. Rudin, M. E., A separable normal, non-paracompact space, *Proc. Amer. Math. Soc.* **7** (1956), 940–1.
332. Rudin, M. E., Interval topology in subsets of totally orderable spaces, *Trans. Amer. Math. Soc.* **118** (1965), 376–89.
333. Rudin, M. E., A Dowker space, *Bull. Amer. Math. Soc.* **77** (1971), 246.
334. Samuel, P., Ultrafilters and compactification of uniform spaces, *Trans. Amer. Math. Soc.* **64** (1948), 100–32.
335. Scarborough, C. T., Minimal Urysohn spaces, *Pacific J. Math.* **28** (1969), 611–17.
336. Shanin, N. A., On special extensions of topological spaces, *DAN USSR* **38** (1963), 6–9.
337. Shapiro, H. L., Extensions of pseudometrics, *Canad. J. Math.* **18** (1966), 981–98.
338. Shapiro, H. L., A note on extending uniformly continuous pseudometrics, *Bull. Belgique Math. Soc.* **18** (1966), 439–41.
339. Shapiro H. L., Closed maps and paracompact spaces, *Canad. J. Math.* **20** (1968), 513–19.

340. Shapiro, H. L., More on extending continuous pseudometrics, *Canad. J Math.* **22** (1970), 984–93.
341. Shiraki, T., On some metrization theorems, *Proc. Japan Acad.* (to appear).
342. Shirota, T., On systems of structures of a completely regular space, *Osaka Math. J.* **2** (1950), 131–43.
343. Shirota, T., On spaces with a complete structure, *Proc. Japan Acad.* **27** (1951), 513–16.
344. Shirota, T., A class of topological spaces, *Osaka Math. J.* **4** (1952), 24–40.
345. Singal, M. K. and Arya, S. P., On *M* paracompact spaces, *Math. Ann.* **181** (1969), 119–33.
346. Slaughter, F., Σ-spaces with point countable separating open covers are σ-spaces, *Research Report*, University of Pittsburgh.
347. Smirnov, Y. M., On normally embedded sets of normal spaces, *Math. Sbornik* **29** (1951), 47–9. (Russian.)
348. Smirnov, Y. M., A necessary and sufficient condition for metrizability of a topological space, *Dokl. Akad. Nauk USSR* (*n.s.*), **77** (1951), 197–200. (Russian.)
349. Smirnov, Y. M., On proximity spaces, *Mat. Sb.* (*n.s.*), **31** (1952), 543–74. (Russian.)
350. Smirnov, Y. M., On metrization of topological spaces, *Amer. Math. Soc.* Transl. No. 91 (1953).
351. Smirnov, Y. M., On completeness of proximity spaces, *Trudy Mosk. Mat. Obsc.* **3** (1954), 271–306. (Russian.)
352. Smith, J. C. and Krajewski, L. L., Expandability and collectionwise normality, *Trans. Amer. Math. Soc.* **160** (1971), 437–51.
353. Sorgenfrey, R. H., On the topological product of paracompact spaces, *Bull. Amer. Math. Soc.* **53** (1947), 631–2.
354. Steen, L. A., Conjectures and counterexamples in metrization theory, *Amer. Math. Monthly* **79** (1972), 113–32.
355. Steiner, E. F., Wallman spaces and compactifications, *Fund. Math.* **61** (1968), 295–304.
356. Steiner, E. F., Graph closures and metric compactifications of **N**, *Proc. Amer. Math. Soc.* **25** (1970), 593–7.
357. Steiner, E. F. and Steiner, A. K., Wallman and *Z* compactifications, *Duke Math. J.* **25** (1968), 269–75.
358. Stoltenberg, R. A., A note on stratifiable spaces, *Proc. Amer. Math. Soc.* **23** (1969), 294–7.
359. Stone, A. H., Paracompactness and product spaces, *Bull. Amer. Math. Soc.* **54** (1948), 977–82.
360. Stone, A. H., Review of 'Extension of coverings, of pseudometrics, and of linear-space-valued mappings', by Richard Arens, *Math. Reviews* **14** (1953), 1108.
361. Stone, A. H., Universal spaces for some metrizable uniformities, *Quart. J. of Math.* **11** (1960), 105–15.
362. Strecker, G. E. and Wattel, E., On semi-regular and minimal Hausdorff embeddings, *Indag. Math.* **29** (1967), 234–7.
363. Suzuki, J., On the metrization and the completion of a space with respect to a uniformity, *Proc. Japan Acad.* **27** (1951), 219–23.
364. Suzuki, J., Some properties of completely normal spaces, *Proc. Japan Acad.* **33** (1957), 19–24.
365. Swaminathan, S., A note on countably paracompact normal spaces, *J. Indian Math. Soc.* **29** (1965), 67–9.

366. Tamano, H., On paracompactness, *Pacific J. Math.* **10** (1960), 1043–7.
367. Tamano, H., On compactifications, *J. Math. Kyoto Univ.* **1** (1962), 161–93.
368. Terasaka, H., On cartesian product of compact spaces, *Osaka Math. J.* **4** (1952), 11–15.
369. Thomas, J. P., Associated regular spaces, *Canad. J. Math.* **20** (1968), 1087–92.
370. Tong, H., Some characterizations of normal and perfectly normal spaces, *Duke Math. J.* **19** (1952), 289–92.
371. Traylor, D. R., Metrizability and completeness in normal Moore spaces, *Pacific J. Math.* **17** (1966), 381–90.
372. Traylor, D. R. and Younglove, J. N., On normality and pointwise paracompactness, *Pacific J. Math.* **25** (1968), 193–6.
373. Urysohn, P., Über die Mächtigkeit der zusammenhängenden Mengen, *Math. Ann.* **94** (1925), 262–95.
374. Urysohn, P., Zum Metrisation problem, *Math. Ann.* **94** (1925), 309–15.
375. Van der Slot, J., Some properties related to compactness, *Mathematical Centre Tract No.* 19, Mathematisch Centrum, Amsterdam (1968).
376. Vidossich, G., Uniform spaces of countable type, *Proc. Amer. Math. Soc.* **35** (1970), 551–3.
377. Wagner, F. J., Notes on compactification, I, II, *Indag. Math.* **19** (1957), 171–81.
378. Wagner, F. J., Normal base compactifications, *Indag. Math.* **26** (1964), 78–83.
379. Wallman, H., Lattices and topological spaces, *Ann. Math.* **39** (1938), 112–26.
380. Warner, S., The topology of compact convergence on continuous function spaces, *Duke Math. J.* **25** (1958), 265–82.
381. Weddington, D. D., On *k* spaces, *Proc. Amer. Math. Soc.* **22** (1969), 635–8.
382. Wenjen, C., Realcompact spaces, *Port. Math.* **25** (1966), 135–9.
383. Whipple, K. E., Cauchy sequences in Moore spaces, *Pacific J. Math.* **18** (1966), 191–9.
384. Whyburn, G. T., Open and closed mappings, *Duke Math. J.* **17** (1950), 69–74.
385. Whyburn, G. T., Compactification of mappings, *Math. Ann.* **166** (1966), 168–74.
386. Willard, S., Absolute Borel sets in their Stone–Čech compactifications, *Fund. Math.* **58** (1966), 323–33.
387. Worrell, J. M., On continuous mappings of metacompact Čech complete spaces, *Pacific J. Math.* **30** (1969), 555–62.
388. Yang, C. T., On paracompact spaces, *Proc. Amer. Math. Soc.* **5** (1954), 185–9.
389. Yuan, S., On paracompact regular spaces, *J. Australian Math. Soc.* **2** (1961), 147–50.
390. Zabrodsky, A., Covering spaces of paracompact spaces, *Pacific J. Math.* **14** (1964), 1489–1503.
391. Zame, A., A note on Wallman spaces, *Proc. Amer. Math. Soc.* **22** (1969), 141–4.
392. Zenor, P., Realcompactifications with projective spectra, *Glasnik Mat.* 5 (25) (1970), 153–6.
393. Zenor, P., Extending completely regular spaces with inverse limits, *Glasnik Mat.* **5** (25) (1970), 157–62.